NANOANTENNA

Pan Stanford Series on the High-Tech of Biotechnology

Founding Series Editor
Robert S. Marks

Avram and Stella Goldstein-Goren Department of Biotechnology Engineering
National Institute for Biotechnology Engineering
Ilse Katz Institute for Nanoscale Science & Technology
Ben Gurion University of the Negev
Israel

Titles in the Series

Published

Vol. 1
Nanoantenna: Plasmon-Enhanced Spectroscopies for Biotechnological Applications
Marc Lamy de la Chapelle and Annemarie Pucci, eds.
2013
978-981-4303-61-3 (Hardcover)
978-981-4303-62-0 (eBook)

Forthcoming

Vol. 2
Viral Diagnostics: Advances and Applications
Amadou Alpha Sall, Leslie Lobel, and Robert S. Marks, eds.
2013
978-981-4364-43-0 (Hardcover)
978-981-4364-44-7 (eBook)

Vol. 3
Electrochemical Biosensors: Functionalization
Serge Cosnier, ed.
2013
978-981-4411-46-2 (Hardcover)
978-981-4411-47-9 (eBook)

Vol. 4
Nanomaterials for Water Management, I: Signal Amplification Using Nanostructures in Biosensing
Ibrahim Abdulhalim and Robert S. Marks, eds.
2013

Vol. 5
Luminescent Biosensors
Gerald Thouand and Robert S. Marks, eds.
2014

Vol. 6
Fibre-Optic Immunosensors and Biosensors
Robert S. Marks, ed.
2014

Pan Stanford Series on the High-Tech of Biotechnology
Volume 1

NANOANTENNA

Plasmon-Enhanced Spectroscopies for Biotechnological Applications

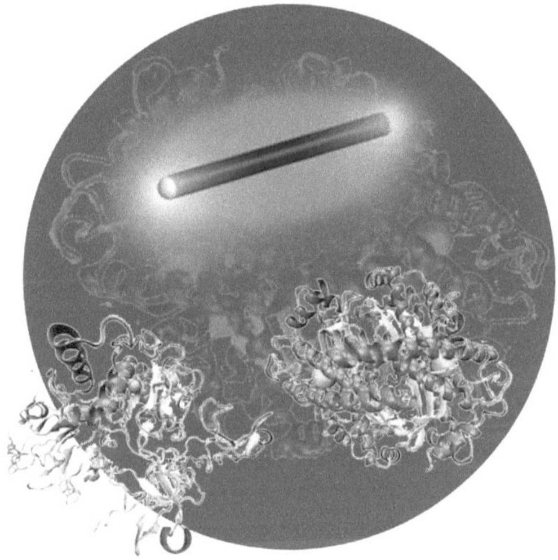

edited by

Marc Lamy de la Chapelle
Annemarie Pucci

Routledge
Taylor & Francis Group
LONDON AND NEW YORK

First published 2013 by Pan Stanford Publishing Pte. Ltd

Published 2019 by Routledge
2 Park Square, Milton Park, Abingdon, Oxon OX14 4RN
52 Vanderbilt Avenue, New York, NY 10017

Routledge is an imprint of the Taylor & Francis Group, an informa business

British Library Cataloguing-in-Publication Data
A catalogue record for this book is available from the British Library.

Nanoantenna: Plasmon-Enhanced Spectroscopies for Biotechnological Applications

ISBN 13: 978-981-4303-61-3 (hbk)

Contents

Foreword

This book is closely related to the EU project Nanoantenna, which is focused on the development of a highly sensitive and specific nanobiosensor based on surface-enhanced vibrational spectroscopy dedicated to protein detection and disease diagnostics.

A. Examples of sensitive assays that biochemistry has developed on its own

1. Quantitative detection of target proteins by immunochemical procedures is very sensitive. The enzyme-linked immunosorbent assay (ELISA) uses an antibody against the protein of interest. The bound enzyme is, for instance, detected by bonding to a fluorescent dye.
2. As most proteins are negatively charged, they can be spatially separated by electrophoresis in, for instance, acrylamide gels by migration from the positive to the negative electrode according to their molecular weight and mobility, yielding characteristic stripes (so-called Western plot), which become visible by various techniques (e.g., by the antibody of a special protein, the antibody being tagged to a fluorophore).

B. Detection of proteins by a combination of optical plasma resonances at silver or gold surfaces and an immunoreaction

1. The oldest and as yet most successful pure optical method uses electromagnetic surface waves (called surface plasmon polaritons [SPPs]) at the interface of a silver film conditioned to bind special proteins or antibodies. The silver film is prepared on a prism. Under illumination from the prism side, the SPP becomes resonant in a small angular range. This resonance is slightly shifted by an immunoreaction, which can thus be recorded with high precision.
2. A promising method to detect immunoreactions are plasma resonances in small gold spheres. So-called reporter molecules with special Raman spectra are adsorbed on the spheres and protected by a very thin glass film. The spheres are conditioned

with a special antibody . A specific reporter molecule is employed for the specific antibody. Only those spheres that bond to the searched-for antisense stick to the assay and are characterized by the Raman spectrum of the special reporter molecule.

C. Label-free detection of proteins

The so-called label-free detection of proteins would observe directly the known specific vibrational spectra by infrared (IR) or Raman spectroscopy without any immunoreaction. Plasmon resonances at the "hot sides" of noble metal nanoparticles would enhance the IR or Raman spectra of small quantities of proteins adsorbed onto the "hot sides." Such spectroscopy is called surface-enhanced infrared spectroscopy (SEIRS) or surface-enhanced Raman spectroscopy (SERS). Surface-enhanced coherent anti-Stokes Raman scattering (CARS) is discussed as a promising method. The goal of the above-mentioned project Nanoantenna is label-free detection of a small number of proteins. Nevertheless, functionalization of gold surfaces to achieve bonding of proteins is necessary.

The book is structured as follows.

C1. *Chapters centered on "bio-spectroscopy"*

Chapter 1 describes Raman spectroscopy in general and its wide use as a method for probing the structure and conformation of native proteins that play an important role in the body. Important structural information can be deduced from specific Raman vibrational bands such as amide I, amide II, and amide III bands. On the other hand, the influence of chemical reactions involving proteins (folding/unfolding, oxidation, reduction, phosphorylation, and polymerization) can be monitored by following the evolution over time of the Raman band related to disulphide bridge stretching, tryptophan and other aromatic ring vibrations, and protein side chain deformation. It is possible to determine reaction-free enthalpy and free enthalpies for the unfolding of protein structure and reaction kinetics in a biological environment.

Chapter 2 focuses on the analysis of biofluids by mid-infrared and Raman spectroscopy. Biological processes strongly depend on the fact that the relevant biochemistry takes place in water, which has a very strong absorption in the mid-infrared spectral range. The chapter describes strategies to mitigate the influence of water. Biomedical vibrational spectroscopy of serum is capable of supporting medical diagnostics by, for instance, simultaneously determining the concentration of multiple analytes or in the form of a direct attribution of a sample's spectrum to a particular disease under investigation. The need for independent validation is pointed out.

Chapter 3 describes the functionalization of gold surfaces by self-assembling films of various thiols. The functional groups attached to the end opposite to the SH group (forming a strong Au–S bond) control the surface properties of the monolayers formed. The flexibility to design different head groups of monolayers using a large number of functional groups makes this functionalization strategy especially useful for the controlled fabrication of structurally ordered assemblies of proteins on surfaces. Different types of thiols such as mono-, di-, tri-, or mixed thiol molecules are discussed in view of their benefits and disadvantages to immobilize special biomolecules.

C2. *Chapters centered on surface plasmonic resonances*

Chapter 6 discusses the resonant excitation of metallic nano-structures by light. Dubbed nanoantennas, these structures yield electromagnetic near-field enhancement, used for sensing applications such as SEIRS and SERS. The interaction between nanoparticles in multimers and arrays of nanoantennas affects the optical properties of the system. These depend on the separation distances, especially between nearest neighbor particles. The interaction effects studied experimentally and theoretically in dimers—consisting of ellipsoids, spheres, nanodisks, nanorods, or nanoantennas—in linear arrays of nanocylinders or two-dimensional nanoparticle arrays are discussed, especially concerning SEIRS.

Chapter 7 gives special attention to nanosculptured thin films, which are assemblies of shaped, parallel, and tilted nanorods, prepared using many variants of the basic oblique angle deposition technique. Comparison is made between different materials such as silver, gold, copper, and silicon and with various shapes such as columns, screws, spheres, and helices that are deposited on different substrates and with different porosities and orientations. Localized surface plasmon resonances in the vis–NIR range lead to surface-enhanced fluorescence and SEIRS.

Chapter 8 presents an extensive discussion of the excitation of surface plasmon polaritons at plane metal surfaces by attenuated total reflectivity. The self-organized templates of anodic porous alumina allow patterning of noble metal structures with plasmon resonances over large areas. Arrays of square-like gold nanoparticles were fabricated by electron beam lithography. Au nanoaggregates were grown in a hole array. The SERS spectra of four different proteins on these samples were recorded. Large arrays of nanoantennas were achieved with reproducibility at the sub-10 nm scale. The sample topography was characterized by scanning electron microscopy and atomic force microscopy equipped with ultrasharp Si tips. Focussed ion beam milling was employed to produce a photonic crystal (PC) with a central cavity, on which a Pt nano-cone was grown by electron beam–induced chemical vapor deposition, and finally it was covered with gold. The adiabatic propagation of surface plasmon polaritons on this conical Au tip and field enhancement at the tip end was achieved by back illumination of the PC. The structure presented an extremely high electromagnetic enhancement at a very small radius of curvature of the tip end demonstrated by SERS of benzene thiol.

Chapter 9 is a general introduction to surface-enhanced infrared absorption (SEIRA) of adsorbates on metal island films at the percolation threshold and surface-enhanced infrared spectroscopy (SEIRS) using infrared resonances of nanoantennas. The metal island films at the percolation threshold have localized electromagnetic resonances throughout a wide spectral infrared range. Therefore, a complete IR spectrum of the adsorbed molecules may be obtained, including some microscopic surface–molecule "chemical" interactions. In SEIRS the resonances of tailored nanoantennas or antenna dimers and arrays are confined to narrow frequency ranges, overlapping only with some IR-active vibrations, directly adsorbed molecules, or molecules grafted by a functionalization layer on the antennas. Very strong enhancements have been observed in this case at the "hot spots" with atto-molar sensitivity. Also, the phonon-polaritons of the silica substrate confined to the hot spots of the antennas become observable.

Chapter 10 gives a short introduction to the history of SERS up to single molecule detection by SERS, followed by a general discussion of electromagnetic field enhancement and of the so-called first layer chemical mechanism. The chapter presents as example the measurement of the SERS enhancement of *trans*-1,2-bis(4-pyridil)ethylene (BPE) on different gold nanoparticle arrays. Various strategies to obtain SERS biosensors and the results achieved till now are described. The application of SERS substrates produced by e-beam nanolithography to detect proteins is demonstrated, especially their conformation change with temperature.

Chapter 11 revisits the theory and experiments of surface plasmon resonance in Kretschmann configuration and the high sensitivity of the resonance that is shifted in angular distribution by very thin cover films. Only 2 nm of chromium and 5 nm of functionalization significantly shift plasmon resonance. Optimization methods may lead to more complex biosensors.

Chapter 12 covers coherent anti-Stokes Raman scattering. SERS requires the presence of a relatively large metal surface, and the intensity in surface plasmon resonances decays rapidly with distance away from the surface. Both linear (fluorescence, Raman, etc.) and nonlinear (second and third harmonic generation) physical processes can be generated from biological molecules near a metal surface. The final objective is to achieve single molecule detection. In the case of Raman processes, CARS generation is stronger than spontaneous Raman scattering. Since CARS is a four-wave mixing process in which the anti-Stokes signal results from the parametric coupling of an incident pump and a Stokes laser, a stronger local enhancement is expected for CARS than for conventional SERS, which would make surface-enhanced (SE-)CARS more challenging than SERS. However, strong field energies are needed in CARS generation, and CARS is not background free. This makes single molecule detection hard.

C3. *Chapters centered on related subjects*

Miniaturization of gold plasmonic antenna structures finally leads to atomic wires made of Au atoms. Chapter 5 discusses dimensionality and electronic correlation effects in relation to the atomic-scale confinement. Atomic wires are prospective materials for supporting plasmonic resonating modes. The dispersion relation of propagating plasmonic modes was measured by electron energy loss spectroscopy. It is quasi-linear with a wavevector (and was therefore named acoustic plasmon in former times). The first examples of measurements of plasmonic resonators in atomic wires are introduced together with nanoscale structure characterization by electron diffraction. The results clearly demonstrate that the frequency of the plasmon in an atomic wire can be fully tuned by controlling its width, separation, and length in the range of one to tens of nanometers.

The aim of chapter 4 is to summarize the recent advances in the field of nanoparticle surface functionalization, especially of magnetic nanoparticles. The intrinsic interaction of magnetic nanoparticles with applied magnetic field gradients makes these particles attractive for a large panel of biomedical applications such as magnetic separation systems for biomolecules and cells, for magnetic resonance imaging contrast enhancement, for therapy such as hyperthermia and drug delivery, as well as for multidetection systems based on biosensors.

Chapter 13 details multi-signal processing biosensors and bioactuators based on biocomputing. Molecular and biomolecular logic gates and their network-processing chemical input signals received high attention and were rapidly developed in the last decade. Networks with computational steps that solely involve biochemical processes are being researched for new technological capabilities: multi-input biosensors with new functionalities as well as approaches that allow removing the batteries from, and generally reducing the need for, inorganic leads and electrical power supply for those stages of information processing that occur during biomedical testing; implantable devices; and other fast decision-making steps in applications such as patient-tailored timely therapy. Most of the activity in such feedback-loop systems is currently being devoted toward the management of diabetes through integration of an electrochemical glucose-sensing element with an insulin-delivery feedback loop for optimal doses of insulin.

All in all, students and researchers in nanotechnology, physics, chemistry, biology, and medicine, especially those with an interest in biosensing technologies, will certainly benefit from this book.

Dr. Andreas Otto
Professor Emeritus
University of Düsseldorf
Germany

Chapter 1

Raman Spectroscopy Applied to Biomolecule Characterization

Catalina David

CSPBAT Laboratory, University Paris 13, 74 rue Marcel Cachin, F-93017 Bobigny, France

catalina.david@univ-paris13.fr

1.1 Introduction

C. V. Raman and K. S. Krishnan were the first researchers to describe, explain, and report (in 1928) experimental observations of the Raman effect.[1–4] But the first theoretical results that predicted the Raman effect belonged to A. Smekel, in 1923.[5] At the same time other research teams, such as G. Landsberg and L. Mandelstam in the Soviet Union, focused their studies on this topic. From then until today, the Raman spectroscopy technique for detection and diagnosis has developed greatly. The use of this technique in many research fields involved exponential development of experimental Raman spectrometer setups. Nowadays, many physico-chemical analysis and biomedical laboratories are equipped with a Raman spectrometer.

Nanoantenna: Plasmon-Enhanced Spectroscopies for Biotechnological Applications
Edited by Marc Lamy de la Chapelle and Annemarie Pucci

ISBN 978-981-4303-61-3 (Hardcover), 978-981-4303-62-0 (eBook)
www.panstanford.com

1.1.1 Principle of the Raman Effect

In Raman spectroscopy, analysis is done by exciting the material. Increased to a virtual energy level by a strong monochromatic light laser source, the material re-emits radiation, which is then collected and analyzed by an adequate detector. This radiation has two types of signals. The first, and the majority, corresponds to Rayleigh scattering: The incident radiation is transmitted elastically without energy and thus without wavelength change. However, in some cases the photons can interact with matter. This matter absorbs (or transfers) energy to the incident photon, thus producing Stokes (or anti-Stokes) radiation. The observed energy variation of the photon can provide useful information about the energy levels of the rotation and the vibration of the studied molecule. In Raman spectroscopy, we are interested in vibrational and rotational energies that are very low compared to those of incident photons. As shown in Fig. 1.1, the energy of the incident photons does not correspond to a direct jump between two levels of energy. Therefore, to explain the Raman phenomenon we must involve some virtual energy levels. Nevertheless, a molecule can move to a more excited state by absorbing a photon only if the energy of the excited state corresponds to the gap between the current energy level of the molecule and that of another "allowed" level. Actually, this is the quantification system of energy levels.

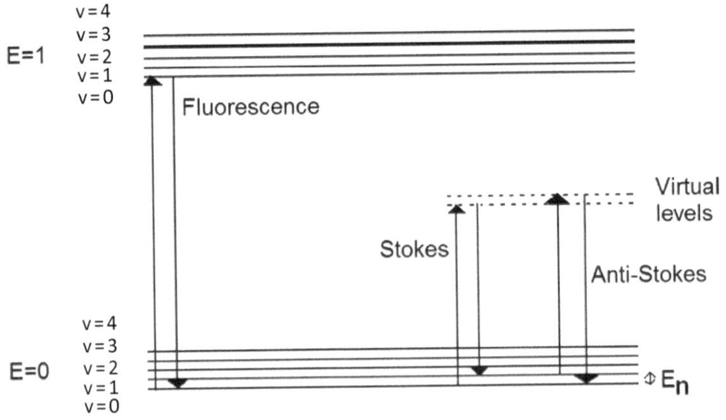

Figure 1.1 Jablonski diagram: evolution of the vibrational energy of an atom.

However, in this case, we observe that the molecule absorbs some incident radiation, just enough to jump to a possible energy level. From the quantum point of view, we can explain this by using the uncertainty relation of Heisenberg, shown in Eq. 1.1:

$$\Delta E * \Delta t \geq \frac{\hbar}{2\pi} \tag{1.1}$$

Because the time during which the molecule acquires the necessary energy is very short, the uncertainty in energy is very large. The molecule can therefore use a small part of the incident energy. The photon, which lost a fraction of its energy, now has the energy of $h\nu - E_n$, where E_n represents a jump of energy between two allowed energy levels (rotational or translational vibration). And it is now characterized by a lower frequency (Stokes lines). The phenomenon leading to the formation of anti-Stokes lines is similar. When the molecule is in an excited state, which is less probable, it may give energy to a photon. Thus the photon energy becomes $h\nu + E_n$. This fact leads to the appearance of spectral lines, which are symmetrical with respect with Stokes lines. However, their amplitude is lower than that of Stokes lines because, according to the Boltzmann distribution law, the population of these excited states is very low.

1.1.2 Raman Applicability

Raman spectroscopy can be applied to different chemical or biological samples to produce a lot of very interesting information:

- Phase transitions and chemical compound identification.
- Material characterization.
- Molecular structure determination.
- Study of amorphous and crystalline systems.
- Direct determination of the progress of chemical reactions in a biological environment.
- Measurement of different reactions kinetics involving proteins in complex environment.
- Determination of physicochemical parameters of great importance in understanding the studied mechanisms: reaction Gibbs free energy, Gibbs free energy of protein unfolding, conformational Gibbs free energy.

- Identification of the conformational changes of biomolecules that accompany their unfolding.

Raman spectroscopy can provide different kinds of information simultaneously. The Raman spectrum of a compound can, for example, indicate both bond type and crystalline structure. Obviously, this technique can quote the most important chemical and biochemical methods. So it is worth it to underline some very important characteristics of Raman spectroscopy:

- It is easy to implement.
- It does not degrade the sample. The method is nondestructive and nonintrusive, and can be applied to real systems.
- It can be used to study a variety of samples (heterogeneous materials, samples with structure that can be oriented or not).
- It can be coupled with other analytical methods and perform *ṅ st u* measurements.
- It can be used on solid samples of very small dimensions.
- It can work in a hostile environment, such as one with very high temperatures, one that is radioactive, or in a controlled atmosphere.
- It is sensitive to small structures, allowing the identification of amorphous systems and analysis of thin films for which diffraction methods are sometimes difficult to apply.

Compared with various other techniques, Raman spectroscopy has many advantages:

- The Raman effect is independent of the excitation wavelength used. Thus it is possible to remove certain undesirable phenomena, such as fluorescence decay or the decomposition of colored substances, by selecting a suitable wavelength.
- No permanent polarization of the molecules is required, making it an efficient method for measuring homonuclear diatomic molecules.
- The presence of water is not a problem because water absorption in the Raman effect is weak. Thus water can be used as a solvent.
- It is possible to use glass cells, because glass is transparent in the relevant spectral regions and its Raman spectrum is very weak.

- Samples can be used in any form, without being diluted or altered.

However, these advantages are offset by the discomfort caused by some phenomena:

- The emission of black body (by heating the sample).
- Fluorescence (which is much more intense than the Raman effect, but can be avoided by changing the wavelength).
- Photochemical reactions (decomposition of colored substances, etc.).
- Multiphoton reactions.
- Decomposition of samples by heating.

A very important aspect that can be studied using Raman spectroscopy is the characterization of biomolecule structures. Indeed, Raman spectroscopy provides information on the chemical composition, molecular structure and molecular interactions between biomolecules in cells and tissues. Therefore, biochemical information relating to proteins, lipids, carbohydrates, and nucleic acids can be obtained, providing information on the health and pathology of the tissue. In addition, polarized Raman spectroscopy can yield information about the alignment and orientation of molecules in tissues.

Proteins represent more than half of the mass of a cell, excluding water. These macromolecules play a crucial role in the functioning of organisms. Knowledge of their native or non-native conformation or of their interactions with the complex biological and chemical environment is important for understanding their role in organisms and more precisely in cells. From its implementation, Raman spectroscopy has been widely used by many research groups to analyze the structures of proteins and their interactions with the environment.[6–30]

Taking Raman measurements on proteins and their assemblies is no easy task. Although Raman spectroscopy is successfully used on organic and inorganic, natural, and synthetic chemical compounds, things are a little more complex in the case of proteins. This is not surprising considering that proteins are large, complex biological systems (10 kDa ÷ 150 kDa). To highlight the structural complexity of proteins in this section, we present a brief overview of their structure.

1.1.3 Protein Structure

Proteins are the most abundant organic components of living cells. They function as structural material in cell walls, cell membranes, and ribosomes. They also function as enzymes, a group of biological molecules that catalyze and regulate most chemical reactions in biological systems. Denaturation of proteins in an organism, caused by heat or chemicals, usually leads to cell death.

Proteins are composed of amino acids that are linked to one another by a peptide (amide) bond (Fig. 1.2). Each free amino acid has a carboxyl group (–COOH) and a free amino group ($-NH_2$) as part of its molecular structure. During protein synthesis, two amino acids can be joined together by a dehydration reaction that combines the carboxyl group of one amino acid to the amino group of another amino acid via a peptide bond (CO–NH).

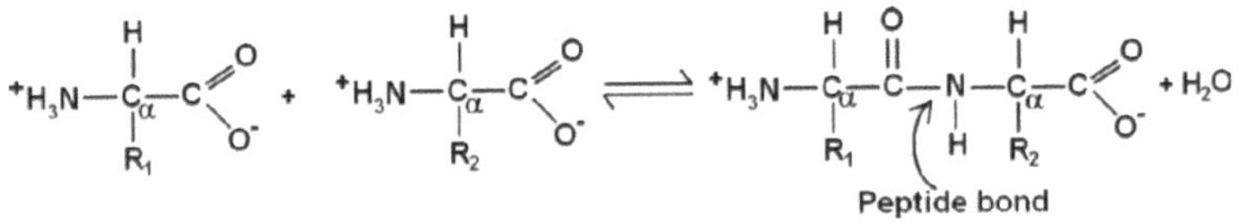

Figure 1.2 Formation of the peptide bond.

There are 20 amino acids used during protein synthesis. They can be neutral and charged residues (positively or negatively), hydrophilic and hydrophobic residues, polar and nonpolar residues, or aromatic and aliphatic residues.

Protein structure is characterized by four levels of organization: primary, secondary, tertiary and quaternary structures, as shown in Fig. 1.3.

The *primary structure* corresponds to the sequence of amino acids in the protein, thus forming the polypeptide chain. Since proteins are polypeptides (many amino acids joined together) consisting of up to hundreds of amino acid molecules, there is unlimited potential in their primary structure.

It is very probable that in this amino acids chain interactions may occur that cause the chain to fold in order to adopt a precise three-dimensional structure. This three-dimensional structure mostly depends on the sequence of amino acids forming the chain and more precisely their order. Actually this aspect is crucial because a little difference in the protein sequence will lead to a completely different protein with a different function. It has been shown that the function of the protein is closely related to its shape (three-dimensional structure).

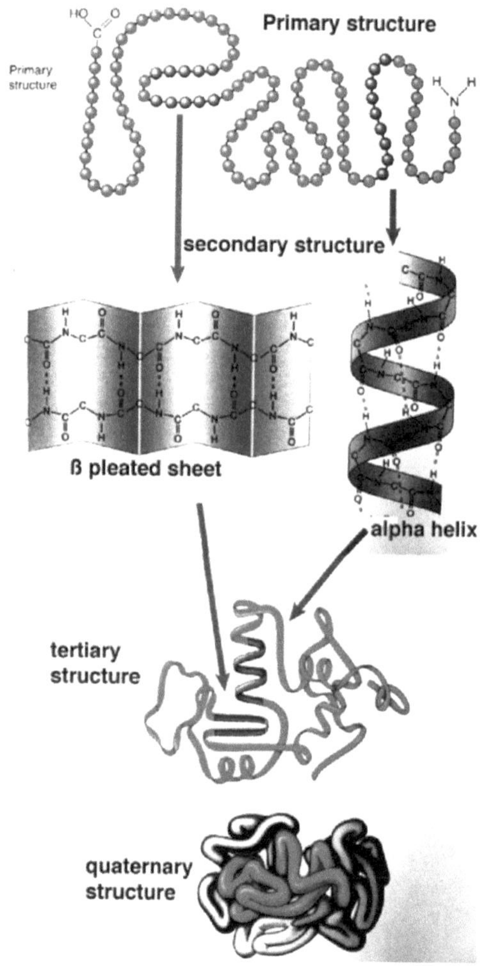

Figure 1.3 Organization level of the protein structure.

The rotation about the two bonds attached to the C_α allows the peptide to fold into certain three-dimensional arrangements. So, proteins can adopt several types of *secondary structures*: α-helix, β-sheet, or random, or all of them.

The α-helix is a spiral arrangement (R groups extending outward) with about 3.6 amino acid residues per turn. This arrangement is stabilized by intra-chain hydrogen bonds (between C=O and N–H groups in the backbone within one chain). Usually the α-helix can be right-handed or left-handed.

The β-sheet is a flat, extended chain arrangement stabilized by the inter-chain hydrogen bonding (between C=O and N–H groups in the backbone of two different chains). Depending on their folding direction, the β-sheets can be parallels or antiparallels.

The random coil is a nonrepeating region with a residual structure in unfolded stats. It has a fixed structure within a given protein and is commonly described as "connecting loop region." Its structure is determined by bonding of side chains (not necessarily hydrogen bonds).

The secondary structures fold and pack together to form the *tertiary structure*. Usually, the tertiary structure has a globular shape and is stabilized by bonds between R groups (i.e., side chains).

Four major types of interactions stabilize the tertiary structure of proteins (Fig. 1.4):

- The hydrogen bonds.
- The covalent disulphide bonds.
- The ionic bonds.
- The hydrophobic effect.

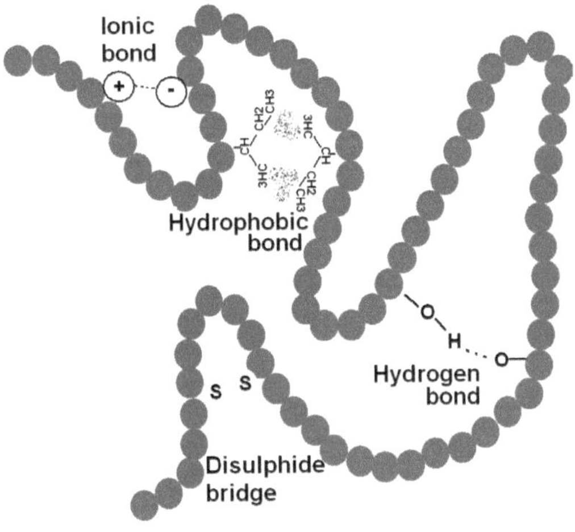

Figure 1.4 Peptide chain folding: types of interactions between amino acids.

The hydrogen bonds formed between chains are weak, thus allowing them to be easily cleaved and reformed. This allows structural change of the proteins that in time will produce functional molecules.

Among the 22 amino acids that serve as building blocks in the construction of proteins, cysteine occupies a special place with its thiol group. Within a protein, the cysteines can be linked together by covalent bonds to form disulfide bridges (S–S). It has been shown that S–S bridges significantly stabilize the tertiary structure of proteins.[31-34] This stabilizing effect has been demonstrated qualitatively by detecting changes in protein conformation induced by the cleavage of S–S bridges, which leads to a global unfolding of the protein.

The tertiary structure can also be stabilized by ionic bonds. Interaction between the positive ammonium group and the negative acid group of the side chain may create salt bridges through these ionic bonds.

The hydrophobic forces within the tertiary structure arise from the close attraction of nonpolar R groups through dispersion forces. The individual interaction is very weak but the collective ones are capable of stabilizing the protein structure on a large area.

The *quaternary structure* of proteins is an arrangement of multiple tertiary structures into a single functional complex. More precisely, is the clustering of at least two polypeptide chains linked, mainly by noncovalent (hydrogen bonds, ionic bonds), but also by S–S bonds.

1.2 Raman Spectrum of Proteins

1.2.1 Introduction

The Raman spectra of most biomolecules have a broadly similar form covering a range of about 3500 cm^{-1} Raman shift from the excitation laser (Fig. 1.5). However, a detailed spectral analysis based on normal mode analysis of all spectral features may reveal important structural and functional characteristics of biomolecules.

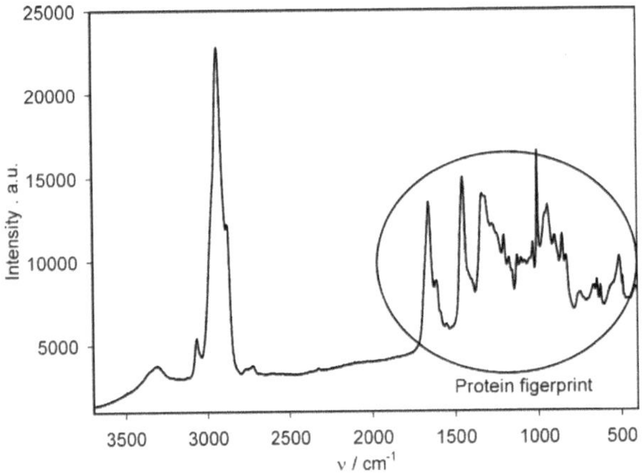

Figure 1.5 Complete Raman spectrum of bovine serum albumin in powder state.

In general, the spectral region between 400 and 1800 cm^{-1} is known as the "fingerprint region" (Fig. 1.6). It contains most of the Raman bands used to identify various structural features of proteins and it seems to be different from one protein to another. This spectral region is very complex and generally cannot be used directly in the identification of biomolecules.

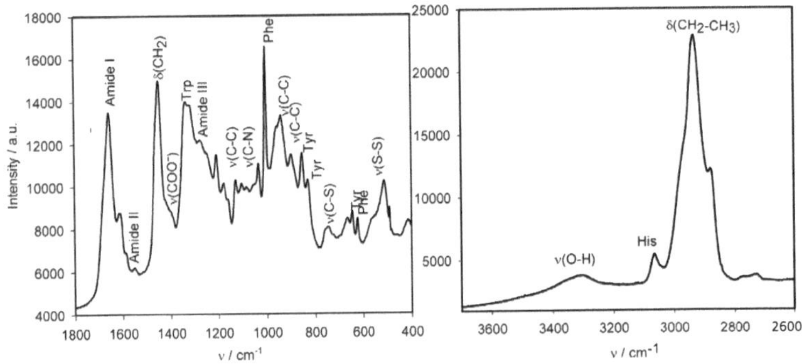

Figure 1.6 The most important Raman vibrational bands of proteins. Two spectral windows are represented: 400–1800 cm^{-1} and 2600–3700 cm^{-1}.

The information obtained from a Raman spectrum is various:

- The band position provides information on the molecular species present in the sample.

- The band width at half height provides information on the structure of the sample.
- The band intensity can be related to the concentration of the species.
- The band shift is a function of several external factors: temperature, pH, hydrogen bonding, and chemical reactions.

1.2.2 Important Raman Bands of Proteins

1.2.2.1 Amides

Nine normal modes are allowed for the amide bands of proteins.[35] These are called A, B, and I–VII in order of decreasing frequency; they have their origin in the vibrations of the different atomic groups of the polypeptide chain (C=O group, N–H group, C–N group). The most important amide bands are the amide I, II, and III bands. Indeed, these vibrational Raman bands are strongly related to protein structure and conformation.

The *amide I band*, located in the spectral range of 1610–1700 cm^{-1} (Fig. 1.6), originated mainly (80%) in the vibrations of carbonyl groups (C=O) of the polypeptide chain.[36-39] Individual vibrations are coupled by a dipole electrostatic interaction,[40-42] which leads to the formation of a broad Raman band called the amide I band. The coupling between the individual C=O vibrations depends on the relative positions in space of the C=O bonds, which, in turn, depend on the secondary structure of the protein.

Therefore, different types of secondary structures (α-helices, β-sheets) are characterized by amide I bands of slightly different position and shape. The exact band position is determined by the backbone conformation and the hydrogen bonding pattern.

The analysis of databases of IR and Raman spectra and X-ray diffraction structures of proteins has established a correlation between different types of secondary structures and the position of the amide I band.[35,43] These results allow us, in principle, to monitor the conformational changes of proteins by detecting and analyzing the Raman amide I band.

The *amide II band* is found in the 1510 to 1580 cm^{-1} (Fig. 1.6) region and it is more complex than amide I.[35] The origin of the amide II band is 40% to 60% in-plane N–H bending and 18% to 40% C–N stretching. Inspection of the amide II band generally provides little help in distinguishing between turn and sheet conformation. It is quite a bit weaker in the case of Raman measurements than in IR measurements.

The *amide III band* has its origin mainly from C–N bending (40%) and from H–N stretching (30%). The structure of an amide III band that has a spectral range of 1200 to 1350 cm^{-1} (Fig. 1.6) can be correlated to the amide I band, giving the complementary structural information on the protein structure and thus making it possible to get additional details about the amide I.[35,44]

1.2.2.2 S–S stretching

Experimental studies performed on model systems show that the frequency of disulfide bridges vibration associated with S–S elongation is in the spectral range 500–545 cm^{-1}.[45–48] For the proteins whose structure contains S–S bridges the S–S Raman bands are also located in the spectral range 500–550 cm^{-1}.

The factors affecting the frequency of vibration are the relative conformation of atoms C_α-C_βS–S′-C'_β-C'_α around C_β-S and C_β-S′ bonds,[49] the mode coupling, and the hydrogen bonds.[50,51]

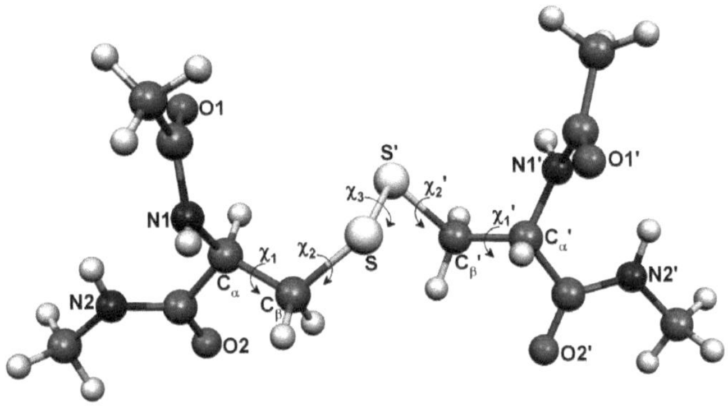

Figure 1.7 The model of an S–S bridge. The conformation of the bridge is characterized by the five dihedral angles.[50]

1.2.2.3 Aromatic amino acids

Some of the vibrational bands of tyrosine (Tyr) or tryptophan (Trp) are sensitive to the microenvironment (Table 1.1).[52,53] Indeed, detailed information about the microenvironment of these aromatic side chains can be obtained from, for example, variations in intensity of the vibrational modes or changes in the intensity ratio of Fermi resonance doublets (835 to 853 cm^{-1}). These bands are very sensitive to the environment—more precisely to the extent of H-bonding of the phenolic hydroxyl.

For instance, the interaction between a protein and a target or another protein is characterized by local changes around the binding site where H-bonds are formed or broken and where dipole-dipole and dipole-polarizability interactions possibly occur. In some instances, it is characterized by additional conformational changes of the protein. If aromatic amino acids are located near a binding site, or if they are exposed to a different environment because of conformational changes in the protein upon binding, it is anticipated that this will be reflected in the Raman spectrum. Hence, these changes have the potential both to allow qualitative detection of drug binding and to enable quantitative study of the nature of the interaction between the protein and its target.

Table 1.1 Important Raman modes of aromatic amino acids within the protein structure[43]

Aromatic residues	Mean frequency (cm^{-1})
Phe	620
Tyr	640
Trp	750
Tyr	830, 850
Phe	1000, 1030
Trp	1011
Tyr, Phe	1170–1200
Trp	1340–1360
Trp	1582
Phe, Trp	1584
Tyr	1590
Phe	1605
Trp, Phe, Tyr	1610–1616
Trp	1618–1621
His	3110–3160

It is very important to properly identify and assign the Raman bands of proteins. Indeed, in most cases the identification of the conformation or structure of proteins by Raman spectroscopy is possible only by studying a specific reaction. This leads us to the well-known controversial history of the assignment of S–S Raman bands of proteins.

As mentioned above, experimental[50,51,54–57] and theoretical[49,51] studies indicate that the Raman bands belonging to S–S stretching are located between 500 and 560 cm^{-1}. Many researcher groups have assigned the differences between the values of these frequencies to different conformations of the dihedral angles -C_α-C_β-S–S-.[45,49] On the other hand, it is known that the rotation about the C–S bond produces *gauche* (g) or *trans* (t) rotamers. Given that there are two C–S bonds, the disulphide bridges conformation may be classified into three types: *gg*, *gt*, and *tt*. Analysis of experimental data indicates that each type is characterized by a specific S–S stretching frequency: 510, 525, and 540 cm^{-1}, respectively.

However, even in the case of model molecules containing disulphide bridges (like cystine[48] or GSSG[50]), the S–S band has a complex structure. This suggests that the environment plays a significant role on the shape and position of these bands.[58,59]

In addition, the Raman spectrum of proteins being rich, it is very likely that in the spectral range characteristic for the S–S bands, there are other vibrational bands. These elements lead to the idea that an analysis of the S–S band that takes into account only the possible conformations of the bridges around the C_β–S bond is too simplistic. It has been shown that there are other factors that influence the structure of the S–S Raman band. This hypothesis was confirmed by

- monitoring changes of the protein S–S bands during reduction of the disulphide bridges
- measuring the depolarization degree of S–S Raman bands
- simulating the molecular Raman spectra of model system of S–S bridges in the spectral range 500–560 cm^{-1}

These three methods based on Raman spectroscopy are complementary and all of them confirm the following:

- The shift to high frequencies of the S–S stretching band is induced by a trans C_α–S conformational.
- Most the bands in this spectral range have a low degree of depolarization even though they are not associated with completely symmetrical vibrational modes.

- The S–S Raman band is complex, since the contributions of the different S–S bridges significantly overlap each other and bands with partial S–S stretching character are also present.
- The S–S bridge can be present even without a dominant S–S stretching character. These bands disappear upon reduction of S–S bridges.

This example shows us clearly that the correct assignment of Raman bands is essential and that the chemical environment plays an important role in the distribution of different vibrational bands.

1.3 Conformational Changes Detected by Raman Spectroscopy

Nowadays, Raman spectroscopy is a common technique in biology and biochemistry applications due to its nondestructive character and because the evolution of conformational changes induced by different factors can be tracked in real time. Therefore, this method allows the study of folding and unfolding processes of proteins. In general, biomolecules and in particular proteins may fold or unfold in the presence of factors such as environmental conditions and/or organic or chemical agents that mediate important reactions in biological organisms.

We now look at examples that highlight the role of Raman spectroscopy in the detection of protein conformational changes due to reactions with external agents and environmental effects.

1.3.1 Environmental Effects

As we have seen, the function of a protein is closely related to its three-dimensional structure. The most important environmental factors that can affect the three-dimensional structure of the protein up to alteration are changes in pH, salt concentration variation, and temperature variation.

1.3.1.1 Changes in pH

Variations in pH induce changes in the ionization of protein side chains, and thus information or breaking of salt bridges. The hydrogen bonds are also disturbed. Usually, this can be reversible, except for the limit values of the pH (too acidic or too basic).[60]

Numerous Raman studies of proteins have emphasized their conformation changes induced by pH variation. For example, Frushour and Koenig[61] studied the Raman spectra of tropomyosin denatured after a pH change. The native structure of this protein is 90% alpha-helix. After the pH was increased up to 9.5, the percentage of alpha-helix type structure decreased. The Raman spectrum of the protein native state showed an amide I band very intense around 1655 cm^{-1}, an amide III band with a relatively low intensity, and a Raman band around 940 cm^{-1} with medium intensity. All three Raman bands correspond to the α-helix contribution. Once the protein was denatured, the Raman spectrum changed. The contribution of the unordered α-helix (~1250 cm^{-1}) to the amide III band became very intense and the intensity of the band located at 940 cm^{-1} decreased. The authors showed that the intensities of these two bands are a sensitive indicator of the amount of alpha-helix and disordered chain.

Another example is the study done by S. W. Ellepola, *et al.*[24] on rice globulin protein, for different pH values. They found a slightly shift of amide I and amide III bands indicating a transition from α-helical structure near neutral pH to antiparallel β-sheets and disordered structures at highly acidic (pH 3) and alkaline (pH 11) conditions. Extreme pH also shifts the C–H vibrations to higher wave numbers, suggesting the formation of random coil structures associated with protein denaturation. Increases in the intensity of amide I, C–H bending, and C–H stretching vibrations at extreme pHs (pH 3 and 11) also indicate protein denaturation.

1.3.1.2 Temperature

As the temperature is increased, a number of bonds in the protein molecule weaken. The first affected are the long-range interactions that are necessary for tertiary structure. As these bonds weaken and break, the protein becomes more flexible and the groups are exposed to solvent. If heating ceases at this stage, the protein should be able to readily refold to its native structure. As heating continues, some of the cooperative hydrogen bonds that stabilize the helical structure will begin to break. As these bonds are broken, water can interact with and form new hydrogen bonds with the amide nitrogen and carbonyl oxygen of the peptide bonds. The presence of water further weakens nearby hydrogen bonds by causing an increase in the effective dielectric constant near them. As the helical structure is broken, hydrophobic groups are exposed to the solvent.

M. C. Chen, *et al.*[62] measured the Raman spectra of lysozyme in aqueous solution in the 32–76°C range. They observed very little quantitative change in the spectra and concluded that any conformational change in the native form of the protein is limited to the side chains. However, if the protein solution is heated to 100°C, denaturation is inevitable. The solution becomes a gel, whose Raman spectrum shows visible changes in amide I and amide III bands with respect with the native form.

1.3.2 Chemical Reactions

The presence of reducing agents in protein solution will cleave the disulphide bridges within the polypeptide chain. Indeed, the native disulfide (S–S) bridges are considered to be a very important key in the folding and structure stabilisation of many proteins. Disulfide bonds significantly limit the configuration space accessible to the protein, thus stabilising its folded conformation.[33] Moreover, during protein folding, intermediate disulfide bridges can have a role in defining the pathway by which a protein reaches its native, active conformation.[63] The reductive breaking of native disulfide bridges is very often used in studying protein folding–unfolding mechanisms. It can be induced by some reducing agents such as tris-2-carboxyethyl phosphine (TCEP)[64] or dithiothreitol (DTT).[50]

Among the experimental techniques that are suitable for monitoring thiol/disulfide exchange reactions (Fig. 1.8), Raman spectroscopy is particularly interesting.

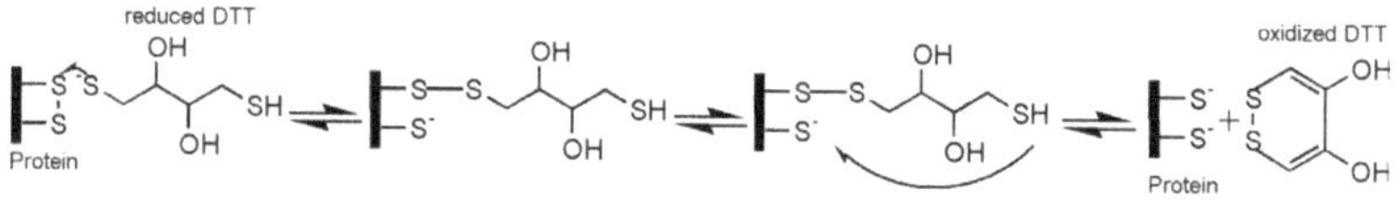

Figure 1.8 Scheme of the thiol/disulphide exchange reaction between protein and DTT.

Proteins containing disulfide bonds usually show complex Raman bands belonging to the S–S stretching mode located between 500 and 545 cm^{-1} that seems to be sensitive to protein conformation

(Fig. 1.9).[54] When combined with *ab initio* calculations, these data contribute to a better understanding of the mechanism of protein reductive unfolding.

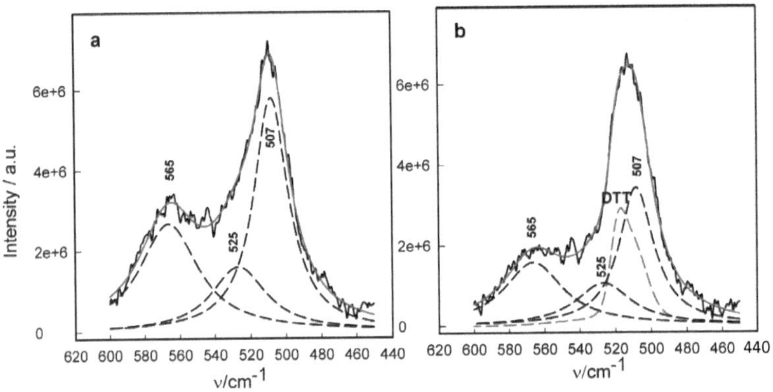

Figure 1.9 Analysis of the S–S Raman band of 1 mM BSA (a) and 1 mM BSA and 20 mM DTT (b), in aqueous solution at pH 7.8. The experimental spectrum is shown by the black solid line, the fitting curve by the red solid line, the protein contributions by the black dashed lines, and the oxidize DTT contribution by the red dashed line.[50] See also Color Insert.

In addition, the protein secondary structure can be characterized by analyzing the protein amide I Raman band.[43] Thus, it is possible to simultaneously follow the reduction of the disulfide bridges (evolution of the S–S Raman bands) and the related protein unfolding, which is revealed by the changes in the protein secondary structure (evolution of the amide I Raman band).

A Raman spectroscopy study of the disulfide bridge exchange reaction between serum albumins and DTT in aqueous solutions is proposed in the literature.[50,51] It has been shown that the gradual reduction in the number of protein S–S bridges due to the addition of DTT results in a spectral evolution by which the protein S–S band decreases while the oxidized S–S band of DTT increases proportionally.

Analysis of the Raman spectrum of the native BSA in the amide I spectral region is shown in Fig. 1.10(a). The first four components (1–4) are ring vibration bands caused by the aromatic residue, while components 5 through 8 effectively belong to the protein amide I band. The peak centered at 1655 cm^{-1} (peak no. 6) is usually attributed to the ordered α-helix, the peak at 1671 cm^{-1} (peak no. 7) to unordered α-helix, and that at 1684 cm^{-1} (peak no. 8) to the vibrations of various turns.[54] The protein amide I band contributions in the 1630 cm^{-1} range are most often assigned to the β-sheet structure. However, in the case of BSA the peak at 1630 cm^{-1} (no. 5) has been attributed to the short-segment chains connecting the α-helical segments.[35] The reason for this assignment is the absence of β-sheet segments in the X-ray diffraction structure of serum albumins.

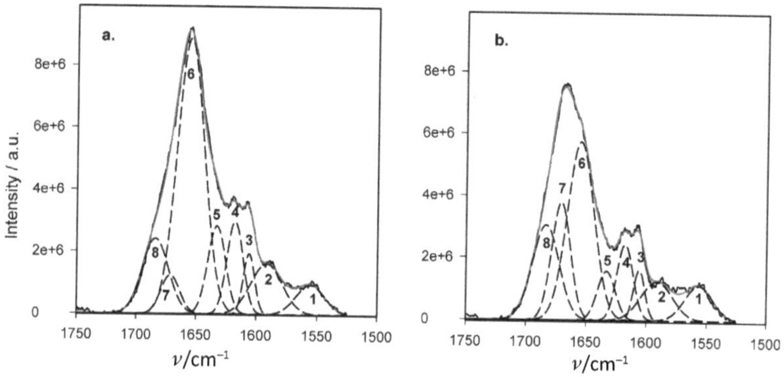

Figure 1.10 Analysis of the amide I Raman band of (a) 1 mM BSA and (b) 1 mM BSA and 30 mM DTT, in aqueous solution at pH 7.8, showing the experimental spectrum (black solid line), the fitting curve (red solid line), and its different contributions (black dashed lines).[50] See also Color Insert.

The addition of DTT to a BSA aqueous solution results in a significant evolution of the protein amide I band (Fig. 1.10b). This evolution consists of a decrease in both peaks 5 and 6 and an increase in peaks 7 and 8. Thus, the changes in the secondary structure may be approximately represented as a decrease in the fraction of the ordered α-helix (peak 6) and an increase in the fraction of unordered α-helix (peak 7).

1.4 Proteins: Spectra Analysis and Contributions to the Baseline

1.4.1 Spectra Analysis and Contributions to the Baseline

Once the protein spectrum is acquired, a first qualitative analysis is possible. This analysis is based on direct observation of different spectral features related to protein conformation and structure. Usually, proteins are studied in solvent. The presence of the solvent is even more important when we want to identify its native, active conformation.

However, there are many results concerning the vibrational study of biomolecules in a solid state.[43] The Raman spectra of molecules in a solid state are quite different from those of molecules in a solvent. First, the intensity of Raman bands in a solid state is more important and, second, the bands are narrower and better resolved. In the case of solutions, the vibrational bands are wider and usually shifted toward higher frequencies. This shift can be assigned to the interaction of the molecule with the solvent and more specifically to a change of coupling of the molecules modes with those of the solvent. Indeed, in aqueous solution the molecule is more flexible, making the modes coupling most efficient. This coupling could also explain the bands enlargement. In the following section, the most common contributions to the protein Raman spectrum are explained.

1.4.1.1 The solvent contribution

The reactions involving the proteins occur mostly in solution. The solvent could be a buffer solution, which maintains the proteins' physiological conditions, or even water. There are also various buffer solutions containing different compounds. Like proteins, these compounds have their own Raman spectrum, composed of one or more bands. Hence, when the spectrum of the mixture is measured, band overlap may occur, which affects the Raman intensity and complicates the quantitative analysis. To overcome the problem, different techniques (fit, simulation of the Raman bands) are used to determine the contributions of the individual bands in the whole band envelope. Another way of tackling the problem is to measure separately the Raman spectrum of all compounds from the mixture as a reference and then subtract it. The situation becomes complex if the Raman bands of the buffer solution change the position, the form, and even relative intensity in the presence of protein.

The most convenient solvent in terms of spectral contribution is water. While water strongly absorbs light at mid-infrared wavelengths, it is a weak Raman scatterer, which means it creates little or no interference in the Raman spectra. Two important vibrational bands are observed on its Raman spectrum: a broad O–H bending band located at 1643 cm^{-1} and the intense O–H stretching band located at ~3400 cm^{-1}.

1.4.1.2 The fluorescence contribution

Another factor that can strongly affect Raman intensity, and thus the protein spectrum, is the fluorescence. Fluorescence from the sample or impurities can cause significant background noise. This contribution may be more or less attenuated. If the fluorescence is an intrinsic property of the molecule, it is possible to attenuate it through the instrumental setting by optimizing the Raman signal. Using the right laser excitation wavelength (usually the laser lines close to the infrared region) will greatly reduce the risk of the appearance of the fluorescence broad band in the spectra of proteins. But if the sample fluorescence is due only to impurities, a biochemical purification of the sample is required. Another alternative is to quench the fluorescence by exposing the sample to the laser beam for a longer time, while the impurities are decomposed. A compromise between the laser power and the spectra acquisition time is necessary, since the longer exposition can affect also the protein. Actually, every time we perform a Raman measurement on biomolecules we must take into account the exposure time and the subsequent thermal degradation of molecules within the sample solution.

1.4.1.3 The signal-to-noise ratio

A quantitative spectral analysis based on Raman spectroscopy requires a Raman spectrum with a minimal signal-to-noise ratio. Such spectra can be obtained after lengthy acquisitions in optimal conditions. Signal averaging is another way to improve signal-to-noise ratio. Thus, n number of spectra, successively measured, under completely identical conditions will be added. The signal-to-noise ratio will be improved by a factor of $\sqrt{n}$. This method can be successfully applied when the measurements are reproducible; hence the sample must be in chemical equilibrium.

The most common limitation of current Raman-based method concerns solvent (even water) and background subtraction. In this case, the subtraction and fit of the background of the protein Raman spectrum were conducted sequentially, and the signal-to-noise ratio of the sample spectrum decreased for each subtraction step. The subtraction of the background containing all inconvenient spectral contributions mentioned above is known as spectrum stripping.

The following example of spectrum stripping looks at the case of protein S–S Raman stretching. The analysis, based on Raman spectroscopy measurements, focuses on the role of the protein structure in the reactivity of the protein S–S bridges; the experimental kinetics of S–S bridge reduction is obtained by monitoring the protein S–S Raman bands. Any quantitative analysis of the S–S bridge reduction based on S–S stretching Raman bands detection requires a correct identification of these bands within the protein Raman spectrum.

First, we examine the spectral analysis of the four lysozyme S–S bridges in reaction with TCEP (tris-2-carboxyethyl phosphine) performed by David, *et al.*[51]

In the spectral domain around the S–S Raman band (450–600 cm^{-1}), the experimental spectrum includes three contributions: the protein S–S Raman bands, the solvent scattering contribution, and a relatively weak contribution from the TCEP Raman spectrum. The background spectrum was obtained by measuring a TCEP aqueous solution of suitable concentration. Then it was multiplied by the appropriate factor and subtracted point by point from the spectrum of the mixture (lysozyme and TCEP in aqueous solution). It is interesting to underline that the TCEP structure does not contain any S–S bridges.

The protein Raman spectrum that results after subtraction of this background is complex and can be fitted using a sum of six Lorentzian functions shown in Eq. 1.2:

$$S(\nu) = a_1L_1(\nu) + a_2L_2(\nu) + \cdots + a_6L_6(\nu) \tag{1.2}$$

Where $L_1 \ldots L_6$ are Lorentzian functions related to the protein Raman spectrum and $a_1 \ldots a_6$ are constants.

The widths and positions of the Lorentzian functions used in the model function were first determined by fitting the Raman spectrum of the native lysozyme (Fig. 1.11), and were then kept constant in the analysis of the reduced samples measured at different reaction times. The only adjustable parameters used to fit the Raman bands were the amplitudes a_1–a_6.

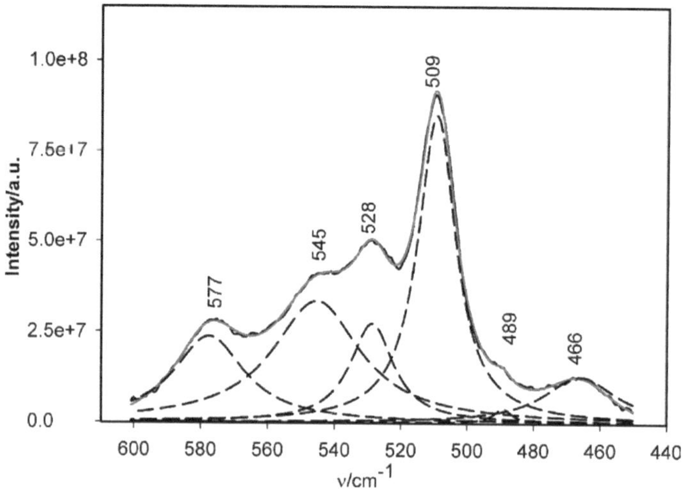

Figure 1.11 Analysis of the lysozyme Raman spectrum in the 450–600 cm^{-1} spectral range using Lorentzian functions. Experimental spectrum in black and simulated spectrum in red (band decomposition in dashed). See also Color Insert.

By analyzing the reaction kinetics of the S–S bridges reduction and at the same time performing QM and QM/MM calculations to simulate the vibrational S–S Raman bands, the researchers concluded that there are three bands belonging to the spectral interval characteristic of the S–S stretching frequency, whose maxima are located at 509 cm^{-1}, 528 cm^{-1}, and 545 cm^{-1}.

In the case of lysozyme, the signal of the fluorescence is very weak, especially in the S–S stretching range, and doesn't affect the shape or the intensity of the S–S bands. But let's see how it is possible to determine a spectrum baseline in the case of proteins that exhibit a strong fluorescence scattering.

One of these proteins is the bovine serum albumin, which has a disulphide bridge.[50] In the spectral range of the S–S Raman band, the experimental spectrum includes four contributions: the S–S bands of the protein (see Fig. 1.10(a)), the S–S band of the oxidized DTT, the solvent scattering contribution, and a fluorescence background. As the fluorescence seemed to quench when the sample was exposed to the laser, spectra were collected within few seconds of laser exposure, and after several minutes of quenching. Because of the significantly different fluorescence background contributions, it was possible to estimate the shape of the fluorescence band by subtracting sets of spectra measured at later times from corresponding spectra measured at earlier times. In general, the shape of the fluorescence follows the form of a polynomial function.

The solvent Raman spectrum was measured directly. The S–S Raman band of oxidized DTT recorded in a separate experiment is complex and was fitted using the sum of three Lorentzian functions. Their position and width were determined by measuring the Raman spectrum of an aqueous solution containing only the native protein. Finally, the model function for fitting the measured spectra in the region of the S–S band is shown in Eq. 1.3:

$$S(\nu) = a_1 L_1(\nu) + a_2 L_2(\nu) + a_3 L_3(\nu) + a_4 f_{\mathrm{DTT}}(\nu) + a_5 w(\nu) + a_6 Fl(\nu) \tag{1.3}$$

Here $L_1(\nu)$, $L_2(\nu)$, $L_3(\nu)$, and $L_4(\nu)$ are Gaussian functions related to the protein Raman spectrum, $f_{\mathrm{DTT}}(\nu)$ is the oxidized DTT S–S band shape, $Fl(\nu)$ is the shape of the fluorescence contribution, and $w(\nu)$ is the solvent Raman spectrum. The S–S band of oxidized DTT is asymmetrical and was previously measured on a pure oxidized DTT solution and fitted using three Gaussian functions. The resulting shape, $f_{\mathrm{DTT}}(\nu)$, was then frozen by fixing the Gaussian functions and their relative amplitudes. The $L_1(\nu)$ to $L_3(\nu)$ Gaussian functions were also frozen. The variable parameters used to fit the Raman bands were the amplitudes a_1–a_6.

1.4.2 Polarization Measurements

When an external electric field, E, will be applied to a molecule, it will induce in the molecule a dipole moment, P, which is proportional to the field, as shown in Eq. 1.4,

$$P = \alpha E \tag{1.4}$$

where α is the polarizability of the molecule. The polarizability measures the easiness with which the electron cloud around a molecule can be changed. From this point of view, Raman scattering occurs because a molecular vibration can change the polarizability. Thus, if a vibration does not change the polarizability, the polarizability derivative will be close to zero, and the intensity of the Raman band will be low.

Raman scatter is partially polarized, especially for molecules in a liquid environment, where they are randomly oriented.

Generally, positioning the sample with respect to the laser beam can greatly influence the intensity of Raman bands. When a polarized laser beam is used, other effects may also arise.

To analyze the polarization properties of Raman scattered radiation, we consider an experimental configuration in which the incident radiation propagates along the *z*-axis. The direction of observation of the scattered light is the *y* direction (perpendicular observation). *x* is the direction of polarization of the electric vector of the beam incident. $I_{||}$ is the intensity of scattered radiation with its electric vector polarized along the *x*-axis while $I_{\perp}$ is the radiation whose polarization is directed along the *z*-axis (see Fig. 1.12). A scattering molecule is located on the system origin.

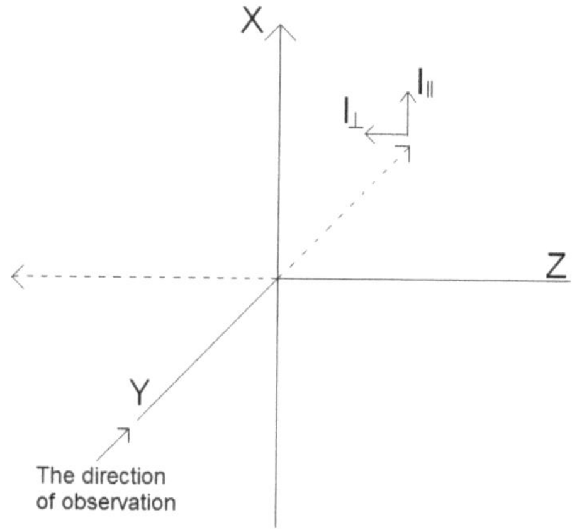

Figure 1.12 The scheme for the experimental conditions to define direction of polarization of Raman scattering.

The relationship between the two intensities ($I_{\perp}$) and ($I_{||}$) is the depolarization factor shown in Eq. 1.5:

$$\rho = \frac{I_{\perp}}{I_{II}} \tag{1.5}$$

The completely symmetrical vibrational modes have a depolarization degree smaller than or equal to ¾. Hence, a polarization measurement is a practical way to identify the totally symmetric modes in the Raman spectra of molecules possessing a certain degree of symmetry.

Furthermore, a mode that would contain a widely dominant contribution to a specific stretching vibration would be, in first approximation, a completely symmetrical mode and therefore easy to identify by a measurement of depolarization degree.

But in the case of molecules without symmetry or a very low symmetry, the things are not so easy. Actually, we can discuss the polarization properties of Raman radiation by considering the local symmetry of the molecule and not the global one. In this case, some modes that are not completely symmetric but have a small depolarization degree. This fact is evident if we compare directly the two Raman spectra of lysozyme, one measured in parallel polarization, the other in perpendicular polarization (with respect to the polarization of the incident beam) (see Fig. 1.13).

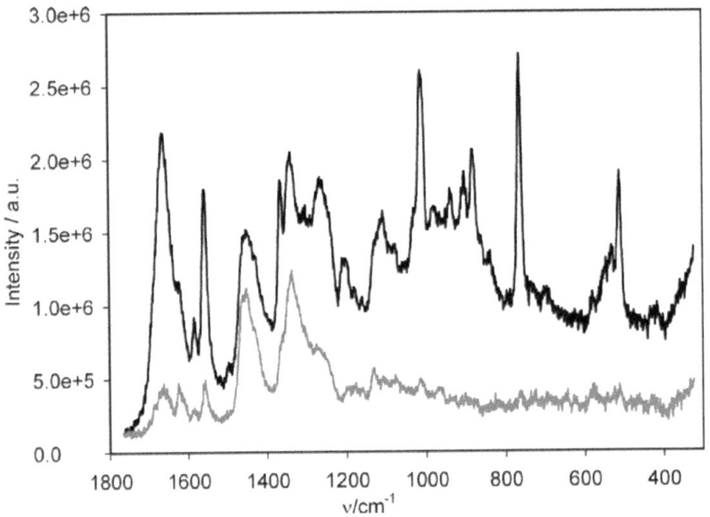

Figure 1.13 Raman spectrum of lysozyme powder measured in parallel (black) and perpendicular (grey) polarization with respect to the monochromator slit.

Apparently, most Raman bands of this spectrum have a very low degree of depolarization. Of course, they do not all belong to totally symmetric modes, but they may belong to modes that are Raman activated by a single coupling, σ–σ^*. This coupling is most likely produced by a bond elongation.

Widely used in Raman spectroscopy method, polarization measurements and, more precisely, the depolarization ratio can facilitate the assignment of unknown bands to molecular vibrations.

1.5 Conclusion

Since its discovery, Raman spectroscopy has proven especially useful in providing important information about biological samples. Its applicability in biology or, more interestingly, in the biomedical field, often provides new information, complementary to other vibrational methods.

At the beginning, Raman spectroscopy was used only for qualitative analysis, because interference with the solvent and the fluorescence background from water, buffers, and other contributions was difficult to eliminate. Now, due to improvements in the measurement systems, quantitative analysis enjoys great success. New types of Raman spectroscopy have been developed, including near-infrared (NIR) and ultraviolet (UV) Raman spectroscopy, which employ wavelengths well away from those of fluorescence.

However, the study of molecules (especially proteins in solution or in solid phase) is no easy task. Their Raman spectra are complex and often difficult to interpret. Some of the difficulties can also be overcome by improvement of Raman spectrometers and of the measurements, by thorough preparation of the sample before measurement, and, not least, through the use of a reliable treatment protocol in order to take into account as many parasite spectral contributions as possible.

Raman spectroscopy has been and still is widely used as a method for probing the structure and conformation of native proteins with an important role in the body. Important structural information can be deduced from specific Raman vibrational bands such as amide I, amide II, and amide III bands. On the other hand, the influence of chemical reaction mechanisms involving proteins (folding and unfolding, oxidation, reduction, phosphorylation, and polymerization) can be monitored by following the evolution over the time of Raman bands related to disulphide bridge stretching, tryptophan and other aromatic ring vibrations, and protein side chain deformation.

The experimental results obtained are usually submitted to complex processing, which may yield information about physical and chemical parameters important to the understanding of the studied mechanisms. Thus, it is possible to determine the reaction free enthalpy and free enthalpies for the unfolding of the protein structure. The study of the reaction kinetics in a biological environment has allowed researchers to determine the rate constants of reaction accurately.

When coupling Raman experimental results with theoretical calculations, we can obtain outstanding resulting feedback.

References

1. Raman, C. V., and Krishnan, K. S. (1928). The negative absorption of radiation, *Nature*, **122**(3062), 12–13.
2. Raman, C. V., and Krishnan K. S. (1928). Polarization of scattered light-quanta, *Nature*, **122**(3066), 169.
3. Raman, C. V., and Krishnan, K. S. (1928). Molecular spectra in the extreme infra-red, *Nature*, **122**(3069), 278.
4. Raman, C. V., and Krishnan, K. S. (1928). A new radiation, *Indü n J. Phys.*, **2**(4), 387–398.
5. Smekal, A. (1923). On the quantum theory of dispersion, *Dė Naturwiṡ senschaften*, **11**, 873–875.
6. Chen, M. C., and Lord, R. C. (1976). Laser-excited Raman spectroscopy of biomolecules. VIII. Conformational study of bovine serum albumin, *J. Am. Chem. Sci.*, **98**(4), 990–992.
7. Carew, E. B., Leavis, P. C., Stanley, H. E., and Gergely, J. (1980). A laser Raman spectroscopic study of Ca^{2+} binding to troponin C, *Bò phys. J.*, **30**, 351–358.
8. Pezolet, M., Pigeon-Gosselin, M., and Nadeau J. (1980). Laser Raman scattering: a molecular probe of the contractile state of intact single muscle fibers, *Bò phys. J.*, **31**, 1–8.
9. Ishikari, H., Balaram, P., Nagaraj, R., Venkatachalapathi, Y. V., and Tu, A. T. (1981). Determination of β-turn conformation by laser Raman spectroscopy, *Bò phys. J.*, **36**, 509–517.
10. Pezolet, M., Pigeon-Gosselin, M., Menard, D., and Caille, J-P. (1988). Raman spectroscopy of cytoplasmic muscle fiber proteins: orientational order, *Bò phys. J.*, **53**, 319–325.
11. Chen, W., Nie, S., Kuck, J. F. R., and Yu, N-T. (1991). Near-infrared Fourier transform Raman and conventional Raman studies of calf γ-crystallins in the lyophilized state and in solution, *Bò phys. J.*, **60**, 447–455.
12. Li, H., and Thomas, G. J., Jr. (1991). Cysteine conformation and sulfhydryl interactions in proteins and viruses. 1. Correlation of

the Raman S–H band with hydrogen bonding and intramolecular geometry in model compounds, *J. Am. Chem. Soc.*, **113**(2), 456–462.

13. Li, H., Wurrey, C. J., and Thomas, G. J., Jr. (1992). Cysteine conformation and sulfhydryl interactions in proteins and viruses. 2. Normal coordinate analysis of the cysteine side chain in model compounds, *J. Am. Chem. Soc.*, **114**, 7463–7469.

14. Tuma, R., Vohnik, S., Li, H., and Thomas, G. J., Jr. (1993). Cysteine conformation and sulfhydryl interactions in proteins and viruses. 3. Quantitative measurement of the Raman S–H band intensity and frequency, *Biophys. J.*, **65**, 1066–1072.

15. Weizhong, K., and Jianzhong, Y. (1994). Influence of the nitrobenzene extraction technique on the protein molecule conformation, *J. Appl. Spectrosc.*, **48**(2), 209–213.

16. Duindam J., Vrensen, G., Otto, C., and Greve, J. (1998). Cholesterol, phospholipid, and protein changes in focal opacities in the human eye lens, *Invest. Ophthalmol. Vis. Sci.*, **39**(1), 94–103.

17. Overman, S. A., and Thomas, G. J., Jr. (1998). Amide modes of the α-helix: Raman spectroscopy of filamentous virus fd containing peptide 13C and 2H labels in coat protein subunits, *Biochemistry*, **37**, 5654–5665.

18. Thomas, G. J., Jr. (1999). Raman spectroscopy of protein and nucleic acid assemblies, *Annu. Rev. Biophys. Biomol. Struc.*, **28**, 1–27.

19. Quinn, E. A., Forbes, R. T., Williams, A. C., Oliver, M. J., McKenzie, L., and Purewal, T. S. (1999). Protein conformational stability in the hydrofluoroalkane propellants tetrafluoroethane and heptafluoropropane analysed by Fourier transform Raman spectroscopy, *Int. J. Pharm.*, **186**, 31–41.

20. Yamamoto, K., Mizutani, Y., and Kitagawa T. (2000). Nanosecond temperature jump and time-resolved Raman study of thermal unfolding of ribonuclease A, *Biophys. J.*, **79**, 485–495.

21. Raso, S. W., Clark, P. L., Haase-Pettingell, C., King, J., and Thomas, G. J., Jr. (2001). Distinct cysteine sulfhydryl environments detected by analysis of Raman S–H markers of Cys-Ser mutant proteins, *J. Mol. Biol.*, **307**, 899–911.

22. Pande, J., Hanlon, E., and Pande, A. (2002). A comparison of the environment of thiol groups in bovine and human γ-crystallins using Raman spectroscopy, *Exp. Eye Res.*, **75**, 359–363.

23. Ngarize, S., Adams, A., and Howell, N. K. (2004). Studies on egg albumen and whey protein interactions by FT-Raman spectroscopy and rheology, *Food Hydrocolloid.*, **18**, 49–59.

24. Ellepola, S. W., Choi, S.-M., Phillips, D. L., and Ma, C.-Y. (2006). Raman spectroscopic study of rice globulin, *J. Cereal Sci.*, **43**, 85–93.
25. Thawornchinsombut, S., Park, J. W., Meng, G., and Li-Chan, E. C. Y. (2006). Raman spectroscopy determines structural changes associated with gelation properties of fish proteins recovered at alkaline pH, *J. Agric. Food Chem.*, **54**(6), 2178–2187.
26. Towler, M. R., Wren, A., Rushe, N., Saunders, J., Cummins, N. M., and Jakeman, P. M. (2007). Raman spectroscopy of the human nail: a potential tool for evaluating bone health? *J. Mater. Sci.: Mater. Med.*, **18**, 759–763.
27. Herrero, A. M. (2008). Raman spectroscopy for monitoring protein structure in muscle food systems, *Crit. Rev. Food Sci. Nutr.*, **48**(6), 512–523.
28. Zhang, D., Jiang, D., Yanney, M., Zou, S., and Sygula, A. (2009). Ratiometric Raman spectroscopy for quantification of protein oxidative damage, *Anal. Biochem.*, **391**(2), 121–126.
29. Ashton, L., Dusting, J., Imomoh, E., Balabani, S., and Blanch, E. W. (2010). Susceptibility of different proteins to flow-induced conformational changes monitored with Raman spectroscopy, *Biophys. J.*, **98**(4), 707–714.
30. Church, J. S., Poole, A. J., and Woodhead, A. L. (2010). The Raman analysis of films cast from dissolved feather keratin, *Vib. Spectrosc.*, **53**(1), 107–111.
31. Goldenberg, D. P., Bekeart, L. S., Laheru, D. A., and Zhou, J. D. (1993). Probing the determinants of disulfide stability in native pancreatic trypsin inhibitor, *Biochemistry*, **32**, 2835–2844.
32. Karlsson, M., Martensson, L. G., Karlsson, C., and Carlsson, U. (2005). Denaturant-assisted formation of a stabilizing disulfide bridge from engineered cysteines in nonideal conformations, *Biochemistry*, **44**, 3487–3493.
33. Wedemeyer, W. J., Welker, E., Narayan, M., and Scheraga, H. A. (2000). Disulfide bonds and protein folding, *Biochemistry*, **39**(15), 4207–4216.
34. Vaz, D. C., Rodrigues, J. R., Sebald, W., Dobson, C. M., and Brito, R. M. M. (2006). Getting specificity from simplicity in putative proteins from the prebiotic Earth, *Protein Sci.*, **15**, 33–44.
35. Schweitzer-Stenner, R. (2006). Advances in vibrational spectroscopy as a sensitive probe of peptide and protein structure: a critical review, *Vib. Spectrosc.*, **42**(1), 98–117.

36. Cho, M. (2003). Local amide I mode frequencies and coupling constants in polypeptides, *J. Phys. Chem. B*, **107**, 9132–9138.
37. Reisdorf, W. C., and Krimm, S. (1996). Infrared amide I band of the coiled coil, *Biochemistry*, **35**, 1383–1386.
38. Overman, S. A., and Thomas, G. J., Jr. (1998). Amide modes of α-helix: Raman spectroscopy of filamentous virus fd containing peptide 13C and 2H labels in coat protein subunits, *Biochemistry*, **37**, 5654–5665.
39. Myshakina, N. S., and Asher, S. A. (2007). Peptide bond vibrational coupling, *J. Phys. Chem. B*, **111**, 4271–4279.
40. Moran, A., and Mukamel, S. (2004). The origin of vibrational mode couplings in various secondary structural motifs of polypeptides, *Proc. Natl. Acad. Sci.*, **101**(2), 506–510.
41. Choi, J.-H., Ham, S., and Cho, M. (2003). Local amide I mode frequencies and coupling constants in polypeptides, *J. Phys. Chem. B*, **107**, 9132–9138.
42. Krimm, S., and Abe Y. (1972). Intermolecular interaction effects in the amide I vibrations of polypeptides, *Proc. Nat. Acad. Sci.*, **69**(10), 2788–2792.
43. Sane, S. U., Cramer, S. M., and Przybycien, T. M. (1999). A holistic approach to protein secondary structure characterization using amide I band Raman spectroscopy, *Anal. Biochem.*, **269**, 255–272.
44. Fu, F-N., DeOliveira, D. B., Trumble, W. R., Sarkar, H. K., and Singh, B. R. (1994). Amide III region of Fourier transform infrared spectroscopy: application to analyze calcium-binding-induced structural changes in calsequestrin, *Appl. Spectrosc.*, **48**(11), 1432–1441.
45. Van Wart, H. E., and Scheraga, H. A. (1977). Stable conformations of aliphatic disulfides: influence of 1,4 interactions involving sulfur atoms, *Proc. Natl. Acad. Sci.*, **74**, 13–17.
46. Sugeta, H., Go, A., and Miyazawa, T. (1973). Vibrational spectra and molecular conformations of dialkyl disulfides, *B. Chem. Soc. Jpn.*, **46**, 3407–3411.
47. Nogami, N., Sugeta, H., and Miyazawa, T. (1975). C-S stretching vibrations and molecular conformations of isobutyl methyl sulfide and related alkyl sulfides, *B. Chem. Soc. Jpn.*, **48**, 2417–2420.
48. Van Wart, H. E., and Scheraga, H. A. (1986). Agreement with the disulfide stretching frequency-conformation correlation of Sugeta, Go, and Miyazawa, *Proc. Natl. Acad. Sci.*, **83**, 3064–3067.
49. Qian, W., and Krimm, S. (1992). Vibrational studies of the disulfide group in proteins. V. Correlation of SS stretch frequencies with the

CCSS dihedral angle in known protein disulfide bridges, *Bio polymers*, **32**, 321–326.

50. David, C., Foley, S., Mavon, C., and Enescu, M. (2008). Reductive unfolding of serum albumins uncovered by Raman spectroscopy, *Bio polymers*, **89**, 623–634.

51. David, C., Foley, S., and Enescu, M. (2009). Protein S–S bridge reduction: a Raman and computational study of lysozyme interaction with TCEP, *Phys. Chem. Chem. Phys.*, **11**, 1–11.

52. Van Dael, H., Lafaut, J. P., and Van Cauwelaert, F. (1987). Tyrosine group behaviour in bovine α-lactalbumin as revealed by its Raman effect, *Eur. Bio phys. J.*, **14**, 409–414.

53. Yu, N. T., and East, E. J. (1975). Laser Raman spectroscopic studies of ocular lens and its isolated protein fractions, *J. Bio l. Chem.*, **250**(6), 2196–2202.

54. Nakamura, K., Era, S., Ozaki, Y., Sogami, M., Hayashi, T., and Murakami, M. (1997). Conformational changes in seventeen cystine disulfide bridges of bovine serum albumin proved by Raman spectroscopy, *FEBs Lett.*, **417**(3), 375–378.

55. Van Wart, H. E., Lewis, A., Scheraga, H. A., and Saeva, F. D. (1973). Disulfide bond dihedral angles from Raman spectroscopy, *PNAS*, **70**, 2619–2623.

56. East, E. J., Chang, R. C. C., Yu, N. T., and Kuck, J. F. R., Jr. (1978). Raman spectroscopic measurement of total sulfhydryl in intact lens as affected by aging and ultraviolet irradiation. Deuterium exchange as a probe for accessible sulfhydryl in living tissue, *J. Bio l. Chem.*, **253**, 1436–1441.

57. Yamamoto, K., Mizutani. Y., and Kitawaga, T. (2000). Nanosecond temperature jump and time-resolved Raman study of thermal unfolding of ribonuclease a, *Bio phys. J.*, **79**, 485–495.

58. Katz, B. A., and Kossiakoff, A. (1986). The crystallographically determined structures of atypical strained disulfides engineered into subtilisin, *J. Bio l. Chem.*, **261**, 15480–15485.

59. Kuwajima, K., Ikeguchi, M., Sugawara, T., Hiraoka, Y., and Sugai, S. (1990). Kinetics of disulphide bond reduction in α-lactalbumin by dithiothreitol and molecular basis of superreactivity of the Cys6-Cys120 disulphide bond, *Bio chemis try*, **29**, 8240–8249.

60. Hseu, T. H., and Chang, H. (1980). Laser Raman studies on the conformation of Pro-Leu-Gly-NH_2, *Bio chim . Bio phys. Acta Protein Struct.*, **624**(2), 340–345.

61. Frushour, B. G., and Koenig, J. L. (1974). Raman spectroscopic study of tropomyosin denaturation, *Biopolymers*, **13**, 1809–1819.

62. Chen, M. C., Lord, R. C., and Mendelsohn, R. (1973). Laser-excited Raman spectroscopy of biomolecules: IV. Thermal denaturation of aqueous lysozyme, *Biochim. Biophys. Acta Protein Struct.*, **328**(2), 252–260.

63. Creighton, T. E. (1997). Protein folding coupled to disulphide bond formation, *Biol. Chem.*, **378**, 731–744.

64. Cline, D. J., Redding, S. E., Brohawn, S. G., Psathas, J. N., Schneider, J. P., and Thorpe C. (2004). New water-soluble phosphines as reductants of peptide and protein disulfide bonds: reactivity and membrane permeability, *Biochemistry*, **43**, 15195–15203.

Chapter 2

Biomedical Vibrational Spectroscopy

Wolfgang Petrich

Kr chhoff Instt ute for Physc s, Unv erst y of Hed elb rg, INF1 1 7, 6911 0 Hed elb rg, Germany

wolfgang.petrich@roche.com

The fundamental vibrations of the molecules that are most relevant to biology or medicine may be investigated using electromagnetic radiation in the mid-infrared spectral range. However, because water constitutes an essential ingredient of biomedical samples and exhibits a strong absorption in the mid-infrared spectral region, means had to be found to mitigate or avoid the pronounced spectral contribution of water. The use of small sample thicknesses, the application of intense light sources, the well-defined drying of the sample, and alternative techniques of vibrational spectroscopy (such as Raman spectroscopy) have all been tried. Today, biomedical vibrational spectroscopy is capable of supporting medical diagnostics by, for example, simultaneously determining the concentration of multiple analytes or by direct attribution of a sample's spectrum to a particular disease under investigation. Since the spectroscopic tools as well as the algorithms for analysis are readily available, this chapter concludes with a note of caution

Nanoantenna: Plasmon-Enhanced Spectroscopies for Biotechnological Applications
Edited by Marc Lamy de la Chapelle and Annemarie Pucci

ISBN 978-981-4303-61-3 (Hardcover), 978-981-4303-62-0 (eBook)
www.panstanford.com

on evaluation processes and stresses the role of independent validation.

2.1 Introduction

When H. H. Mantsch inaugurated a series of conferences on "biomedical vibrational spectroscopy" in 1998,[1] strong technical advances had already enabled the sensitive and reproducible study of biomedical samples by means of the spectroscopy of molecular vibrations. These advances have nowadays found their continuation in, for example, the availability of mid-infrared detector arrays or the commercialization of quantum cascade lasers. Many review articles, special issues, conference proceedings, and books testify to the growing importance of this vivid field of research.[2-11] In order to elucidate the key factors of biomedical vibrational spectroscopy, this chapter focuses on the analysis of biofluids. Most of the concepts and approaches are also being used for tissue diagnostics and breath analysis.

2.2 On the Role of Water

Since the spring constants of typical bonds in organic molecules such as C–C, C–H, or O–H are on the order of 500 to 2000 N/m, the fundamental frequencies of simple diatomic stretch vibrations correspond to wavelengths of electromagnetic radiation around, say, 3 to 12 μm, which lies in the mid-infrared spectral range.

Biological processes strongly depend on the fact that the relevant biochemistry takes place in water. For example, water contributes approximately 60% to the total mass of the human body. Thus, the spectroscopy of molecular vibrations in biological or medical samples is strongly related to the spectroscopic properties of water. The fundamental vibrations of water correspond to wavenumbers (i.e., inverse wavelengths) around 3700 cm^{-1} and 1600 cm^{-1} such that electromagnetic radiation with wavelengths in the mid-infrared region is strongly absorbed by the molecular vibrations of water (see Fig. 2.1). Given the abundance of water, the spectroscopy of fundamental vibrations of biomolecules in cells, tissue, microorganisms, body fluids, and so on will be dominated by the absorption of water, while the biomolecules of interest will hide their smaller absorption signals in the "water background." Since the mid-infrared spectroscopy of molecules is based on changing the dipole moment of the molecule, the fairly high dipole moment of water worsens the issue. An example is shown in Fig. 2.2, which shows the relative transmission of mid-infrared laser light through a thin microcavity filled with air (a) and which illustrates the 80% decrease of the transmission if the cavity is filled with water (b, solid line). If glucose is added to the water at a concentration of 500 mg/dL (which would be a pathologically high glucose level in blood, five times above the normal glucose concentration), only a very small further signal change is observed.[12]

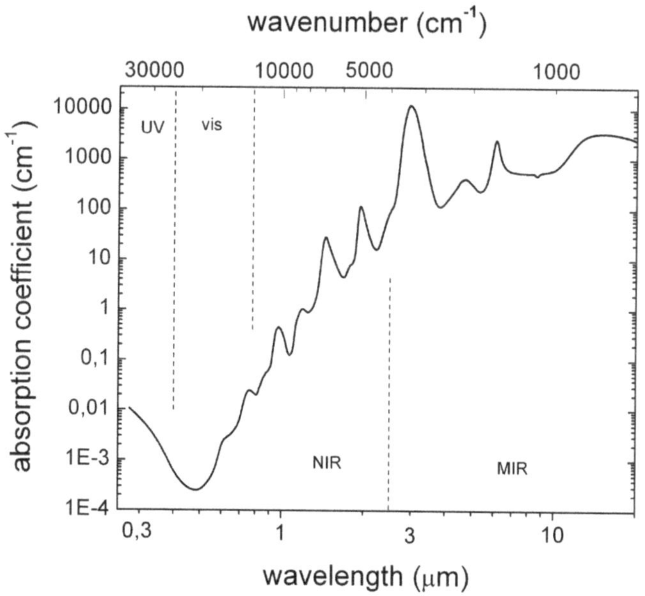

Figure 2.1 Absorption coefficient of water as a function of the wavelength of electromagnetic radiation from the ultraviolet (UV) radiation and visible (vis) light to the near-infrared (NIR) and mid-infrared (MIR) part of the spectrum. The inverse of the wavelength, that is, the wavenumber, is shown in units of cm^{-1} on the top axes for reference.

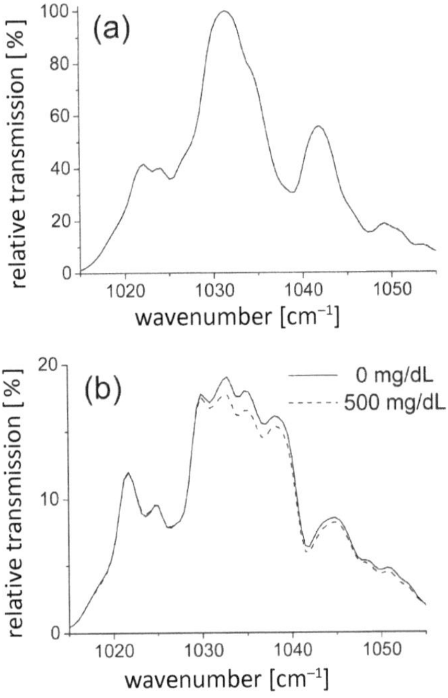

Figure 2.2 Relative transmission signal of mid-infrared laser radiation through a thin microcavity filled with (a) air and (b) (solid line) water and (dashed line) aqueous glucose solution (500 mg/dL). The spectral shape of the transmitted radiation is dominated by the emission profile of the laser. The signal is normalized to the maximum transmission in (a). Note the change in scale between (a) and (b). See also Vrančić *et al.*[12]

2.3 Strategies to Mitigate the Impact of Water

Various attempts have been made to overcome the limitations imposed by water absorption. Briefly, the potential of biomedical vibrational spectroscopy is exemplified in the scope of using small sample thicknesses, applying novel, intense light sources, drying the sample, and Raman spectroscopy.

2.3.1 Reducing Sample Thickness

According to Lambert–Beer's law, absorption depends on the extinction coefficient, concentration, and path length. Since in biomedical samples only the latter can be controlled by the experimenter', thin sample cuvettes have been produced that make short path lengths possible and, in this way, may leave sufficient light for detection despite the strong absorption of water. Disadvantages of this method are first that the analyte signal is reduced in the same manner as is the water signal, second that the convective filling of a narrow cuvette with aqueous solution requires high pressure, and third that cellular debris may cause clotting.

2.3.2 Increasing Light Power

If sufficiently high reproducibility of a measurement is guaranteed, the mid-infrared spectroscopy of biological samples is limited by the signal-to-noise ratio. Most often, this ratio can be increased by using more intense light sources, since the signal increases linearly with the number of photons while the noise increases merely with the square root of the number of photons. When compared to a usual thermal radiation source, such as a Globar, the photon flux may be increased by at least two orders of magnitude by means of an infrared synchrotron radiation beamline. Thanks to pioneering work in France, Germany, and Britain in the 1970s, more than twenty dedicated infrared synchrotron beamlines are now in operation, and researchers worldwide can take advantage of the better signal-to-noise ratio, shorter measurement times, and/or high spatial resolution.[11] Alternatively, the spectral power density of the radiation source may be enhanced by at least one order of magnitude as compared to a Globar by means of quantum cascade lasers. It has recently been shown that transmission spectroscopy with these light sources is capable of delivering superior signal-to-noise ratios in the quantification of analytes in aqueous solution.[12] The data displayed in Fig. 2.2 has been obtained with this type of spectroscopy, showing the capability to measure analyte concentrations (here, glucose) with a high degree of sensitivity.

2.3.3 Drying the Sample

Another approach to analyzing biomedical samples by means of infrared spectroscopy is through drying the sample. This method is particularly useful for analysis of bodily fluids such as blood, plasma, serum, or urine.[13–17] After drying, these samples tend to form a ~2–10 μm thin film of concentrated biological material, so the technical skill rests in the reproducible creation of these thin films under controlled conditions.[18] As an example, Fig. 2.3 depicts the Fourier-transform mid-infrared spectrum of 1 μL of serum after it has been dried on a silicon sample carrier.

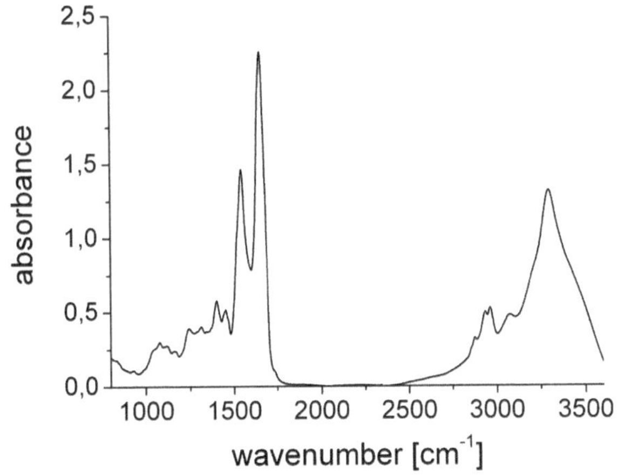

Figure 2.3 Absorbance spectrum of 1 μL of serum after drying. The pronounced mid-infrared spectral signatures of proteins (amide I and amide II bands around 1500–1700 cm^{-1}) and lipids (CH_2 and CH_3 vibrations around 2700-3000 cm^{-1}) are evident.

2.3.4 Raman Spectroscopy

In contrast to infrared spectroscopy, which samples the oscillations of a molecule's permanent dipole moment, Raman spectroscopy is based on the inelastic scattering of light, and its amplitude depends on the induced dipole moment, that is, the polarizability. Technically, Raman spectroscopy is usually performed using ultraviolet, visible or near-infrared radiation with wavelengths readily accessible from standard light sources, optics, and detection technology. Even more important, though, is the fact that for biomedical vibrational spectroscopy the wavelengths may be chosen to be near the minimum of the absorption of water (see Fig. 2.1). However, this advantageously low absorption of water may be partly counteracted by the low cross section of

Raman spectroscopy, by the occurrence of fluorescence as a signal background, and by possible photo-induced damage to the sample by high light intensities.

2.3.5 Further Methods

Numerous other techniques are available for the spectroscopy of the vibrations of biomolecules. Among these are near-infrared spectroscopy, attenuated total reflection spectroscopy, and, more recently, terahertz spectroscopy. In addition, improvements have been made in Raman spectroscopy and mid-infrared spectroscopy, including surface-enhanced Raman (SERS) or infrared (SEIRS) spectroscopy, coherent anti-Stokes Raman spectroscopy (CARS), and tip-enhanced Raman spectroscopy (TERS). These techniques are discussed in depth in the corresponding chapters of this book. Further information can also be found in Lasch and Kneipp.[8]

2.4 Applications

From a diagnostic standpoint, the application of vibrational spectroscopy may be categorized into two groups. While the first category aims at the quantification of the concentration of the molecule of interest, the second category targets the overall interpretation of the shape of a spectrum in relation to the disease of interest. It is the weakness of this latter classification scheme that only very limited systematic insight into the biochemistry of the disease under investigation is obtained, while at the same time it is the strength of the method that no such insight into the biochemistry is needed for mathematical classification.

2.4.1 Quantitative Analysis

In contrast to what happens in many other spectroscopic methods, with vibrational spectroscopy the biomolecule of interest usually contributes to multiple absorption peaks in the vibrational spectrum of a sample rather than to a single signature because spectral signatures are primarily determined by the bond strengths and modes of vibration. Thus, multivariate (rather than univariate) analysis of the spectra is closely interrelated with vibrational spectroscopy.

One of the goals of biomedical vibrational spectroscopy is the quantitative determination of the concentration of analytes. After preprocessing (e.g., baseline correction or background compensation), the spectra are subjected to quantitative multivariate methods such as partial least square analysis or principal component regression. A direct comparison between Raman spectroscopy and mid-infrared spectroscopy has been performed under identical conditions in the view of a clinical laboratory.[19] Aliquots of serum samples originating from 247 volunteers underwent both types of spectroscopy as well as standard clinical laboratory chemical analysis to determine the concentrations of total protein content, cholesterol, high-density lipoproteins, low-density lipoproteins, triglycerides, urea, uric acid, and glucose. While the mid-infrared spectroscopy of dried films of serum had been considered superior in terms of the quantitative analysis of serum prior to this study, it could be shown that it is the reproducibility of mid-infrared spectroscopy rather than the (excellent) signal-to-noise ratio that limits its performance. The study revealed thatd– espite its low signal-to-noise ratioR– aman spectroscopy was able to provide similar results in terms of the quantitative analysis of analytes as compared to mid-infrared spectroscopy.

These results have been obtained with an average turnaround time of 5 minutes per serum sample, which would be acceptable for clinical laboratory analysis with regards to the simultaneous determination of eight analytes in parallel. Besides this advantage of deriving multiple parameters from a single spectroscopic measurement, the spectra may also be stored and, possibly, analyzed retrospectively if at a later point in time some further analytes need to be assessed, perhaps due to changes in the health of the sample donor. With an accuracy around ~10% (see Fig. 2.4), the spectroscopic quantification reaches the clinically acceptable range, although, of course, the conceivable commercialization of biomedical vibrational spectroscopy for the quantification of serum analytes would benefit from further improvement in accuracy. Finally, the relative error appears to increase with decreasing mean analyte concentration (in mass units), such that one might infer a lower limit for quantification of analytes of approximately 0.1 mmol/L at best.

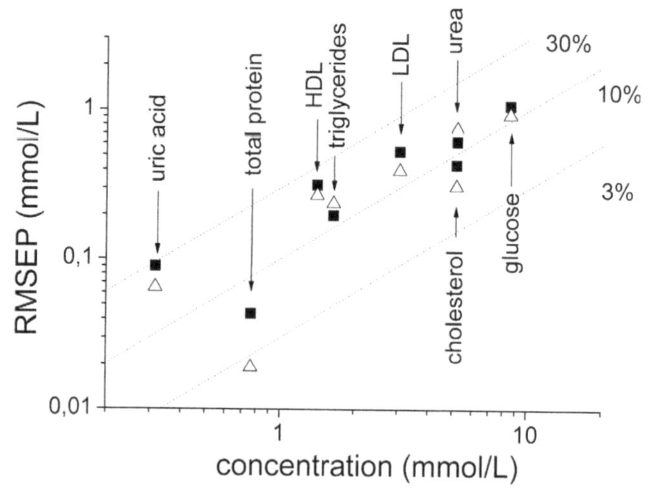

Figure 2.4 Results of the quantitative analysis of serum sample by mid-infrared spectroscopy of dried films of serum (squares) and the Raman spectroscopy of native serum samples (triangles). RMSEP denotes the root mean squared error of prediction calculated on the basis of 99 independent serum samples used for validation.

Beyond these *in vitro* applications, a lot of interest in vibrational spectroscopy originates from its potential to allow continuous surveillance of blood glucose concentration *in vivo*. In preparation for such a sensor system, laboratory experiments have been carried out using a fiber-based mid-infrared quantum cascade laser sensor in combination with a thin, miniaturized cavity. First results in aqueous glucose solutions (i.e., without drying) deliver a noise level that would correspond to glucose concentrations around 1% of its average concentration.[12]

2.4.2 Classification

From a scientific standpoint it is always desirable to identify and, if possible, quantify molecules that give rise to particular diseases. However, biomedical vibrational spectra are very complex, individual molecules contribute to multiple peaks, and individual peaks can usually not be attributed to a single specific molecule.

Pragmatically, all the accessible information must somehow be contained in the overall spectral signatures. Hence it seems legitimate to apply methods for the analyses of spectra that exceed the quantification of individual molecules. Indeed, pattern recognition tools are now being used to analyze biomedical vibrational spectra on a routine basis. These tools try to identify a correlation between the input parameters, such as the absorbance values at each individual wavenumber, and the target value. Many mathematical tools are available for grouping the spectra into two or more classes. This may be performed on the basis of an unsupervised classification algorithm such as cluster analysis or supervised algorithms, which requires a data set for training of the algorithm plus an independent data set for validating the algorithm.

While this approach is also being used for infrared or Raman hyperspectral images, here the fundamental principles are once again illustrated using non-imaging examples such as the spectroscopy of serum: Small but reproducible differences in the spectra of serum have been found in various applications, ranging from the detection of bovine spongiform encephalopathy ("mad cow disease") in cattle serum[20,21] through the metabolic syndrome[22] or rheumatoid arthritis[23] all the way to supporting acute diagnostic quests such as myocardial infarction.[24] The classification of spectra originating from bodily fluids is often referred to as "diagnostic pattern recognition (DPR)."

For example, in one of the largest and most comprehensive studies on mid-infrared spectroscopy in medicine, a total of 1,429 serum samples from 389 consecutive patients with acute chest pain were analyzed with a goal of aiding the rapid diagnosis of acute myocardial infarction. A supervised classification algorithm, namely the robust linear discriminant analysis, was trained using about two-thirds of the sample spectra and was subsequently applied to the spectra of the remaining 421 samples. Within these independent validation samples, 100 out of 113 samples originating from patients with acute myocardial infarction were identified correctly, as well as 262 out of the remaining 308 samples originating from patients, who suffered from chest pain of another origin. This corresponds[a] to a sensitivity (fraction of samples correctly classified

[a] The 95% confidence interval for sensitivity ranges from 81% to 94%; for specificity it is 81% to 89%.

as "acute myocardial infarction") of 88.5% and a specificity (fraction of samples correctly classified as "chest pain of other origin") of 85.1%.

By choosing the appropriate cut-off value for the classification algorithm, it is possible to tune the algorithm for high sensitivity at the expense of lower specificity (and vice versa). This flexibility is often displayed in forms of the receiver operating characteristics (ROC) curve in which pairs of sensitivity and specificity are plotted under variation of the cut-off value (see Fig. 2.5). The area under this curve frequently serves as an indicator for the quality of a marker. As described in detail in Petrich, *et al.* (2009),[24] the diagnostic pattern recognition approach performs equally well or even better than the present cardiac markers of clinical laboratory diagnostics.

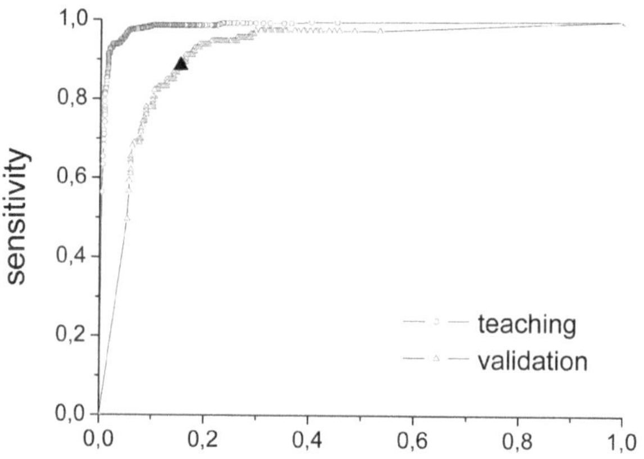

Figure 2.5 Receiver operating characteristics (ROC) of the teaching and the independent validation set as derived from "diagnostics pattern recognition," that is, mid-infrared spectroscopy of serum samples in combination with mathematical classification methods. In this study on acute chest pain, the term *senstivity* denotes the fraction of samples correctly classified as acute myocardial infarction, while the term *specificity* enumerates the fraction of samples correctly classified as chest pain of other origin. The area under the ROC curve amounts to 0.921 for the independent validation set.

In another study, four supervised classification algorithms were blindly and independently compared to each other using identical data sets.[21] A standard linear discriminant analysis, a robust discriminant analysis, a neural network, and a support vector machine model were applied to the classification of serum spectra. It turns out that all of the algorithms were able to distinguish between spectra originating from BSE-positive samples versus BSE-negative samples with similar accuracy when validated with the blinded, independent part of the data set. Using a combination of the four classification algorithms in forms of a decision tree, an accuracy of 94% could be achieved.

2.5 Analysis Methods: The Need for Independent Validation

It is well beyond the scope of this chapter to elaborate on the various mathematical tools for analyzing vibrational spectra with respect to the biomedical application. However, all supervised schemes have one concern in common, which is related to the number of fit parameters versus the number of spectra or samples. In a crude simplification one may say that it is always possible to find a solution for 100 coupled linear equations, if one allows for 100 free fit parameters.

This concern of "overfitting" may be illustrated using the above example of the quantification of glucose in serum. Here 148 samples were used for training a partial least square algorithm for various numbers of fit parameters (which in partial least square regression may be given by the number of latent variables). The trained algorithm was then applied to 99 further, independent samples for validation. The result is shown in Fig. 2.6. The root mean square error of calibration (RMSEC) is used to assess the quality of the classification algorithm during training. Very low RMSECs can be obtained for large numbers of latent variables (Fig. 2.6a, filled squares). When the algorithms are applied to an independent validation set, the root mean square error of prediction (RMSEP) first also decreases with an increasing number of latent variables (Fig. 2.6b, filled squares). Beyond a certain number of latent variables, the algorithm more and more tends to fit the random noise of the teaching data and consequently fails to enhance the prediction accuracy of an independent validation sett— hat is, the data had been overfitted.

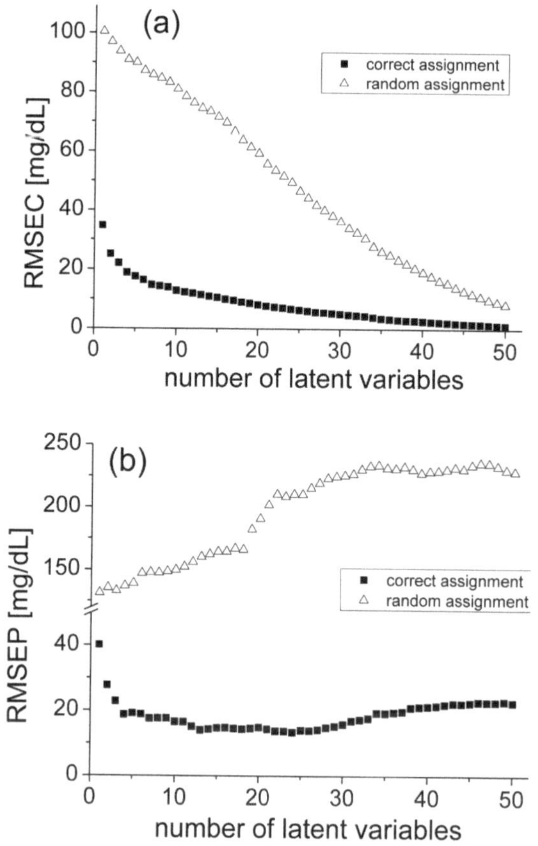

Figure 2.6 The root mean square of (a) calibration (RMSEC) and (b) validation (RMSEP) are plotted as a function of the number of latent variables used in a partial least square regression of mid-infrared spectra of serum samples with glucose as the target analyte. The correct assignment between spectra and glucose concentration is symbolized by the filled squares, whereas the results of the intentionally flawed, random assignment is shown by the open triangles. Note the change in scale and the axes break in Fig. 2.6b (see also Petrich (2008)[25]).

It is thus extremely important to provide independent validation data. In order to illustrate the dramatic effect of overfitting, the same data was now used for training, but the relations between spectra and glucose concentrations were first deliberately randomized. Again, the RMSEC can deliver good values for large numbers of latent variables (Fig. 2.6, open triangles) while the independent validation and the RMSEP show the severe failure of the algorithm due to overfitting (Fig. 2.6, open triangles).

2.6 Conclusion

As the Hungarian scientist Albert Szent-Gyorgyi has pointed out, "Water is life's matter and matrix, mother and medium." As a consequence, the spectroscopy of the fundamental vibrations of biologically relevant molecules is hampered by the strong absorption of the omnipresent water in the mid-infrared region. However, multiple options are available nowadays to get around this problem. Advances in spectroscopic technologies, such as intense light sources or highly sensitive Raman spectrometers, in combination with sophisticated mathematical algorithms for the analysis of the spectra have enabled tremendous progress in fundamental as well as applied research in biomedical vibrational spectroscopy.

References

1. Mantsch, H. H., and Jackson, M. (eds) (1998). *Infrared Spectroscopy: New Tool ṅ Medċ ṅ e*, Proceedings of the Society of Photo-optical Instrumentation Engineers, Vol. 3257, SPIE Press, San Jose, CA.
2. Gremlich, H.-U., and Yan, B. (eds) (2001). *Infrared and Raman Spectroscopy of Bȯ logċ al Materȧ ls*, Marcel Dekker, New York.
3. Naumann, D. (2001). FT-infrared and FT-Raman spectroscopy in biomedical research, *Appl. Spectrosc. Rev.*, **36**, 238–298.
4. Petrich, W. (2001). Mid-infrared and Raman spectroscopy for medical diagnostics, *Appl. Spectrosc. Rev.*, **36**, 181–237.
5. Naumann, D., Petrich, W., and Schmitt, J. (eds) (2007). Special section on biomedical vibrational spectroscopy, *Anal. Bȯ anal. Chem.*, **387**, 1589–1829.
6. Mahadevan-Jansen, A., Petrich, W., Alfano, R. R., and Katz, A. (eds) (2008). *Bȯ medċ al Optċ al Spectroscopy*, Proceedings of the Society of Photo-optical Instrumentation Engineers, Vol. 6853, SPIE Press, San Jose, CA.
7. Diem, M., Griffiths, P., and Chalmers, J. (eds) (2008). *Vb atȯ nal Spectroscopy for Medċ al Dȧ gnosṡ* , Wiley, Hokboken, NJ.

8. Lasch, P., and Kneipp, J. (eds) (2008). *Biomedical Vibrational Spectroscopy*, Wiley, Hoboken, NJ.

9. Copsey, M. (ed) (2010). Themed issue on optical diagnostics, *Analyst*, **135**, 3027–3272.

10. Mahadevan-Jansen, A., and Petrich, W. (eds) (2010). *Biomedical Vibrational Spectroscopy VI: Advances in Research and Industry*, Proceedings of the Society of Photo-optical Instrumentation Engineers, Vol. 7560, SPIE Press, San Jose, CA.

11. Moss, D. (ed) (2011). *Biomedical Applications of Synchrotron Infrared Microspectroscopy*, RSC Analytical Spectroscopy Monographs No. 11, Royal Society of Chemistry, Cambridge, UK.

12. Vrančić, C., Fomichova, A., Gretz, N., Herrmann, C., Neudecker, S., Pucci, A., and Petrich, W. (2011). Continuous glucose monitoring by means of mid-infrared transmission laser spectroscopy *in vitro*, *Analyst*, **136**, 1192–1198.

13. Werner, G. H., Boecker, D., Haar, H. P., Kuhr, H. J., Mischler, R. (1998). Multicomponent assay for blood substrates in human sera and hemolysed blood by mid-infrared spectroscopy, in *Infrared Spectroscopy: New Tool in Medicine* (ed. Mantsch, H. H., and Jackson, M.), SPIE Press, San Jose, CA, pp. 91–100.

14. Shaw, R. A., Kotovich, S., Leroux, M., and Mantsch, H. H. (1998). Multianalyte serum analysis using mid-infrared spectroscopy, *Ann. Clin. Biochem.*, **35**, 624–632.

15. Naumann, D. (2001). FT-Infrared and FT-Raman spectroscopy in biomedical research, in *Infrared and Raman Spectroscopy of Biological Materials* (ed. Gremlich, H.-U., and Yan, B.), Marcel Dekker, New York, pp. 323–378.

16. Petrich, W., Dolenko, B., Früh, J., Ganz, M., Greger, H., Jacob, S., *et al.* (2000). Disease pattern recognition in infrared spectra of human sera with diabetes mellitus as an example, *Appl. Opt.*, **39**, 3372–3379.

17. Heise, H. M., Voigt, G., Lampen, P., Küpper, L., Rudloff, S., and Werner, G. (2001). Multivariate calibration for the determination of analytes in urine using mid-infrared attenuated total reflection spectroscopy, *Appl. Spectrosc.*, **55**, 434–443.

18. Moecks, J., Kocherscheidt, G., Köhler, W., and Petrich W. (2004). Progress in diagnostic pattern recognition, *Proc. SPIE*, **5321**, 117–123.

19. Rohleder, D., Kocherscheidt, G., Gerber, K., Kiefer, W., Köhler, W., Möcks, J., and Petrich, W. (2005). Comparison of mid-infrared and Raman spectroscopy in the quantitative analysis of serum, *J. Biomed. Opt.*, **10**, 031108.

20. Lasch, P., Schmitt, J., Beekes, M., Udelhoven, T., Eiden, M., Fabian, H., *et al.* (2003). Ante-mortem identification of bovine spongiform encephalopathy from serum using infrared spectroscopy, *Anal. Chem.*, **75**, 6673–6678.

21. T. Martin, T., Moecks, J., Belooussov, A., Cawthraw, S., Dolenko, B., Eiden, M., *et al.* (2004). Classification of infrared signatures in serum specific to bovine spongiform encephalopathy, *Analyst*, **129**, 897–901.

22. Früh, J., Jacob, S., Dolenko, B., Häring, H.-U., Mischler, R., Quarder, O., *et al.* (2002). Diagnosing the predisposition for diabetes mellitus by means of mid-infrared spectroscopy, *Proc. SPIE*, **4614**, 63–69.

23. Staib, A., Dolenko, B., Fink, D. J., Früh, J., Nikulin, A. E., Otto, M., *et al.* (2001). Disease pattern recognition testing for rheumatoid arthritis using infrared spectra of human serum, *Clin. Chim. Acta.*, **308**, 79–89.

24. Petrich, W., Lewandrowski, K. B., Muhlestein, J. B., Hammond, M. E. H., Januzzi, J. L., Lewandrowski, E. L., *et al.* (2009). Mid-infrared spectroscopy can aid the triage of patients with acute chest pain, *Analyst*, **134**, 1092–1098.

25. Petrich, W. (2008). From study design to data analysis, in *Biomedical Vibrational Spectroscopy* (ed. Lasch, P., and Kneipp, J.), Wiley, Hoboken, NJ, pp. 315–332.

Chapter 3

Bioreceptor Functionalization of Gold-Coated Sensor Surfaces

Daria Prilutsky,[a,b,c] Lev Tsapovsky,[d] and Robert S. Marks[e,f]

[a]*Department of Virology, Faculty of Health Sciences, Ben-Gurion University of the Negev, Beer-Sheva, Israel*
[b]*Department of Information Systems Engineering, Faculty of Engineering Sciences, Ben-Gurion University of the Negev, Beer-Sheva, Israel*
[c]*National Institute of Biotechnology in the Negev, Ben-Gurion University of the Negev, Beer-Sheva, Israel*
[d]*Unit of Environmental Engineering, Faculty of Engineering Sciences, Ben-Gurion University of the Negev, Beer-Sheva, Israel*
[e]*Department of Biotechnology Engineering, Faculty of Engineering Sciences, Ben-Gurion University of the Negev, Beer-Sheva, Israel*
[f]*The Ilse Katz Center for Meso and Nanoscale Science and Technology, Ben-Gurion University of the Negev, Beer-Sheva, Israel*
rsmarks@bgu.ac.il

3.1 Biosensors as Analytical Tools

Biosensors are analytical self-contained devices for monitoring and detecting various analytes in a concentration-dependent manner. Biosensors combine three principal components broken down into physical, chemical, and biological parts. The first core physical part involves signal transduction and configuration of the system in order to translate the biological detection into an interpretable one. The second, chemical part concerns mainly immobilization

Nanoantenna: Plasmon-Enhanced Spectroscopies for Biotechnological Applications
Edited by Marc Lamy de la Chapelle and Annemarie Pucci

ISBN 978-981-4303-61-3 (Hardcover), 978-981-4303-62-0 (eBook)
www.panstanford.com

techniques and modification of sensor transducer inorganic surfaces with biospecific elements. The last, biological part is the biorecognition element or biomolecule, which can be an enzyme, antibody, nucleic acid, bioreceptor, ligand, whole cell, or something else. This biological element will interact with a specific target analyte and will transfer the molecular recognition signal from a chemical to the physical environment of a biosensor.[1]

In recent years, greater investment has made biosensors increasingly important in environmental monitoring, medicinal diagnostics, and food industries, and extensive research has been done to optimize the technology. Biosensors offer a lot of advantages in comparison to conventional analytical methods: they are cheaper, simpler to operate, more sensitive and specific in providing an alert to potential danger, and in some cases may be connected to some form of actuation technology. The chemical functionalization of the surface and immobilization of the biomolecule are particularly important aspects of the fabrication of biosensors, as both of them must reproducibly keep the biorecognition molecule close to the transducer surface, while retaining their biological activity.[2]

This chapter explores the chemical part of a biosensor whose surface is made of gold and introduces techniques for its functionalization via the conjugation of antibodies and DNA strands.

3.2 Physical Properties of Gold Surfaces

Noble metals are called "noble" because of their inert nature; they have several applications due to their catalytic, chemical, optical, magnetic, mechanical, and electrical properties. Many of the noble metals are highly resistant to oxidation, even at elevated temperatures.[3] Gold, being a noble metal, is reasonably stable, easy to prepare, and easy to pattern using a combination of lithographic tools. Gold readily binds certain selective chemistries, which can then be modified through conjugation so the biomolecular reaction of target analytes can be detected without undergoing any unusual reaction. The chemistry we focus on in this chapter is a gold-thiol bond. In line with the advantages of gold as a stable, easily operable material, thiols linked to gold provide reliable conjugation systems. It should be pointed out that thin gold films are common substrates in several analytical techniques, such as surface plasmon resonance (SPR) spectroscopy, quartz crystal microbalances (QCMs), ellipsometry, nanostructured surfaces such as nanoantennas, and electrochemical methods.[4]

SPR spectroscopy is a powerful tool for qualitatively and quantitatively analyzing the specific binding of a mobile reactant to a binding partner immobilized on the sensor surface of a noble metal film. Optical biosensors based on SPR use the characteristic evanescent field under total internal reflection conditions.[5] Furthermore, a thin layer of gold deposited on a highly refractive index glass surface will absorb laser light, producing electron waves (surface plasmons) on the gold surface. The surface plasmon is a collective oscillation of free electrons (free electron metallic property of gold), which propagates along the surface of a metal with the magnetic field vector parallel to the metal surface. The total internal reflection of light is used to excite nonradiative surface plasmons in the metal film.[6] This phenomenon occurs only at a specific angle and wavelength of incident light and is highly dependent on the surface of the gold, such that binding of a target analyte to a receptor on the gold surface will generate a measurable signal.[5,7–8]

Another very popular biosensing methodology is piezoelectric QCM, which has been employed as a highly sensitive mass detector, enabling the measurement of real-time changes in mass.[9] A QCM is an acoustic wave resonator, in which a thin quartz crystal disk is sandwiched between two metal electrodes that are vapor deposited on either side of the crystal.[10–12] Gold electrodes have been most commonly used in QCM studies, because of the inert nature of the metal and ease with which gold is evaporated and manipulated. However, Cu, Ni, Pt, and other metals have also been employed. There is a linear proportionality between the deposited mass and the frequency decrease of an oscillating quartz crystal, as was demonstrated by Sauerbrey in 1959.[13] As a consequence and ease of its use, QCM attained a significant popularity as an analytical device in recent years.[14]

The next sections mainly review the immobilization process on bare and chemically modified gold surfaces, and discuss the suitability of this metal as an immobilization platform.

3.3 Chemical Characterization and Modification of Gold Surface Contacting the Biological Media

When the construction of a biosensor device is discussed, there are a lot of prerequisites to consider in order to design a functional platform: the immobilized biomolecules should retain their full biological activity and stability after the immobilization process, their accessibility after immobilization should be ensured, and compatibility and chemical inertness toward the host environment should be guaranteed.[15-16] The preservation of the biomolecule activity emerges as one of the most important issues in the field, as some biomolecules can lose part of their activity due to their sensitive nature (especially enzymes) and conformational instability, while immobilized on surfaces.[4] The immobilization of proteins is considered to be more complicated when compared to nucleic acids, since proteins are much more complex than nucleic acids from both a chemical and a structural point of view.[17-19] An immobilized biomolecule has many operational advantages over free biomolecules, such as

- reusability (in case of expensive enzymes),
- focusing the measurement at a given location and thus economizing reagents and ensuring greater signal capture,
- overall reduced cost of operation,
- continuous operational mode in flow systems, and
- enhanced stability and shelf life in biosensors.[20]

Immobilization of biomolecules can be performed using many different methods, including weak noncovalent binding or physical adsorption, electrostatic and hydrophobic interactions, and strong covalent bonding.[4,15,21]

Covalent linking of the protein to a reactive surface through surface-accessible residues often lacks region specificity and can potentially induce conformational changes within the inner structure; hence, an immobilized protein may not be correctly oriented, altering its functionality. Furthermore, the reactive site of a protein can be blocked by the immobilization procedure, resulting in the reduction or loss of protein activity. Therefore, one of the paramount challenges for a viable technology is to achieve controlled immobilization of the proteins on surfaces in such a way that they maintain their biological features and properties, integrity, and native conformation, while avoiding nonspecific protein binding. Among the different available strategies for biomolecule immobilization, the simplest known one is direct adsorption by depositing the protein solution onto in our specific case a gold surface.[9] Proteins physically adsorb to the gold surface through nonspecific interactions mainly via $-NH_2$ groups of lysine, –SH groups of cysteine[22] or –OH groups of serine or threonine.[23] In such conjugation methodologies, there is no possible control over the orientation of the attached molecule, leading to a random distribution of protein adsorption geometries. In the case of antibodies, a share of them is either denatured (less likely) or linked via their variable regions, resulting in the loss of binding activity of their specific target antigen.[22]

Another popular and powerful surface modification strategy is self-assembly: it is easy to apply, which makes it possible to change or tune the surface character by modifying the end groups of the layer accurately and predictably. The self-assembly strategy, which forms surfaces with stimuli-responsive properties, is also known as smart surfaces and is discussed in the following section.

3.3.1 Thiol-Gold Coordination: Self-Assembled Monolayers (SAMs)

The process of self-assembly involves the spontaneous non-covalent arrangements of atoms and molecules to form a functional unity toward an energetically stable form, whose novel structure and properties are determined by their nature.[24] Natural and prominent examples of self-assembly are provided by double-helical[25] and triple-helical DNA,[26] multichromophore structures of the photosynthetic reaction center,[27] and light-harvesting antennae apparatus.[28] Among self-assembly strategies, the generation of a self-assembled monolayer (SAM) is one of the most elegant ways to create a film with specific surface properties.[4]

A self-assembled monolayer is a layer of molecular thickness formed by self-organization of active surfactant molecules in an ordered manner on a solid surface. Molecular structures forming SAMs can be divided into three parts (Fig. 3.1):[29–31]

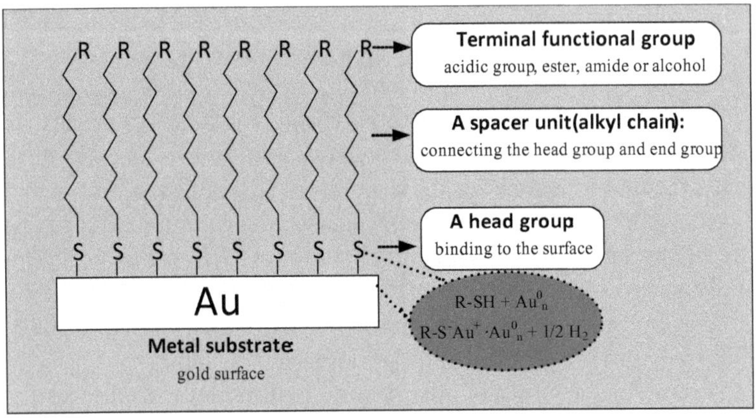

Figure 3.1 Model for self-assembled monolayer of alkanethiols on gold.

- A surface-active head group, which strongly binds to the solid substrate.
- An end group that allows the introduction of a variety of organic functionalities (e.g., acidic group, ester, amide, or alcohol) and plays an important role in terms of coupling the biomolecule to a monolayer.
- A spacer unit (alkyl chain) that connects the head group and end group, giving stability to the assembly by van der Waals interactions, affecting the intermolecular separation and molecular orientation in the film, while ensuring that the reactive groups targeting the analytes are far enough from the surface so as to reduce any putative deleterious influence.

Each those constituents can be easily manipulated and controlled by various chemistries, making them compatible for desired applications. Through proper selection of terminal functional groups, specific surface interactions can be exploited to immobilize molecules at an interface. The most common strategy is based on the use of alkanethiol molecules, which attach strongly to gold surfaces via their sulfur atoms, forming SAMs. In fact, this forms good, well-packed SAMs and is historically the most studied strategy. In general, sulfur compounds are well known for their reactivity to noble metals. Gold surfaces offer an advantage of being easily functionalized with SAMs and have been extensively used, although a variety of other substrates were possible candidates for the deposition of sulfur moieties.[32] Because the formation of SAMs on such surfaces depends on the crystalline morphology of the metal, gold yields SAMs with the highest density and degree of regularity.[19] In addition, the method to modify gold surfaces with self-assembled thiol derivatives is extremely simple and can be achieved in a laboratory just by immersing the desired slides into a thiol solution for a specified time followed by thorough washing with the same solvent and drying, often using a jet of dry argon.[33]

SAM formation, pioneered by Nuzzo[34] in 1983, demonstrated that dialkyl disulfides generate oriented monolayers on gold. This opened a new era for immobilization, and many studies have been performed since then, showing that alkanethiols adsorb from solution onto gold, forming S–Au bonds.[35–36] This bond forms spontaneously, when the thiol end of the alkanethiol approaches the gold, resulting in the weakening of the sulfur-hydrogen bond and release of hydrogen (Fig. 3.1):[37–40]

$$R\text{–}SH + Au_n^0 \rightarrow R\text{–}S\text{–}Au^+ \cdot Au_n^0 + 1/2\ H_2$$

The thiolate molecule acts as nucleophile and donates a pair of electrons to the gold surface, forming a strong covalent bond. Despite its zero valence, inert gold binds sulfur with longer Au–S bonds than Au–Au, demonstrating a weak but partial covalent bonding with –H the leaving group, which tends to combine into H_2 molecules.[41,42] Long chain alkanethiols ($X(CH_2)_nSH$, $n = 11 - 18$) connect to a gold surface by forming densely packed, ordered, oriented monolayer films and are considered to be thermally more stable than films formed from short chain thiols.[43] Such crystalline-like structures, with all-trans conformation alkyl chains, exhibit average tilt angles of the chain axis in a range of 28° to 40° from the surface normal, and an approximate 55° twist (rotation angle) of the chain axis away from a configuration with the plane perpendicular to the surface.[44,45] The dense packing of SAMs results mainly from van der Waals interactions between the neighboring carbon chains.[31] The separation distance between neighboring alkanethiols is 0.5 nm, representing a $(\sqrt{3} \times \sqrt{3})$ R 30° overlayer structure on the gold surface.[31,46–48] In order to stabilize the SAMs even more, internal substitutions can be performed. It has been reported that increasing the stability of such a layer can be achieved by incorporation of internal hydrogen bonding cross-links, such as amide-containing chains instead of methylene chains.[49,50] Because the resulting hydrogen bond between the amide groups of neighboring chains are stronger than van der Waals interactions, it will deform at higher temperatures.[51] Porter *et al.*[52] have demonstrated the importance of the alkyl chain length on the packing density, intermolecular environment, and geometry of the monomolecular assemblies, stating that with a decrease of the chain length ($n < 8$), the layer becomes increasingly disordered. Therefore, chains containing mainly $n > 10$ carbons can form well-organized structures.

Another property affecting the packing and molecular orientation in SAMs is the type of terminating groups and their mutual interactions. It has been reported that differently functionalized end groups of *n*-alkanethiols (COOH–, CH_3–, and OH– terminated thiols) with the same chain length will not form the same packing, thus influencing the structure of alkanethiolate monolayers on gold substrates.[53] The crystalline-like monolayer can be destabilized by different environmental conditions, such as elevated temperature or extreme pH.[54,55] Other factors influencing the fabrication of SAMs are the nature of the solvent and the time of immersion of gold surfaces in concentrated thiol solutions, while longer periods may result in the formation of multilayers.[56]

All the aforementioned intrinsic properties of alkanethiol-based SAMs on a golden surface enable reliable construction of sensing systems. The stability, the uniform surface structure, and the relative ease of their functional modification make SAMs an ideal choice for immobilizing sensing molecules on surface. The design and use of SAMs for application in biomolecular recognition, including different kinds of modifications and their dynamic interactions with biomolecules, are reviewed in the following sections.

3.3.1.1 Monothiol layers

The vast majority of alkanethiols derivatives contain a single –SH group and are anchored to gold via single S–Au bonds. These molecules differ by their size and by the functional terminating group for attachment of biomolecules, as discussed above. The functional terminal group can be acidic (COOH-terminated thiols), amide

(NH_2-terminated thiols), alcohol (OH-terminated thiols), or phosphonic ($-PO_3H_2$).

3.3.1.1.1 *Alkanethiol with carboxylic terminus (–COOH)*

The most popular molecules for immobilization are alkanethiols with a carboxylic acid tail, varying by chain size. In comparison to their simple alkanethiol analogues, the organization of acid SAMs is complicated by the presence of the carboxyl head groups. When protonated, strong hydrogen bonding between the -COOH head groups can influence the functionality of the film, whereas in the deprotonated state, electrostatic repulsions between neighboring carboxylate anions can disrupt the film's packing density. Therefore, both existing and future applications of these surfaces rely on a thorough understanding of the state of the carboxylic acid group from which further surface functionality can be derived.[57]

The molecule with the shortest chain is a thioglycolic acid (Fig. 3.2y), comprising two carbons, one attached to -COOH for bio-conjugation and the other to -SH, to form a S–Au bond with gold.[58–60] The most common chain length is 11-mercaptoundecanoic acid (11-MUA) (Fig. 3.2q),[61,62] 16-mercaptohexadecanoic acid (MHDA) (Fig. 3.2l),[63] and 3-mercaptopropionic acid (MPA) (Fig. 3.2o).[64,65] The popularity of the carboxylic-terminated molecules is mainly because of the relatively simple protocol needed to modify them for biomolecule immobilization, usually with ethyl-3-(3-dimethyl-amino-propyl)-carbodiimide (EDC) and *N*-hydroxy-succinimide (NHS). Those long-chain alkanethiols form highly ordered and densely packed monolayers suitable for bioconjugation.[4,36] Nevertheless, other studies show that the -COOH terminus of the chains form strong hydrogen bonds with other -COOH groups leading to a dimerization of the COOH groups, thus preventing the formation of highly oriented films.[53] Such dimerization can be prevented by adding acetic acid to the thiol solution, which will protonate the COO– tails.[66] The monolayer can achieve a significantly higher degree of order by using a more rigid backbone chain, consisting of aromatic rings such as mercaptohexadecanoic acid and mercapto-methyl-terphenyl-carboxylic acid, instead of a CH_2 backbone.[67]

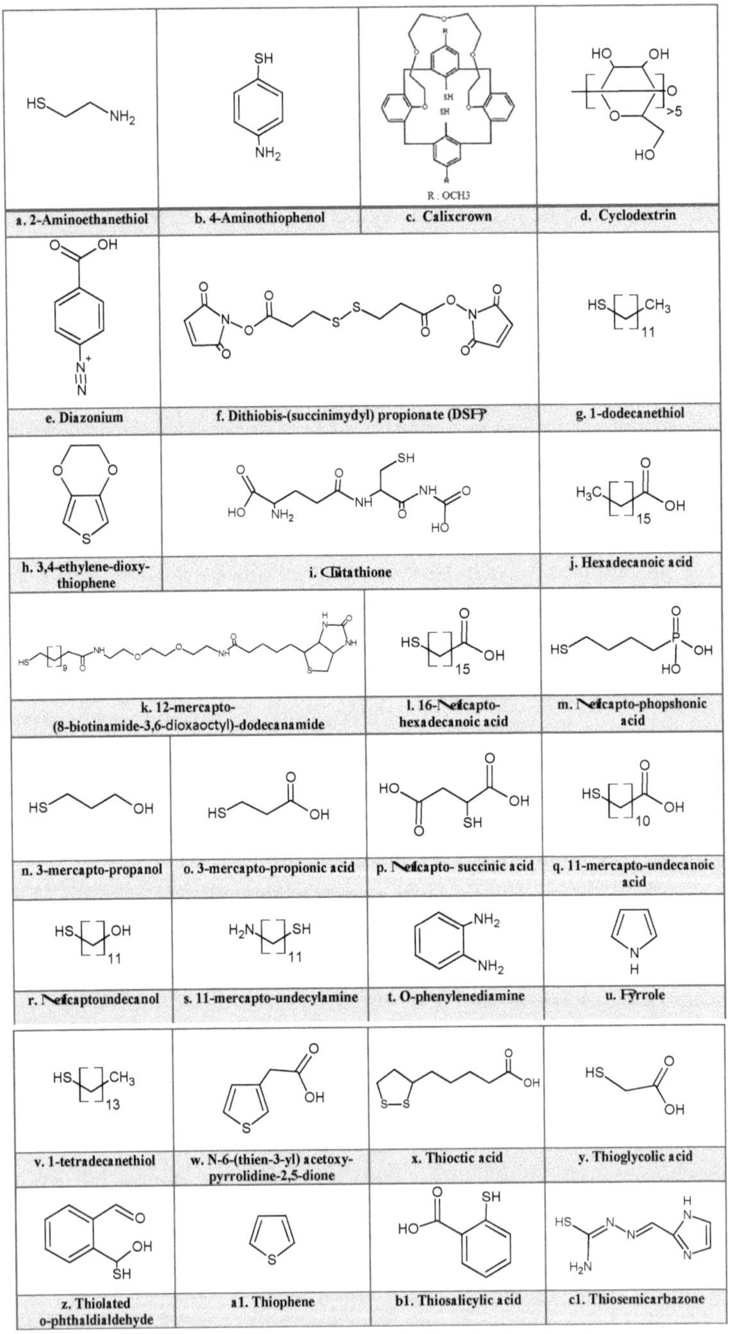

Figure 3.2 Chemical structure of the molecules.

3.3.1.1.2 *Alkanethiol with amine terminus (–NH_2)*

Immobilization of alkanethiols with an amino terminus is less common when compared to those with carboxylic acid termini, mainly because the activation of the amine tail requires a much more cumbersome procedure than the COOH tail does. Amine is usually activated by a bifunctional crosslinker such as glutaraldehyde.[68] Other activation techniques are based on the use of N-succinimidyl S-acetylthiopropionate to create a sulfhydryl-terminated alkanethiol[69] or protein A with glutaraldehyde.[70] The common chain length used for immobilization with amino-terminated alkanethiols are 2-aminoethanethiol (cystamine) (Fig. 3.2a)[68,70–73] and 11-mercaptoundecylamine (Fig. 3.2s).[69]

3.3.1.1.3 *Mercaptosuccinic acid (MSA)*

Mercaptosuccinic acid (Fig. 3.2p), also known as thiomalic acid is a four-carbon chain containing a thiol group and two carboxylic groups. Surface coverage with MSA provides a charged surface. Such a surface is suitable for nonionic interactions,[74] electrochemical characterization,[75] and antibody immobilization.[76] The disadvantage of this molecule for monolayer construction is the networked structures formed due to hydrogen bonding between their carboxylic acid terminus, which can prevent immobilization of biomolecules.[77,78]

3.3.1.1.4 *Mercaptophosphonic acid*

The use of mercaptophosphonic acid (Fig. 3.2m) is mainly for immobilization of charged molecules due to the phosphonic terminus of the molecule (–PO_3^{2-}), which is negatively charged. Another reason for a growing interest in using phosphonic termini is mainly due to the high biocompatibility of the phosphorus atom, which is one of the important elements in biological systems due to its pervasive existence in physiological organs and tissues.[79] The charging current of the electrode modified with phosphonic acid is much lower than a carboxylic terminated acid with the same carbon length, thus enhancing the sensitivity of the electrochemical analysis. Xu *et al.*[80] modified the gold surface with 4-mercaptobutylphosphonic acid and created an aluminum alkanebisphosphonate film for ds-DNA or ss-DNA immobilization. A well-ordered film exposed many aluminum binding sites, which would interact strongly with the phosphate groups of DNA. The examples of various phosphoalkanethiol chain lengths, which were investigated for the adsorption onto gold surfaces, are 3-mercaptopropylphosphonic[81] and 10-mercaptodecanylphosphonic acid.[79]

3.3.1.1.5 *Aromatic molecules*

Monolayers formed from aromatic thiols, although less thoroughly studied, are interesting as surface modification agents because they possess relatively high electrical conductivities due to their delocalized electrons. Moreover, since literature studies indicate that some long-chain normal alkanethiols desorb from gold surfaces,[36] the desorption of the chelating thiols should be at least partially retarded by the ring strain created from the cyclic internal structure.[82] SAMs formed of aromatic thiols, as compared to aliphatic chains, are less packed, even though there is some evidence that a close-packed structure might be possible for longer aromatic thiols.[83] It has been demonstrated that aromatic SAMs are more stable and desorb more slowly in the conditions of elevated temperatures. This is due to the π-stacking interactions that lower the ground state energy[82] and the steric repulsion between adjacent aromatic rings, raising the transition state barrier.

Aminothiophenol (ATP)

Many studies have established the exact structure of aminothiophenol (Fig. 3.2b) that is the most appropriate to form densely packed and well-defined SAMs. In general, the molecule of p-aminothiophenol demonstrated strong covalent bond formation with a gold layer, because of the high concentration of electron density between S and the bonded Au atom[84] and, therefore, was proven to be a suitable candidate for biomolecule immobilization.[85–87] Batz *et al.*[88] showed that 4-aminothiophenol (4-ATP), a small aromatic thiol derivative, forms the densest and well-ordered monolayers on gold, as compared to other p-aminothiolphenols. The organization of the molecules is presumably governed by their intermolecular and surface interactions. Therefore, 4-aminothiophenol is always preferred in immobilization processes over other aminothiol-phenols. Halamek *et al.*[89] and Horacek *et al.*[90] investigated three different thiol compounds used for SAMs for attachment of anti-bodies: cystamine, 4-aminothiophenol, and 3,3′-dithio-bis (propionic acid

N-hydroxysuccinimide ester) (DTSSP). Similar to previous studies, Horacek *et al.* achieved the best immobilization of antibodies on a gold surface with 4-aminothiophenol. However, Halamek *et al.* found that the binding capacity of gold crystals covered with DTSSP is almost two times higher than those covered with cystamine and 4-aminothiophenol, despite the fact that 4-aminothiophenol is more stable than DTSSP for antibody attachment. Pita *et al.*[91] compared the adsorption of the fungal laccase enzymes on bare and thiol-modified gold electrodes. Gold-modified electrodes with 4-aminothiophenol conjugated to laccase showed high reproducibility and stable electrochemistry. Li *et al.*[92] demonstrated diazotization to immobilize on a 4-aminothiophenol–covered gold film, followed by coupling of thionine with the diazonium group (Fig. 3.2e) to form a covalent diazo bond.[93]

Thiosalicylic acid

Thiosalicylic acid (TSA) (Fig. 3.2b1), also known as a 2-mercaptobenzoic acid (MBA), is an aromatic molecule, containing carboxylic as well as a thiol group. p-MBA molecules adsorb strongly on a gold surface. The binding sites and their orientation with respect to the surface are strongly dependent on the concentration of the p-MBA solution and the pH.[94]

2-carboxythiophenol or 2-MBA is the most common molecule from the p-MBAs used for immobilization, even though 4-carboxythiophenol (4-MBA) has been also reported for the construction of SAMs monolayers. Pattabi *et al.*[95] reported the synthesis of water-soluble gold nanoparticles stabilized with thiosalicylic acid/ 2-carboxythiophenol. This study showed that H-bonding and consequent aggregation between the 2-carboxythiophenol/TSA stabilized gold nanoparticles was minimal compared to that reported for 4-carboxythiophenol (4-MBA).

Vaughan *et al.*[96,97] investigated different thiol derivatives and checked their ability as a method to covalently bind antibodies and antigens to the surface of piezoelectric crystals. Thiosalicylic acid was observed to give the most reproducible results for antigen and gave a large, fairly reproducible reduction for antibody binding. Rosendahl and Burgess[57] used electrochemical and infrared spectroscopy to study SAMs of 4-MBA formed on gold surfaces and checked how the state of the terminal carboxylic head group of 4-MBA can influence the functionality of SAMs.

IR measurements have provided valuable information concerning the state of the –COOH head group under various conditions, and it has been shown that the carboxylic acid moiety of 4-MBA SAMs can be protonated and deprotonated by varying the potential applied to the electrode surface.

Thiophene

Thiophene (Fig. 3.2a1) is a molecule in which sulfur is part of a heteorocyclic, aromatic, flat, five-membered ring. Thiophene adsorption onto gold films raises some interesting questions about sulfur-based monolayers, because the sulfur atom in thiophene is part of a stable aromatic ring. Unlike alkanethiols and dialkyl disulfides, which can in principle react with a gold surface in a number of ways, thiophene is relatively inert; it is difficult to reduce, and its ring structure is difficult to cleave.[98] Nevertheless, it has been demonstrated in several studies that sulfur-containing molecules of thiophene spontaneously adsorb from an ethanol solution onto the gold surface and form a well-ordered SAM.[98,99] An alternative molecule for thiophene is constructed by introduction of long, flexible hydrocarbon chains to the 3 position of the thiophene ring, yielding high molecular weight poly-(3-alkyl-thiophenes). This offers a good opportunity for the construction of SAMs, as such soluble poly-(3-alkylthiophenes) possess all the needed elements for SAM manipulation.[100]

An array of various thiophene derivatives has been studied as candidates for immobilization processes. Peng *et al.*[101] synthesized a new thiophene-based molecule: *N*-[6-(thien-3-yl)acetoxy]-pyrrolidine-2,5-dione (Fig. 3.2w), which can self-assemble on a gold surface and react with amino groups, immobilizing bioelements such as DNA on gold electrodes. Bornemann *et al.*[102] modified oligonucleotides with thiophene molecules and demonstrated the suitability of such a conjugate for immobilization onto gold substrates. Kim *et al.*[103,104] reported the preparation of a polythiophene derivative with pendant NHS ester moieties for immobilization of biomolecules. Another thiophene derivative candidate for biomolecule immobilization is poly-(3,4-ethylenedioxythiophene) (PEDOT) (Fig. 3.2h), which was developed by Bayer AG research laboratory in Germany in the late 1980s,[105] and brought PEDOT and its derivatives to the forefront of the field of conductive polymers due to their attractive properties: high stability, high conductivity, and high transparency.[106] Xu *et al.*[105] reported a modified gold electrode with PEDOT-poly(styrene sulfonic acid) (PEDOT-PSS/Au) nanocomposite for horseradish peroxidase (HRP) immobilization, demonstrating that the soluble nanocomposite constructed is an excellent biomaterial for enzymes immobilization and bioelectrochemical device fabrication.

3.3.1.1.6 *Additional thiols*

Among the less common thiol derivatives that are barely exploited, but can still be efficient for biomolecule immobilization, are the thiosemicarbazones (Fig. 3.2c1),[107] glutathione (Fig. 3.2i),[108] 3-mercaptopropyl-trimethoxysilane,[109] and o-phthaldialdehyde modified with thiol (Fig. 3.2z).[110]

3.3.1.2 Disulfides

Because the poor stability of normal SAMs limits their usefulness in real-world applications, many researchers have explored strategies for enhancing SAM stability.[111] Some of these strategies include the use of adsorbates that have multiple sulfur gold interactions, such as disulfides. Several studies have compared SAMs formed from dialkyl disulfides with those from alkanethiols, demonstrating that they both give rise to the same chemical species on a surface, characterized by one S–Au bond.[112] SAMs from alkanethiols and corresponding dialkyl disulfides are indistinguishable by many physical criteria such as thickness and rate of formation. Alkanethiols and dialkyl disulfides differ, however, in their ability to replace thiolate groups present in formed SAMs. It has been demonstrated that dialkyl disulfides are slower to interchange in the competitive adsorption between solution and the SAM than alkanethiols by a factor of ~75, probably due to steric hindrance.[113]

3.3.1.2.1 *Thioctic acid (TA)*

Thioctic acid (TA) (Fig. 3.2x), also known as a lipoic acid (LA), is an organosulfur compound derived from octanoic acid. It is a naturally occurring substance, synthesized by humans as well as by some plants and animals. It is found inside every cell of the body, where it helps to generate energy in the metabolism of glucose molecules. Thioctic acid has received great attention in recent years in the area of gold functionalization for a variety of applications, ranging from the immobilization on gold electrodes and the investigation of metal ion chelation, to the probing of nanoparticle surface adsorption immobilizing various molecules for bioanalytical applications.[114]

Most of the studies show that cyclic disulfide binds particularly strongly to the surface of gold and results in a densely packed thiol layer. The greater thermodynamic stability probably arises from the ring strain generated in the formation of cyclic disulfide desorption products.[111] The main reason for the high stability of cyclic disulfides results, in part at least, from anchoring each molecule to gold through two sulfur atoms, which is more difficult to detach as compared to only one S–Au bond.[115] Nevertheless, some controversial studies have demonstrated the opposite effects of double-bonded thiols on SAMs stability. Langry *et al.* [116] showed with atomic force microscopy that thioctic acid, a spatially constrained system with two sulfur linkages to gold, is less stable to tensile stress than a thiolate with a single attachment to gold.

Regardless of the double-bond stability disadvantages demonstrated, thioctic acid has been reported to be a popular conjugating molecule for DNA[114,117] and antibody immobilization[118] through the carboxylic acid terminus. Addition of a thioctic acid solution onto a gold surface results in dissociation of the disulfide bond and spontaneous interaction between the two sulfurs and gold surface with tilt angle of 38° from the surface normal. Thioctic acid has a very short alkyl chain (four CH_2 units) with weak van der Waals interactions between neighboring molecules. A successful method for constructing reproducibly well-ordered thioctic acid monolayers is achieved by adding the acetic acid into solution, resulting in a high degree of interaction and hydrogen bonding between carboxyl groups.[119] Dougan *et al.*[114] modified oligonucleotides with thioctic acid and demonstrated that gold nanoparticles conjugated to thioctic acid-oligonucleotides are more stable compared to monothiol or dialkyl disulfide modifications.

3.3.1.2.2 *Dithiobis-(succinimidyl propionate) (DSP)*

Dithiobis-(succinimidyl) propionate (DSP) (Fig. 3.2f) is a cleavable and membrane-permeable crosslinker that contains an amine-reactive NHS-ester at each end. Adsorption of this molecule on the gold surface results in a breakage of the disulfide bond and formation of two mono-thiol bonds. There has been some controversy concerning the order of the monolayer formed from

DSP. Some studies show that it is unlikely that such a molecule will form a highly ordered monolayer on a gold surface.[120] However, others showed that, for example, attaching DNA to gold via DSP, instead of directly through a chemisorbed sulfur, allows a better defined surface density of DNA to be prepared, providing an improved framework for hybridization.[121,122]

3.3.1.2.3 *Calixcrown*

Calixcrowns or calixarenes (Fig. 3.2c) are cyclic, cup-shaped oligomers based on a hydroxyalkylation product of a phenol and an aldehyde. The word *calixarene* is derived from *calyx* or *chalice*, because this type of molecule resembles a vase, and, from the word *arene*, which refers to the aromatic building block. Calixcrowns are fixated to the gold surfaces through their thiol moieties to form SAMs, allowing tight binding of capture proteins to the crown moiety of the molecule.[123,124] Oh *et al.*[125] showed that calixcrowns interact with proteins containing extra arginine residues, immobilizing them with the correct orientation. The vertical orientation of the proteins resulted in a high-density attachment of protein, leading to a high specificity and reproducibility.

3.3.1.3 Trithiol molecules

Trithiol molecules are characterized with a three S–Au bonds formation, thus further improving the attachment strength of the molecule when compared to single-thiol group-containing molecules. The stable immobilization of oligonucleotide probes was demonstrated by the use of synthesized tripodal thiol derivatives, because they could be bound to the gold surface through three thiol groups. In addition, oligonucleotide probes could be immobilized perpendicular to the surface of the substrate based on the regular tetrahedron structure of tripodal derivatives. This kind of immobilized probe orientation makes hybridization much more efficient, even though the density of monolayers formed by molecules containing one thiol group is higher than the molecules with tripodal thiol.[126] The density is much lower in the case of tripodal thiol molecules due to the structural difference, because they take up more space on the gold surface.

One of the well-known issues in the long-term storage of gold covered with SAMs is that their storage in buffer solution, such as PBS, results in deterioration of SAM monolayer.[127] An additional problem with SAM layers is their instability at high temperatures as a result of which the S–Au bond dissociates. Sakata *et al.*[126] investigated the S–Au thermal bond strength of tripodal and single thiols. He found that at high temperatures tripodal thiols were tightly bound to the gold surface and remained connected to the gold surface at one-point contact, even if the other two S–Au bonds detached. Phares *et al.*[128] has demonstrated that a trihexylthiol molecule deposited on the gold surface improves solution-phase storage stability, thermostability, and robustness. The use of trithiol anchors to improve the stability of electrochemical biosensors represents an advance over earlier efforts to improve sensor stability by increasing SAM thickness. Fox *et al.*[129] used aminotrithiol to form a monolayer onto a gold surface and demonstrated high yields of surface coupling and significant local order within the two-dimensional array. Li *et al.*[130] reported the synthesis of trithiol-terminated nucleotides and gold nanoparticles conjugates prepared from it. These DNA-gold nanoparticle conjugates appeared to be very stable and can be used in different types of assays. Other research, such as that of Siiman *et al.*[131] showed conjugation of antibodies to aminotrithiols. Zheng *et al.*[132] introduced a new amplification strategy for very sensitive electrochemical detection of thrombin through the formation of network-like thiocyanuric acid (TCA)/gold nanoparticles. This star-shaped molecule can form bonds with three nanoparticles, creating a network that can enhance significantly the sensitivity of the system.

To conclude, trithiol molecules significantly improve the stability of the SAM layers without sacrificing electron transfer efficiency. In general, shorter SAMs support efficient electron transfer, but they pack poorly and are thus relatively unstable. Longer SAMs are more stable but suffer from less efficient electron transfer, thus degrading sensor performance. Taken together, this suggests that the flexible trithiol linkage chemistry may be of significant general utility in the fabrication of biosensors and efficient immobilization of capture proteins on solid matrixes.

3.3.1.4 Mixed SAMs

In recent years the construction of mixed self-assembled monolayers (mixed SAMs) has been investigated intensively owing to its potential application for biosensors. A controlled chemistry and structure can be introduced at the mixed SAMs interfaces by co-adsorbing thiols of different chain lengths or end group functionality.[133–135] Mixed SAMs consist of two different functionalities simultaneously incorporated on the gold surface, which allows for conjugation of bioreceptor molecules on one hand and induces specific interaction of the immobilized biomolecules instead of a nonspecific adsorption of undesired biological entities, on the other hand.[136,137] Mixed SAMs using alkanethiols of different chain length or end-group functionality appear to form a two-layer structured monolayer. Consequently, an efficient biomolecule immobilization can be achieved due to enhancement in a specific interface area and reduction in steric hindrance inherent to SAM interface, which was brought out by using alkanethiols of different chain length and end-group functionality.[133,138] Bain *et al.*[134,135] investigated the formation of monolayers by the coadsorption of thiols on gold as a function of the length of the alkyl chain, head group, tail group, and solvent present. The main conclusion from his studies is that long-chain alkanethiols form ordered, oriented monolayers on gold and are adsorbed preferentially over molecules containing a wide range of other functional groups. In general, the ratio of the concentration of the two components in a mixed monolayer is not the same as in solution, but reflects the relative solubility of the components in solution and interactions between the tail groups in the monolayers. Owing to a combination of all the mentioned positive properties, mixed layers can be an extremely useful tool for immobilization of biomolecules on metal surfaces.

Nevertheless, one of the issues when dealing with mixed monolayers is the formation of islands composed from one type of thiol.[139,140] Bertilsson[141] and Bain[135] demonstrated that the islands that do form are too small and insignificant to influence the monolayer integrity. Different combinations of the molecules have been composed in an effort to improve even more the mixed thiol layers properties: a mixture of 3-MPA and 11-MUA;[142] of mercaptoundecanol (Fig. 3.2r) and 12-mercapto-(8-biotinamide-3,6-dioxaoctyl) dodecanamide (Fig. 3.2k);[143] of 1-hexadecanoic acid (Fig. 3.2j) for antibody immobilization and 11-mercapto–1-undecanol (Fig. 3.2r) (for avoiding nonspecific interaction with gold surface;[136] of dithiobis-*N*-succinimidyl propionate (DTSP) (Fig. 3.2f) and 1-tetradecanethiol (Fig. 3.2v);[144] of thioctic acid (Fig. 3.2x) and 1-dodecanethiol (Fig. 3.2g);[145] and of 3-mercaptopropanol (Fig. 3.2n) and 11-MUA (Fig. 3.2q).[146] All of these definitely show improved properties of the thiol layer in comparison to a single-component layer and, therefore, can fulfill the necessary requirements for forming stable and well-ordered mixed SAMs.

3.4 Immobilizing Bioreceptors on the Surface

SAMs combine several properties that make them the best and the most studied available chemistries for modeling biological surfaces. Nonetheless, there is an array of different immobilization techniques that are not based on thioled surface bonding chemistry with their advantages and disadvantages. This section discusses the immobilization techniques that are not based on thiolated SAMs, as well as those that involve thiol-metal bonds. The section also investigates insight onto the modification and chemical properties of the biomolecule: what kind of modifications can be performed on them and which kind of interactions can those molecules exhibit?

Various types of immobilization procedures are available to couple the biomolecules to the substrate surface. These procedures can be divided to two main kinds of interactions or coupling: noncovalent interactions and covalent binding of the biomolecule to the linker or to the surface itself. Noncovalent coupling between the monolayers and the biospecies can be represented via electrostatic, hydrophobic/hydrophilic interactions. Covalent attachment, on the other hand, should involve the participation of functional groups from both sides of the linker and the molecule. The most prominent example is the attachment of enzymes or antibodies to the surface of the monolayer containing some free amine terminal groups through amide linkage formations or, probably, the attachment through some crosslinker with the surrounding functional groups. Other kinds of interactions such as bioaffinity, supramolecular interactions, physical adsorption, sol-gel entrapment, polymeric membranes, use of intercalators, and spin-on-glass are also gaining a lot of interest in the immobilization of biomolecules. In the following, the techniques and methods usually employed for immobilization are discussed according to the kind of interactions exploited.

3.4.1 Noncovalent Interactions

Noncovalent immobilization methods are widely employed, involving simple passive physical adsorption onto surfaces (e.g., hydrophobic character), weak interactions such as hydrogen bonding and van der Waals forces, ionic bonds, or electrostatic interactions with charged surfaces. Which of these intermolecular forces dominate the interaction will depend on the particular biomolecule and surface involved.

3.4.1.1 Physical adsorption

The simplest way to conjugate protein on gold is by physical adsorption, depositing the protein solution on the substrate surface. Protein physically adsorbs to the gold surface mainly via an $-NH_2$ group of lysine, but also via an –SH group of cysteine,[22] and an –OH group of serine or threonine.[23] As it mentioned above, a high affinity of cysteine residues in the protein toward the gold surface can be used for efficient chemisorptions of the proteins.[147,148] With this type of adsorption, no control over the orientation of the attached molecule is achieved: the attachment is nonspecific. Some of the antibodies that will immobilize in this way will be denaturated and, consequently, will lose the ability to bind the specific antigen.

Adsorption strategies are mainly used with three-dimensional porous materials that can entrap the bioreceptor, such as gels and membranes. Polypropylene membranes modified with polyaniline allow an adhesion mechanism of combined electrostatic and hydrophobic interactions and demonstrate high affinity and compatibility toward different proteins.[149] However, the most-used polymer substrates for immobilization on gold are hydrogels, mainly due to the widespread success of Biacore SPR read-out technology.[150] Among hydrogels, sulfate-modified dextrans have performance better than aspartic-modified dextrans and commercial carboxymethyl cellulose, when used to adsorb proteins contained in a crude extract of *E. coli*.[151]

Drawbacks of the physical adsorption mechanism are random orientation and weak attachment, since proteins may be removed by some buffers or detergents when assays are performed.[17]

3.4.1.2 Hydrophobic interactions

Immobilization via hydrophobic interactions on alkanethiol SAMs has been reported to be suitable for some proteins.[152] Such interactions were achieved through coupling between the hydrophobic area of the molecule and the akanethiol methyl heads. Hydrophobic interactions of alkanethiol SAM can be used to assemble a second phospholipid layer, creating a so-called hybrid bilayer membrane (HBM): one layer of such a membrane includes the hydrophobic thiol molecules and another is located on the top of the first one.[153] This system resembles a true biological membrane, at least regarding the second layer and the internal hydrophobicity. In particular, the SAM turns the bilayer into a highly stable structure. HBMs are well suited for investigating reactions occurring at the membrane interface, such as protein surface interactions, binding of ligands to membrane-confined receptors, and embedding of some redox proteins, which make these interfaces very promising for biorecognition and biosensing applications.[4,154,155]

3.4.1.3 Electrostatic interactions

Biomolecule immobilization can also be achieved through electrostatic interactions. Surfaces that are modified with positively charged amine groups or negatively charged carboxyl groups are most suitable for this approach. The resulting adsorbed protein layer is likely to be heterogeneous and randomly oriented, since each molecule can form many contacts in different orientations, minimizing repulsive interactions with the surface and previously adsorbed proteins.[19] Although electrostatic interactions are not directive and are difficult to control, they are often used when patterning antibodies onto negatively charged nanoparticles that are deposited on surfaces in specific patterns.[156]

Wrobel *et al.* investigated the physical adsorption of [32] P-labeled DNA oligomers onto SAMs formed by carboxy-modified alkylthiols on gold.[157] Immobilization of proteins has been also studied using phosphate and phosphonate-terminated SAMs as a negatively charged substrate,[158,159] indicating that such SAM-functional interfaces provide a favorable biocompatible environment for the protein and fabrication of biosensors in general.

The major advantage of electrostatic-based immobilization is that neither additional coupling reagents nor further modification of the protein is required. Nevertheless, noncovalent immobilization typically involves relatively weak and reversible interactions. This has several implications for the reusability of systems, particularly when they are used in analytical assay and sensor devices.[4] During adsorption, the proteins often undergo conformational changes and consequent denaturation that can significantly reduce their activity.[160] Furthermore, since the packing density of the immobilized proteins is uncontrolled, their activity may be further reduced by steric hindrance and nonspecific adsorption.[161]

3.4.2 Covalent Interactions

Methods for immobilization that rely on covalent coupling of biomolecules to surfaces are inherently more controllable and result in layers of molecules that are unable to dissociate easily from the surface. Covalently bonded proteins are stable and are not usually subject to solvent mediated desorption.[162] This type of coupling also allows the controlled orientation of immobilized antibody, which can be crucial for antibody-antigen bonding.[163] A variety of surface chemistries and SAM modifications were discussed above. In general, many functional groups of biomolecules can be used for direct chemical coupling with suitable surface via well-known bioconjugate reactions: the peptide bond between amine and carboxylic acid termini, aldehydes reacting with amine to form imine, thiol chemistry such as maleimide functional group, or disulfides reacting with cysteine groups.

Chemical binding via functional groups, mainly side chains of amino acids, is often random, because it is based on residues typically present on the exterior of the protein. Therefore, the attachment may occur simultaneously through many residues, thereby restricting degrees of conformational freedom and enhancing heterogeneity in the population of immobilized proteins. Chemical attachment can also be guided in an orderly manner to attain oriented immobilization.[19] Well-defined immobilization procedures provide reproducible and oriented immobilization, avoiding denaturation, due to a reaction with amino acid functional groups not involved in binding. Site-specific immobilization requires functionalization of the molecules blocking some specific residues or tailoring of the surface. In some cases, oriented immobilization can be obtained if the protein possesses a single reactive amino acid in the structure. Theoretically, site-specific immobilization should be optimal, overcoming the insufficient exposure of functional residues due to unpredictable orientation. Consequently, proteins can be attached to surfaces in a well-ordered manner, allowing reproducibility and conformational stability.[17,164]

3.4.2.1 Amine chemistry

By far the most common way to attach proteins to surfaces covalently uses the amine group of the lysine side residue, typically present on the exterior of the protein. However, the abundance of those groups can lead to heterogeneity and restricted conformational flexibility due to the multipoint attachment to the substrate. Immobilization efficiency depends on several parameters, such as pH, concentration, ionic strength, and reaction time, and in specific cases the conditions need to be optimized for each type of protein. Nevertheless, the NHS ester is sufficiently reactive and stable toward hydrolysis if standard reaction procedures are applied, such as phosphate-buffered saline at pH 7.3 and at room temperature.[19] Another widely used strategy employs surface-bound aldehyde groups that can react with amine groups of proteins to form labile imine linkages, which can be subsequently stabilized by reduction with sodium borohydride to form secondary amines.[165–167]

3.4.2.2 Carboxyl chemistry

An alternative to amine-based immobilization is immobilization through carboxyl groups of aspartic and glutamic acids of surface groups on proteins. Therefore, instead of adsorbing proteins via multipoint interactions onto aminated surfaces, mild coupling methods may be used to attain covalent immobilization.

Fernandez-Lafuente *et al.* proposed an effective strategy to covalently bind proteins via carboxyl groups using carbodiimide (CDI) as an activator of COOH groups. This method, however, has the disadvantage of rapid cross-linking induction of the proteins, because of the reaction of the activated carboxylic ester with the amine groups of lysine residues.[168]

3.4.2.3 Supramolecular interactions

An additional strategy for modifying the surface composition through noncovalent interactions providing the highest degree of specificity is based on the use of designed supramolecular interactions.[169] One of the most intriguing molecules in the field of supramolecular interactions is cyclodextrin (Fig. 3.2d). Cyclodextrins are a family of cyclic oligomers of sugar molecules bound together in a ring. As a consequence of this peculiar structure, the molecule features a conical cavity that is essentially hydrophobic in nature. The cavity allows the inclusion of several types of guest molecules of appropriate size to form inclusion complexes. Moreover, cyclodextrins can be chemically or enzymatically modified, which makes them of high interest for immobilization on surfaces. Monolayers based on supramolecular interactions of cyclodextrin SAMs, also called "molecular printboards" (surfaces functionalized with receptor molecules), offer unique properties for immobilizing proteins through small multivalent, orthogonal linker molecules.[170] This method allows control both over the binding strength, by varying the valence of the linker at the "printboard," and over the orientation of the protein, by the bioengineering of a specific binding site at a predetermined location in the protein. These monolayers have recently been applied to study the specific binding of some model proteins.[171,172]

3.4.2.4 Bioaffinity interactions

One of the current challenges for mimicking biological interfaces, proteomics, and medicine screening is the control over protein adsorption at surfaces and interfaces. In most cases, this control relies on a "key-lock system" for specifically attaching a particular protein to an engineered surface. Biochemical affinity reactions offer a gentle oriented immobilization of proteins, providing an important advantage over other immobilization techniques. Moreover, it is also possible to detach proteins and to use the same surface over and over again.[4,17]

Avidin-biotin system

One of the most well-known and extensively studied strategies, which was adapted from affinity chromatography, is the specific noncovalent binding of biotin to the proteins avidin or streptavidin. Biotin, also known as the B_7 vitamin, is a coenzyme that takes part in the metabolism of fats and amino acids. Streptavidin, a tetrameric protein purified from a bacterium *Streptomyces avidini*, possesses four binding sites for a biotin molecule. Owing to the high binding affinity and specificity between streptavidin and biotin (with an association constant of 10^{15} M^{-1} and a binding free energy of −18.3 kcal/mol), the formation of this complex can be regarded as nearly an irreversible and the strongest known bond in biochemistry, on a scale nearly comparable to a covalent bond.[173–174]

Biotin-streptavidin bond formation is very rapid, thermostable, and resistant to extreme pH, organic solvents, and other denaturating agents. Streptavidin is usually preferred over avidin, because glycol-amino acids occurring on avidin can potentially cause unwanted nonspecific adsorption. Consequently, this high-affinity biotin-streptavidin binding system and its intrinsic properties have been found to have many surface applications in immobilization processes.

The biotin-streptavidin design principle was first demonstrated by Muller, *et al.*,[175] who conjugated biotinylated concanavilin. A and biotinylated antibody Fab fragments to biotin-streptavidin monolayers, producing multilayers that present uniformly oriented proteins.

The simplest procedure is to immobilize alkanethiols conjugated with streptavidin onto the gold surface. Afterward, biotinylated DNA or protein can be firmly attached to those layers, thereby forming the biotin-streptavidin linkage.[176,177] However, such a design principle yields a lower degree of organization. Therefore, another procedure of stacked composition biotin/streptavidin/biotin instead of direct streptavidin immobilization is generally preferred.[19]

Protein A-mediated immobilization

Protein A is an immunoglobulin-binding bacterial protein found in the cell wall of the bacterium *Staphylococcus aureus*. This protein has gained increasing importance as a tool in both quantitative and qualitative immunological techniques because of its ability to interact with immunoglobulins, mainly IgG, from mammalian species. It can be divided into two main structurally and functionally different regions: an N-terminally located region consisting of four globular highly homologous immunoglobulin-binding units and one C-terminally located nonimmunoglobulin binding region that is covalently linked to the cell wall.[178] An advantage of using protein A for immobilization is its stability in high temperatures, low acidic pH, and after exposure to denaturating agents. Therefore, a gold surface coated with protein A can stay stable for weeks, without any signs of deterioration.[179] Another advantage of the noncovalent bond between protein A and gold is the relative ease with which the molecule can be detached from the surface, allowing the re-use of the surface for other assays in experimental setups.

Unlike thiols, protein A binds to gold via adsorption. The major force of attraction of protein A to gold is considered to be the van der Waals force, and the major force of repulsion is thought to be the electrostatic double-layer force. Horisberger and Clerc[180] have investigated a protein A-gold bondage extensively. They showed that the adsorption of protein A can be divided into two regions: low and high coverage. At low gold surface coverage, it is thought that the molecules lie mostly flat at the interface, resulting in a better attachment to a substrate, but binding sites of protein A will become inaccessible to the Fc portion of IgG. At a high coverage, protein A molecules become attached by a shorter segment, decreasing stability. However, because they are attached by a shorter segment, flexible molecules such as protein A can extend into the solution. Thus one or both binding sites of protein A become available for combining with IgG.

The immobilization of antibodies using protein A relies on the specific interaction with the Fc constant region of IgG molecules.[181] Free protein A can bind two IgG molecules, while the immobilized protein binds only one IgG molecule, due to a steric hindrance. Harboe *et al.*[182] showed that some fractions of IgM, IgA, and IgE interact with protein A as well.

Numerous studies have been performed to investigate protein A attachment to gold and antibodies. Anderson *et al.*[183] investigated the effectiveness of protein A as an IgG binding protein on the surface of fiber optics. He demonstrated that no difference was observed in the binding of IgG, whether the capture antibody was bound to the probe surface via protein A or covalently attached. In the study done by Kanno *et al.*,[184] the B-domain, which is one of IgG-binding domains of staphylococcal protein A, was repeated five times and a cysteine residue was introduced at its C-terminus using a genetic engineering technique. The newly designed protein retained the same IgG-binding activity as native protein A, utilizing a strong affinity between thiol of cysteine and a gold surface.

A disadvantage of protein A-mediated immobilization is the lack of control on the orientation of protein A itself. Therefore, different approaches have been presented in order to achieve higher orientation control upon immobilization. Shen *et al.*[185] tried to improve immobilization via protein A by indirectly binding protein A onto sensors through a three-dimensional hyperbranched polymer, synthesized from *p*-phenylenediamine (Fig. 3.2t) and trimesic acid. Such a three-dimensional structure and its characteristics, including orientation control and biocompatibility of protein A, led to highly efficient immunoreactions and enhanced detection system performance. In the work of Johnson *et al.*[186] a hexahistidine tag was added at the C-terminus of the excised B-domain of a truncated version of protein A. The tag orients protein A on a surface and lets the binding site be free to interact with the Fc region of the antibodies. Lee *et al.*[187] fabricated an ordered layer-by-layer architecture, with a self-assembled *N*-succinimidyl-3-(2-pyridyldithio)propionate layer, attached to the gold surface through an –SH bond, reacting with the primary amine of protein A.

In addition to protein A, other immunoglobulin-binding bacterial proteins such as protein G and protein L are commonly used to purify, immobilize, or detect immunoglobulins. Each of these immunoglobulin-binding proteins has a different antibody binding profile in terms of the portion of the antibody that is recognized and the species and type of antibodies it will bind.

3.4.2.5 Electrochemical immobilization

Although conventional procedures for deposition of biomolecules onto solid surfaces, such as physical adsorption, covalent bondage, and entrapment have been used extensively, they suffer from a low reproducibility and have a poor spatially controlled deposition. Recently, the focus of immobilization has been shifted toward the entrapment of biomolecules in the layers of electrochemically synthesized polymers. Aizawa *et al.*[191] pioneered the application of PPy and PANI to the immobilization of enzymes. Since then, hundreds of papers have been published and patents have been granted in this area.[188–191]

An electrically conducting polymer is generally comprised of a conjugated polymer chain with electrons delocalized along the backbone, yielding a semiconducting polymer that can be reversibly tuned through doping, an oxidation/reduction process in which charge carriers are introduced to the polymeric backbone either chemically or electrochemically.[30] Electrically conducting polymers have considerable flexibility in the available chemical structure, which can be modified as required, enabling modulation of specific electronic and mechanical properties. Electrochemical immobilization methods have four major advantages:

1. The deposition on the gold electrode is direct, only one step, and faster than in all immobilization procedures.

2. The distribution of the immobilized biomolecule is spatially controlled.
3. The electrogenerated film thickness can be precisely controlled.
4. It allows the reproducible and precise formation of a polymer coating over surfaces.[16,192]

Grafting conducting polymers onto gold surfaces can be an effective way to immobilize biomolecules. The immobilization of biomolecules onto conducting polymers is generally achieved by incorporating them as functional dopant, capable of imparting molecular recognition to the polymer structure.[193] This does not require a multistep immobilization process, as using thiol derivatives does. Moreover, the conducting polymers can efficiently transfer the electric charge produced by the biochemical reactions.[194] These properties make conductive polymers popular linkers in the construction of amperometric biosensors.

Different types of conducting and nonconducting polymers are used for biomolecule immobilization. Si *et al.*[195] compared polymers containing NH_2 groups, such as o-aminophenol, o-phenylene-diamine (oPD), and m-phenylene-diamine (mPD) and demonstrated that they can be used as a functional coating for antibody immobilization on gold. Loyprasert, *et al.*[196] used polytyramine (nonconducting polymer) to coat a gold electrode. Li[197] and Huang[198] used *O*-phenylene diamine, constructing a semi-insulating membrane that directly adsorbed the antibodies, and that can be used for electrochemical sensors.

One of electroconducting polymers, which became an extremely popular molecule in the polymerization process, is polypyrrole (Fig. 3.2u). The electrochemical oxidization of pyrrole gives, in one step, a solid polypyrrole film laid on the surface of the electrode. Copolymerization of pyrrole and protein-linked pyrrole is an efficient grafting process, which immobilizes molecules at defined positions on a gold surface.[199] DNA strands tethered on polymerized pyrrole form a solid layer of grafted DNA.[201,210] Livache, *et al.*[202] introduced an electro-spotting process, in which the spots are 500 μm in diameter, 10 nm thick, covalently linked to oligonucleotides, and compatible for SPR measurements. Such surfaces coated with pyrrole can be reused for numerous monitoring procedures, saving money and time.

3.4.2.6 Other examples of biomolecule immobilization

Intercalators

Conjugation of molecules to a DNA strand can be performed in any of three general ways: (1) by covalent attachment; (2) through bonding to the groove by weak forces, including hydrophobic, electrostatic, and hydrogen bonding; (3) or with intercalators, which are hetero-aromatic moieties stacked between DNA base pairs. The intercalating surface is tightly packed and rigidly held between the base pairs and stabilized electronically in the helix by π–π stacking and dipole-dipole interactions. Intercalation produces an extension, unwinding, and stiffening of the DNA helix. These changes are a consequence of the untwisting of the base pairs and helical backbone needed to accommodate the intercalator.[203] Letsinger[115] used epiandrosterone as a steroidic intercalator that was connected to cyclic disulfide to anchor to gold nanoparticles oligonucleotides. Gold nanoparticle/oligonucleotide conjugates made using this cyclic disulfide linker served as effective probes for detecting specific oligonucleotide sequences, and they exhibit much greater stability than corresponding conjugates prepared with the conventional mercaptohexyl group or an acyclic disulfide unit. Nakamura *et al.*[204] used anthryl moieties attached to the terminal OH groups in SAM of 1-Mercapt–11-undecanol to immobilize DNA to a gold surface. Sato *et al.*[205] used naphthalene diimide as an intercalating molecule between double stranded DNA. The intercalator was synthesized with two dithiolane moieties for the attachment to gold surfaces. All of these manipulations can be carried out under mild conditions and hence the bound DNA retains functional integrity as revealed by the higher efficiency of hybridization with target DNA. Wang *et al.*[206] investigated an ethidium intercalation as a means of inducing binding of gold nanoparticles to DNA. An acridine derivative containing an alkyl chain whose end is modified with disulfide bond was studied by Higashi[207] to create a novel acridine derivative connected with a disulfide bond through a long methylene and its interactions with DNA in solutions at a self-assembled monolayer surface on gold.

The main disadvantage of work with intercalators is their ability to cause deleterious effects to human health, including different types of cancer, because they interfere with human DNA; therefore, working with those materials needs to be done in extreme caution.

Diazonium

Using diazonium derivatives to develop electrochemical biosensors have been a subject of great interest due to their excellent conductivity. Diazonium group-functionalized electrodes are of special interest owing to their ability to further react with phenolic, imidazole, or amino groups to form covalent diazo bonds for the achievement of different types of surface derivatization.[208] Corgier *et al.*[209] introduced the use of diazonium-modified antibodies for the direct electrically addressable immobilization of proteins. The main advantages of such an immobilization strategy are the ease of preparation, the stability of the covalent surface chemistry bond, and the ability to address the protein spatially onto microelectrode arrays.[210] Civit *et al.*[211] studied the thermal stability of a series of modified gold surfaces with mono- and dithiol molecules, as compared with diazonium salts with one and two diazo-groups for grafting on a gold electrode. Diazonium salt–derived layers were found to be thermally stable up to significantly higher temperatures than alkanethiol SAMs on gold surfaces. Li, *et al.*[208] conjugated 4-aminothiophenol to a gold surface using diazoaminobenzene, subsequently immobilizing a thiolated ss-DNA. This biosensor displayed high sensitivity, good selectivity, regeneration ability, and an improved electron transfer ability.

Spin-on-glass

Spin-on-glass is a type of glass that can be applied as a liquid and curved to form a layer of glass having characteristics similar to those of SiO_2. Spin-on-glass is mainly used for planarization and as a dielectric material. If applied with reactive ends it can be highly suitable to optic biosensors, such as SPR. Kessler *et al.*[212] synthesized a poly (methylsilsesquioxane)-poly(pentafluorophenyl acrylate) (PMSSQ-PFPA) polymer to immobilize protein on a thin and reactive gold surface. PMSSQ are being deposited on the surface by spin-coating, and strong adhesion of those polymers to the surface is being established. By grafting PFPA from PMSSQ, reactive coating is established and biomolecules can be immobilized.

3.5 Conclusion

The aim of this chapter was to introduce the reader to some of the chemical platforms of biosensors and the myriad strategies for modifying gold surfaces in order to immobilize biomolecules, such as proteins and DNA. As shown, thiols have a strong tendency to be chemisorbed onto gold surfaces, such as when they are used to form self-assembled monolayers. Immobilization through thiol derivatives is the most popular way of bioconjugation, since the early work of Nuzzo in 1983. The simple protocol for manufacturing, dense and oriented layer formation, and relative ease of bio-molecules attachment make thiols, undoubtedly, leading molecules in the biosensor field. Thiols are among the most successfully employed chemicals for modification of metal surfaces because they react chemically with gold, silver, and copper, forming very stable metal-sulfur bonds.

Additional, functional groups attached to the end opposite to the SH group control the surface properties of formed monolayers. The flexibility to design different head groups of monolayers using a large number of functional groups makes this functionalization strategy especially useful for the controlled fabrication of structurally ordered assemblies of proteins on surfaces. Therefore, one of the most important applications of monolayers of functionalized alkanethiols is the binding of many different molecules, which cannot be directly attached to an unmodified metal surface. Such an intrinsic feature of strong S–Au bond formation has been exploited in numerous applications, including the immobilization of DNA on gold surfaces for the generation of DNA microarrays or gold electrodes for measuring charge migration mediated by double stranded DNA, immobilization of antibodies or antigens.

Understanding the benefits and disadvantages of using various types of thiols, such as mono-, di-, triple, or mixed thiol molecules, can help the researcher choose the right molecule for the specific purpose of immobilization. Other methods, not including the modification of gold through thiol groups, were also presented. Possible interactions of biomolecules, such as noncovalent, covalent, bioaffinity, and supramolecular bonding have been mentioned in the context of use during immobilization onto chemically modified surfaces. Functionalization of a gold surface with alkanethiols to bind bioreceptors through different immobilization techniques is summarized in Fig. 3.3.

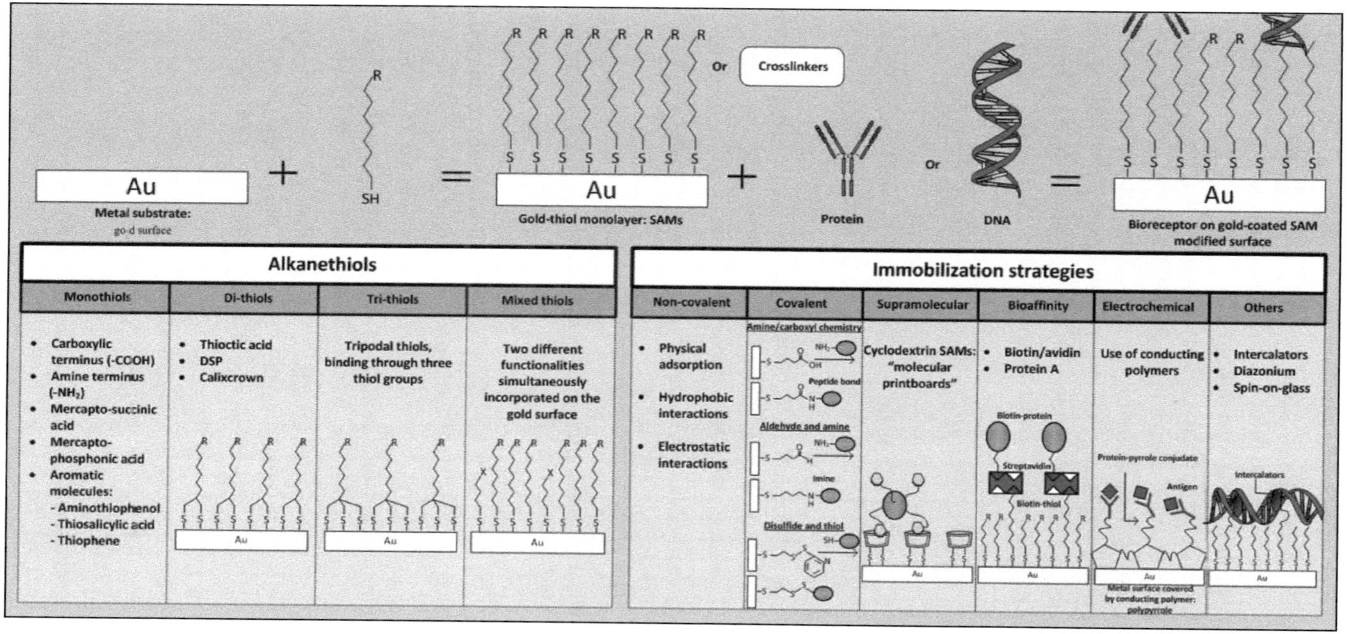

Figure 3.3 Bioreceptor functionalization of gold-coated sensor surfaces. Different alkanethiols are immobilized through thiol linkage onto a gold surface. Afterward, the bioreceptor (protein or DNA) can be linked to the modified surface through various immobilization strategies: noncovalent, covalent, supramolecular, electrochemical interactions, bioaffinity, use of intercalators, and various chemicals and techniques.

SAMs of alkanethiols on gold are one of the most promising surfaces available to understand interactions of proteins and other biomolecules with artificial surfaces. The ease with which a conjugation complex can be presented in a controlled environment, combined with simple methods that can pattern the formation of SAMs in the plane of the monolayer, make these surfaces well suited for studies of fundamental aspects of biointerfacial science. Numerous other thiolated strategies do bring the biosensor researcher quite a number of possible conformations to be adapted in biomolecule immobilization protocols.

References

1. Marks, R. S., Lowe, C. R., Cullen, D. C., Weetall, H. H., and Karube, I. (eds) (2007). *Handbook of Biosensors and Biochips*, Wiley-Inter Science, Hoboken, NJ.
2. Eltzov, E., and Marks, R. S. (2009). Parameters to consider in the construction of fiber-optic biosensors as alternative bioanalytical tools, *IEEE Instrum. Meas. Mag.*, **12**(5), 10–16.
3. Lide, D. R. (ed.) (2001). *CRC Handbook of Chemistry and Physics*, 82 edn, CRC Press, Boca Raton, FL.
4. Frasconi, M., Mazzei, F., and Ferri, T. (2010). Protein immobilization at gold-thiol surfaces and potential for biosensing, *Anal. Bioanal. Chem.*, **398**(4), 1545–1564.
5. Ekgasit, S., Thammacharoen, C., Yu, F., and Knoll, W. (2005). Influence of the metal film thickness on the sensitivity of surface plasmon resonance biosensors, *Appl. Spectrosc.*, **59**(5), 661–667.
6. Schuck, P. (1997). Use of surface plasmon resonance to probe the equilibrium and dynamic aspects of interactions between biological macromolecules, *Annu. Rev. Biophys. Biomol. Struct.*, **26**, 541–566.
7. Abanulo, J. C., Harris, R. D., Sheridan, A. K., Wilkinson, J. S., and Bartlett, P. N. (2002). Waveguide surface plasmon resonance studies of surface reactions on gold electrodes, *Faraday Discuss.*, **121**, 139–152; discussion, 229–251.
8. Gryczynski, I., Malicka, J., Gryczynski, Z., and Lakowicz, J. R. (2004). Surface plasmon-coupled emission with gold films, *J. Phys. Chem. B*, **108**(33), 12568–12574.
9. Casero, E., Vázquez, L., Parra-Alfambra, A. M., and Lorenzo, E. (2010). AFM, SECM and QCM as useful analytical tools in the characterization of enzyme-based bioanalytical platforms, *Analyst*, **135**(8), 1878–1903.

10. Ferreira, G. N., da-Silva, A. C., and Tome, B. (2009). Acoustic wave biosensors: physical models and biological applications of quartz crystal microbalance, *Trends Biotechnol.*, **227**(12), 689–697.

11. Kurosawa, S., Park J.-W., Aizawa, H., Wakida, S.-I., Tao, H., and Ishihara, K. (2006). Quartz crystal microbalance immunosensors for environmental monitoring, *Biosens. Bioelectron.*, **22**(4), 473–481.

12. Dixon, M. C. (2008). Quartz crystal microbalance with dissipation monitoring: enabling real-time characterization of biological materials and their interactions, *J. Biomol. Tech.*, **19**(3), 151–158.

13. Sauerbrey, G. (1959). Verwendung von Schwingquarzen zur wägung dünner schichten und zur mikrowägung, *Z. Phys.*, **155**, 206–222.

14. Cooper, M. A., and Singleton, V. T. (2007). A survey of the 2001 to 2005 quartz crystal microbalance biosensor literature: applications of acoustic physics to the analysis of biomolecular interactions, *J. Mol. Recognt.*, **20**(3), 154–184.

15. Ahuja, T., Mir, I. A., Kumar, D., Rajesh, (2007). Biomolecular immobilization on conducting polymers for biosensing applications, *Biomaterials*, **28**(5), 791–805.

16. Cosnier, S. (2003). Biosensors based on electropolymerized films: new trends, *Anal. Bioanal. Chem.*, **377**(3), 507–520.

17. Rusmini, F., Zhong, Z., and Feijen, J. (2007). Protein immobilization strategies for protein biochips, *Biomacromolecules*, **8**(6), 1775–1789.

18. Sarma, A. K., Vatsyayan, P., Goswami, P., and Minteer, S. D. (2009). Recent advances in material science for developing enzyme electrodes, *Biosens. Bioelectron.*, **24**(8), 2313–2322.

19. Jonkheijm, P., Weinrich, D., Schröder, H., Niemeyer, C. M., and Waldman, H. (2008). Chemical strategies for generating protein biochips, *Angew. Chem. Int. Ed.*, **47**(50), 9618–9647.

20. Boyukbayram, A. E., Kiralp, S., and Toppare, L. (2006). Preparation of biosensors by immobilization of polyphenol oxidase in conducting copolymers and their use in determination of phenolic compounds in red wine, *Bioelectrochemistry*, **69**(2), 164–171.

21. Mrksich, M. (2005). Dynamic substrates for cell biology, *MRS Bull.*, **30**(3), 180–184.

22. Lu, B., Smyth, M. R., and O'Kennedy, R. (1996). Oriented immobilization of antibodies and its applications in immunoassays and immunosensors, *Analyst*, **121**(3), 29R–32R.

23. Sarikaya, M., Tamerler, C., Jen, A. K.-Y., Schulten, K., and Baneyx, F. (2003). Molecular biomimetics: nanotechnology through biology, *Nat. Mater.*, **2**(9), 577–585.

24. Tecilla, P., Dixon, R. P., Slobodkin, G., Alavi, D. S., Waldeck, D. H., and Hamilton, A. D. (1990). Hydrogen-bonding self-assembly of multichromophore structures, *J. Am. Chem. Soc.*, **112**(25), 9408–9410.

25. Saenger, W., and Egli, M. (1983). *Princp les of Nuclei Acid Structure*, Springer-Verlag, New York.

26. Moser, H. E., and Dervan, P. B. (1987). Sequence-specific cleavage of double helical DNA by triple helix formation, *Scie nce*, **238**(4827), 645–650.

27. Deisenhofer, J., Epp, O., Miki, K., Huber, R., and Michel, H. (1984). X-ray structure analysis of a membrane protein complex. Electron density map at 3 A resolution and a model of the chromophores of the photosynthetic reaction center from Rhodopseudomonas viridis, *J. Mol. Bio l.*, **180**(2), 385–98.

28. Zuber, H. (1986). Structure of light-harvesting antenna complexes of photosynthetic bacteria, cyanobacteria and red algae, *TIBS*, **11**(10), 414–419.

29. Arya, S. K., Solanki, P. R., Datta, M., and Malhotra, B. D. (2009). Recent advances in self-assembled monolayers based biomolecular electronic devices, *Bio sens. Bio electron.*, **24**(9), 2810–2817.

30. Mendes, P. M. (2008). Stimuli-responsive surfaces for bioapplications, *Chem. Soc. Rev.*, **37**(11), 2512–2529.

31. Ulman, A. (1996). Formation and structure of self-assembled monolayers, *Chem. Rev.*, **96**(4), 1533–1554.

32. Davis, F., and Higson, S. P. (2005). Structured thin films as functional components within biosensors, *Bio sens. Bio electron.*, **21**(1), 1–20.

33. Chaki, N. K., and Vijayamohanan, K. (2002). Self-assembled monolayers as a tunable platform for biosensor applications, *Bio sens. Bio electron.*, **17**(1–2), 1–12.

34. Nuzzo, R. G., and Allara, D. L. (1983). Adsorption of bifunctional organic disulfides on gold surfaces, *J. Am. Chem. Soc.*, **105**(13), 4481–4483.

35. Weisbecker, C. S., Merritt, M. V., and Whitesides, G. M. (1996). Molecular self-assembly of aliphatic thiols on gold colloids. *Langmui r*, **12**(16), 3763–3772.

36. Nuzzo, R. G., Fusco, F. A., and Allara, D. L. (1987). Spontaneously organized molecular assemblies 3. preparation and properties of solution adsorbed monolayers of organic disulfides on gold surfaces. *J. Am. Chem. Soc.*, **109**(8), 2358–2368.

37. Whitesides, G. M., and Laibinis, P. E. (1990). Wet chemical approaches to the characterization of organic-surfaces– elf-assembled mono-

layers, wetting, and the physical organic-chemistry of the solid liquid interface, *Langmuir*, **6**(1), 87–96.

38. Mrksich, M., and Whitesides, G. M. (1996). Using self-assembled monolayers to understand the interactions of man-made surfaces with proteins and cells, *Annu. Rev. Biophys. Biomol. Struct.*, **25**, 55–78.

39. Dubois, L. H. (1992). Self-assembled monolayers as models for organic surfaces, *Abstracts of Papers of the American Chemical Society*, **204**, 92–Phys.

40. Dubois, L. H., and Nuzzo, R. G. (1992). Synthesis, structure, and properties of model organic-surfaces, *Annu. Rev. Phys. Chem.*, **43**, 437–463.

41. Jadzinsky, P. D., Calero, G., Ackerson, C. J., Bushnell, D. A., and Kornberg, R. D. (2007). Structure of a thiol monolayer-protected gold nanoparticle at 1.1 A resolution, *Science*, **318**(5849), 430–3.

42. Mankefors, S., Grigoriev, A., and Wendin, G. (2003). Molecular alligator clips: a theoretical study of adsorption of S, Se and S-H on Au(111), *Nanotechnology*, **14**(8), 849–858.

43. Bain, C. D., Troughton, E. B., Tao, Y. T., Evall, J., Whitesides, G. M., and Nuzzo, R. G. (1989). Formation of monolayer films by the spontaneous assembly of organic thiols from solution onto gold. *J. Am. Chem. Soc.*, **111**(1), 321–335.

44. Nuzzo, R. G., Dubois, L. H., and Allara, D. L. (1990). Fundamental studies of microscopic wetting on organic-surfaces 1. formation and structural characterization of a self-consistent series of polyfunctional organic monolayers, *J. Am. Chem. Soc.*, **112**(2), 558–569.

45. Laibinis, P. E., Whitesides, G. M., Allara, D. L., Tao, Y. T., Parikh, A. N., and Nuzzo, R. G. (1991). Comparison of the structures and wetting properties of self-assembled monolayers of normal alkanethiols on the coinage metal-surfaces, Cu, Ag, Au, *J. Am. Chem. Soc.*, **113**(19), 7152–7167.

46. Strong, L., and Whitesides, G. M. (1988). Structures of self-assembled monolayer films of organosulfur compounds adsorbed on gold single-crystals—electron-diffraction studies, *Langmuir*, **4**(3), 546–558.

47. Widrig, C. A., Alves, C. A., and Porter, M. D. (1991). Scanning tunneling microscopy of ethanethiolate and normal-octadecanethiolate monolayers spontaneously adsorbed at gold surfaces, *J. Am. Chem. Soc.*, **113**(8), 2805–2810.

48. Alves, C. A., Smith, E. L., and Porter, M. D. (1992). Atomic scale imaging of alkanethiolate monolayers at gold surfaces with atomic force microscopy, *J. Am. Chem. Soc.*, **114**(4), 1222–1227.

49. Clegg, R. S., Reed, S. M., and Hutchison, J. E. (1998). Self-assembled monolayers stabilized by three-dimensional networks of hydrogen bonds, *J. Am. Chem. Soc.*, **120**(10), 2486–2487.
50. Tam-Chang, S. W., Biebuyck, H. A., Whitesides, G. M., Jeon, N., and Nuzzo, R. G. (1995). Self-assembled monolayers on gold generated from alkanethiols with the structure RNHCOCH 2 SH, *Langmuṙ* , **11**(11), 4371–4382.
51. Valiokas, R., Östblom, M., Svedhem, S., Svensson, S. C. T., and Liedberg, B. (2002). Thermal stability of self-assembled monolayers: influence of lateral hydrogen bonding, *J. Phys. Chem. B*, **106**(40), 10401–10409.
52. Porter, M. D., Bright, T. B., Allara, D. L., and Chidsey, C. E. D. (1987). Spontaneously organized molecular assemblies. 4. Structural characterization of *n*-alkyl thiol monolayers on gold by optical ellipsometry, infrared spectroscopy, and electrochemistry, *J. Am. Chem. Soc.*, **109**(12), 3559–3568.
53. Dannenberger, O., Weiss, K., Himmel, H.-J., Jäger, B., Buck, M., and Wöll, C. (1997). An orientation analysis of differently endgroup-functionalised alkanethiols adsorbed on Au substrates, *Thṅ Solḋ Fl ms*, **307**(1–2), 183–191.
54. Delamarche, E., Michel, H. Kang, H., and Gerber, C. (1994). Thermal stability of self-assembled monolayers, *Langmuṙ* , **10**(11), 4103–4108.
55. Bensebaa, F., Ellis, T. H., Badia, A., and Lennox, R. B. (1998). Thermal treatment of *n*-alkanethiolate monolayers on gold, as observed by infrared spectroscopy, *Langmuṙ* , **14**(9), 2361–2367.
56. Kim, Y. T., Mccarley, R. L., and Bard, A. J. (1993). Observation of *n*-octadecanethiol multilayer formation from solution onto gold, *Langmuṙ* , **9**(8), 1941–1944.
57. Rosendahl, S. M., and Burgess, I. J. (2008). Electrochemical and infrared spectroscopy studies of 4-mercaptobenzoic acid SAMS on gold surfaces, *Electrochṁ . Acta*, **53**(23), 6759–6767.
58. Krolikowska, A., Kudelski, A., Michota, A., and Bukowska, J. (2003). SERS studies on the structure of thioglycolic acid monolayers on silver and gold, *Surf. Sci.*, **532**, 227–232.
59. Ye, Y. K., Zhao, J. H., Yan, F., Zhu, Y. L., and Ju, H. X. (2003). Electrochemical behavior and detection of hepatitis B virus DNA PCR production at gold electrode, *Bḃ sens. Bḃ electron.*, **18**(12), 1501–1508.
60. Ju, H. X., Ye, Y. K., Zhao, J. H., and Zhu, Y. L. (2003). Hybridization biosensor using di(2,2'-bipyridine)osmium (III) as electrochemical indicator for

detection of polymerase chain reaction product of hepatitis B virus DNA, *Anal. Biochem.*, **313**(2), 255–261.

61. Bae, Y. M., Oh, B.-K., Lee, W., Lee, W. H., and Choi, J.-W. (2004). Immunosensor for detection of Yersinia enterocolitica based on imaging ellipsometry, *Anal. Chem.*, **76**(6), 1799–803.

62. Kausaite, A., van Dijk, M., Castrop, J., Ramanaviciene, A., Baltrus, J. P., Acaite, J., and Ramanavicius, A. (2007). Surface plasmon resonance label-free monitoring of antibody antigen interactions in real time, *Biochem. Mol. Biol. Edu.*, **35**(1), 57–63.

63. Su, X. L., and Li, Y. B. (2004). A self-assembled monolayer-based piezoelectric immunosensor for rapid detection of Escherichia coli 0157 : H7. *Bio sen. Bio electron.*, **19**(6), 563–574.

64. Fung, Y. S., and Wong, Y. Y. (2001). Self-assembled monolayers as the coating in a quartz piezoelectric crystal immunosensor to detect salmonella in aqueous solution, *Anal. Chem.*, **73**(21), 5302–5309.

65. Susmel, S., Guilbault, G. G., and O'Sullivan, C. K. (2003). Demonstration of labeless detection of food pathogens using electrochemical redox probe and screen-printed gold electrodes, *Bio sen. Bio electron.*, **18**(7), 881–889.

66. Willey, T. M., *et al.* (2004). Chemically transformable configurations of mercaptohexadecanoic acid self-assembled monolayers adsorbed on Au(111), *Langmuir*, **20**(7), 2746–2752.

67. Arnold, R., *et al.* (2002). Preparation, modification, and crystallinity of aliphatic and aromatic carboxylic acid terminated self-assembled monolayers, *Langmuir*, **18**(10), 3980–3992.

68. Jiang, D., *et al.* (2003). Covalently coupling the antibody on an amine-self-assembled gold surface to probe hyaluronan-binding protein with capacitance measurement, *Bio sens. Bio electron.*, **18**(9), 1183–91.

69. Smith, E. A., *et al.* (2001). Formation, spectroscopic characterization, and application of sulfhydryl-terminated alkanethiol monolayers for the chemical attachment of DNA onto gold surfaces, *Langmuir*, **17**(8), 2502–2507.

70. Boujday, S., *et al.* (2008). Detection of pathogenic Staphylococcus aureus bacteria by gold based immunosensors., *Microchim ica Acta*, **163**(3–4), 203–209.

71. Ge, C., *et al.* (2003). Electric potential control of DNA immobilization on gold electrode, *Bio sens. Bio electron.*, **18**(1), 53–8.

72. Ruan, C. M., *et al.* (1998). A reagentless amperometric hydrogen peroxide biosensor based on covalently binding horseradish peroxidase

and thionine using a thiol-modified gold electrode, *J. Electroanal. Chem.*, **455**(1–2), 121–125.

73. Ferreira, A. A. P., *et al.*, (2009). Electrochemical and spectroscopic characterization of screen-printed gold-based electrodes modified with self-assembled monolayers and Tc85 protein, *J. Electroanal. Chem.*, **634**(2), 111–122.

74. Lisdat, F., *et al.* (2000). Electrochemical behavior and nitric oxide interaction of immobilized cytochrome c′ from Rhodocyclus gelatinosus, *Electroanalysis*, **12**(12), 946–951.

75. Chirea, M., *et al.* (2005). Electrochemical characterization of polyelectrolyte/gold nanoparticle multilayers self-assembled on gold electrodes, *J. Phys. Chem. B*, **109**(46), 21808–21817.

76. Gan, N., *et al.* (2008). Electrochemical Immunosensor for human immunodeficiency virus p24 antigen based on mercapto succinic acid hydrazide copper monolayer modified gold electrode, *Chinese J. Anal. Chem.*, **36**(9), 1167–1171.

77. Chen, S. H., and Kimura, K. (1999). Synthesis and characterization of carboxylate-modified gold nanoparticle powders dispersible in water, *Langmuir*, **15**(4), 1075–1082.

78. Wang, S. H., *et al.* (2004). Inclusion-water-cluster in a three-dimensional superlattice of gold nanoparticles, *J. Am. Chem. Soc.*, **126**(24), 7438–7439.

79. Tsai, M. Y., and Lin, J. C. (2001). Surface characterization and platelet adhesion studies of self-assembled monolayer with phosphonate ester and phosphonic acid functionalities, *J. Biomed. Mater. Res.*, **55**(4), 554–565.

80. Xu, X. H., *et al.* (1994). Immobilization of DNA on an aluminum(iii) alkanebisphosphonate thin-film with electrogenerated chemiluminescent detection. *J. Am. Chem. Soc.*, **116**(18), 8386–8387.

81. Chen, Y., *et al.* (2009). 3-mercaptopropylphosphonic acid modified gold electrode for electrochemical detection of dopamine, *Bioelectrochemistry*, **75**(1), 26–31.

82. Garg, N., Carrasquillo-Molina, E., and Lee, T. R. (2002). Self-assembled monolayers composed of aromatic thiols on gold: Structural characterization and thermal stability in solution, *Langmuir*, **18**(7), 2717–2726.

83. Hayes, W. A., and Shannon, C. Electrochemistry of surface-confined mixed monolayers of 4-aminothiophenol and thiophenol on Au, *Langmuir*, **12**(15), 3688–3694.

84. Maniu, D., *et al.* (2007). Density functional theory investigation of p-aminothiophenol molecules adsorbed on gold nanoparticles, *J. Optoelectron. Adv. Mater.*, **9**(3), 733–736.

85. Bardavid, Y., Kotlyar, A. B., and Yitzchaik, S. (2006). Conducting polymer-coated DNA, *Macromol. Symp.*, **240**, 102–106.

86. Valerio, E., Abrantes, L. M., and Viana, A. S. (2008). 4-aminothiophenol self-assembled monolayer for the development of a DNA biosensor aiming the detection of cylindrospermopsin-producing cyanobacteria, *Electroanalysis*, **20**(22), 2467–2474.

87. Billah, M. M., *et al.* (2010). Directed immobilization of reduced antibody fragments onto a novel SAM on gold for myoglobin impedance immunosensing, *Bioelectrochemistry*, **80**(1), 49–54.

88. Batz, V., *et al.* (2000). Electrochemistry and structure of the isomers of aminothiophenol adsorbed on gold, *J. Electroanal. Chem.*, **491**(1–2), 55–68.

89. Halamek, J., Hepel, M., and Skladal, P. (2001). Investigation of highly sensitive piezoelectric immunosensors for 2,4-dichlorophenoxyacetic acid, *Biosen. Bioelectron.*, **16**(4–5), 253–260.

90. Horacek, J., and Skladal, P. (1997). Improved direct piezoelectric biosensors operating in liquid solution for the competitive label-free immunoassay of 2,4-dichlorophenoxyacetic acid, *Anal. Chim. Acta*, **347**(1–2), 43–50.

91. Pita, M., *et al.* (2006). Direct heterogeneous electron transfer reactions of fungal laccases at bare and thiol-modified gold electrodes, *Electrochem. Commun.*, **8**(5), 747–753.

92. Li, F., *et al.* (2010). Electrochemical sensing platform based on covalent immobilization of thionine onto gold electrode surface via diazotization-coupling reaction, *Talanta*, **83**(1), 205–209.

93. Lin, H., *et al.* (2011). Thionin attached to a gold electrode modified with self-assembly of Mo(6)S(9-X)I(X) nanowires for amplified electrochemical detection of natural DNA, *Biosens. Bioelectron.*, **26**(5), 1866–70.

94. Michota, A., and Bukowska, J. (2003). Surface-enhanced Raman scattering (SERS) of 4-mercaptobenzoic acid on silver and gold substrates, *J. Raman Spectrosc.*, **34**(1), 21–25.

95. Pattabi, R. M., and Pattabi, M. (2009). Synthesis and characterization of thiosalicylic acid stabilized gold nanoparticles, *Spectrochim. Acta A*, **74**(1), 195–199.

96. Vaughan, R. D., O'Sullivan, C. K., and Guilbault, G. G. (1999). Sulfur-based self-assembled monolayers (SAM's) on piezoelectric crystals for immunosensor development, *Fresen. J. Anal. Chem.*, **364**(1–2), 54–57.

97. Vaughan, R. D., O'Sullivan, C. K., and Guilbault, G. G. (2001). Development of a quartz crystal microbalance (QCM) immunosensor for the detection of Listeria monocytogenes, *Enzyme Mic rob. Tech.*, **29**(10), 635–638.

98. Dishner, M. H., Hemminger, J. C., and Feher, F. J. (1996). Formation of a self-assembled monolayer by adsorption of thiophene on Au(111) and its photooxidation, *Langmuir* , **12**(26), 6176–6178.

99. Matsuura, T., Nakajima, M., and Shimoyama, Y. (2001). Growth of self-assembled monolayer of thiophene on gold surface: an infrared spectroscopic study, *Jpn. J. Appl. Phys.* 1, **40**(12), 6945–6950.

100. Gao, Z. Q., Siow, K. S., and Chan, H. S. O. (1995). Self-assembled conducting polymer monolayers of poly(3-octylthiophene) on gold electrodes, *Synthetic Met.*, **75**(1), 5–10.

101. Peng, T. Z., Cheng, Q., and Cheng, Q. (2002). Determination of short DNA oligomers using an electrochemical biosensor with a conductive self-assembled membrane, *Electroanalysis* , **14**(6), 455–458.

102. Bornemann, B., *et al.* (2008). Thiolated nucleotides for immobilisation of DNA oligomers on gold surfaces, *Chemphyschem.*, **9**(9), 1241–1244.

103. Kim, H. C., *et al.* (2008). Detection of C-reactive protein on a functional poly(thiophene) self-assembled monolayer using surface plasmon resonance, *Ultramic roscopy*, **108**(10), 1379–1383.

104. Kim, H. C., *et al.* (2009). A reactive polythiophene for protein immobilization, *Polym. Advan. Technol.*, **20**(3), 298–302.

105. Xu, J. J., *et al.* (2010). A highly soluble poly(3,4-ethylenedioxythiophene)-poly(styrene sulfonic acid)/Au nanocomposite for horseradish peroxidase immobilization and biosensing, *Talanta*, **82**(4), 1511–1515.

106. Rozlosnik, N., (2009). New directions in medical biosensors employing poly(3,4-ethylenedioxy thiophene) derivative-based electrodes, *Anal. Bio anal. Chem.*, **395**(3), 637–645.

107. Rodriguez-Arguelles, M. C., *et al.* (2010). A copper(II) thiosemicarbazone complex built on gold for the immobilization of lipase and laccase, *J. Colloid Interf. Sci.*, **348**(1), 96–100.

108. Damrongchai, N., *et al.* (1997). Self-assembling of glutathione S-transferase/calmodulin fusion protein on chemically modified gold surface, *J. Bio tech.*, **55**(2), 125–133.

109. Bendikov, T. A., *et al.* (2008). Biological sensing and interface design in gold island film based localized plasmon transducers, *Anal. Chem.*, **80**(19), 7487–7498.

110. Kyprianou, D., *et al.* (2010). The application of polythiol molecules for protein immobilisation on sensor surfaces, *Bio sen. Bio electron.*, **25**(5), 1049–1055.

111. Shon, Y. S., and Lee, T. R. (2000). Desorption and exchange of self-assembled monolayers (SAMs) on gold generated from chelating alkanedithiols, *J. Phys. Chem. B*, **104**(34), 8192–8200.

112. Biebuyck, H. A., Bian, C. D., and Whitesides, G. M. (1994). Comparison of Organic monolayers on polycrystalline gold spontaneously assembled from solutions containing dialkyl disulfides or alkenethiols, *Langmuir*, **10**(6), 1825–1831.

113. Bain, C. D., Biebuyck, H. A., and Whitesides, G. M. (1989). Comparison of self-assembled monolayers on gold— oadsorption of thiols and disulfides, *Langmuir*, **5**(3), 723–727.

114. Dougan, J. A., *et al.* (2007). Enhanced oligonucleotide-nanoparticle conjugate stability using thioctic acid modified oligonucleotides, *Nucleic Acids Res.*, **35**(11), 3668–3675.

115. Letsinger, R. L., *et al.* (2000). Use of a steroid cyclic disulfide anchor in constructing gold nanoparticle-oligonucleotide conjugates, *Bio conjugate Chem.*, **11**(2), 289–291.

116. Langry, K. C., *et al.* (2005). The AFM measured force required to rupture the dithiolate linkage of thioctic acid to gold is less than the rupture force of a simple gold-alkyl thiolate bond, *Langmuir*, **21**(26), 12064–12067.

117. Pang, D. W., *et al.* (1996). Modification of glassy carbon and gold electrodes with DNA, *J. Electroanal. Chem.*, **403**(1–2), 183–188.

118. Smith, A. M., Ducey, M. W., and Meyerhoff, M. E. (2000). Nature of immobilized antibody layers linked to thioctic acid treated gold surfaces, *Bio sen. Bio electron.*, **15**(3–4), 183–192.

119. Willey, T. M., *et al.* (2004). Surface structure and chemical switching of thioctic acid adsorbed on Au(111) as observed using near-edge X-ray absorption fine structure, *Langmuir*, **20**(12), 4939–4944.

120. Yue, M., *et al.* (2008). Label-free protein recognition two-dimensional array using nanomechanical sensors, *Nano Lett.*, **8**(2), 520–524.

121. Rudzinski, W. E., and Francis, K. (2010). Evaluating the surface density and heterogeneity of a dithiobis (succinimidylpropionate) self-

assembled monolayer on gold and its coupling with DNA embedded within a matrix, *Appl. Surf. Sci.*, **256**(17), 5399–5405.

122. Liss, M., *et al.* (2002). An aptamer-based quartz crystal protein biosensor, *Anal. Chem.*, **74**(17), 4488–4495.

123. Amiri, A., Choi, E. Y., and Kim, H. J. (2010). Development and molecular recognition of Calixcrownchip as an electrochemical ALT immunosensor, *J. Incl. Phenom. Macro.*, **66**(1–2), 185–194.

124. Chen, H. X., *et al.* (2010). Surface plasmon resonance spectroscopic characterization of antibody orientation and activity on the calixarene monolayer, *Sensor. Actuat. B-Chem.*, **147**(2), 548–553.

125. Oh, S. W., *et al.* (2005). Calixarene derivative as a tool for highly sensitive detection and oriented immobilization of proteins in a microarray format through noncovalent molecular interaction, *Faseb J.*, **19**(8), 1335–1337.

126. Sakata, T., *et al.* (2007). Stable immobilization of an oligonucleotide probe on a gold substrate using tripodal thiol derivatives, *Langmuir*, **23**(5), 2269–2272.

127. Flynn, N. T., *et al.* (2003). Long-term stability of self-assembled monolayers in biological media, *Langmuir*, **19**(26), 10909–10915.

128. Phares, N., White, R. J., and Plaxeo, K. W. (2009). Improving the stability and sensing of electrochemical biosensors by employing trithiol-anchoring groups in a six-carbon self-assembled monolayer, *Anal. Chem.*, **81**(3), 1095–1100.

129. Fox, M. A., Whitesell, J. K., and Mckerrow, A. J. (1998). Fluorescence and redox activity of probes anchored through an aminotrithiol to polycrystalline gold, *Langmuir*, **14**(4), 816–820.

130. Li, Z., *et al.* (2002). Multiple thiol-anchor capped DNA-gold nanoparticle conjugates, *Nucleic Acids Res.*, **30**(7), 1558–1562.

131. Siiman, O., *et al.* (2000). Tris(3-mercaptopropyl)-*N*-glycylaminomethane as a new linker to bridge antibody with metal particles for biological cell separations, *Bioconjugate Chem.*, **11**(4), 549–556.

132. Zheng, J., *et al.* (2007). A new amplification strategy for ultrasensitive electrochemical aptasensor with network-like thiocyanuric acid/gold nanoparticles, *Biosen. Bioelectron.*, **23**(3), 341–347.

133. Xu, Y. Y., *et al.* (2006). A microelectronic technology based amperometric immunosensor for alpha-fetoprotein using mixed self-assembled monolayers and gold nanoparticles, *Anal. Chim. Acta*, **561**(1–2), 48–54.

134. Bain, C. D., Evall, J., and Whitesides, G. M. (1989). Formation of monolayers by the coadsorption of thiols on gold—variation in

the head group, tail group, and solvent, *J. Am. Chem. Soc.*, **111**(18), 7155–7164.

135. Bain, C. D., and Whitesides, G. M. (1989). Formation of monolayers by the coadsorption of thiols on gold—variation in the length of the alkyl chain, *J. Am. Chem. Soc.*, **111**(18), 7164–7175.

136. Frederix, F., *et al.* (2003). Enhanced performance of an affinity biosensor interface based on mixed self-assembled monolayers of thiols on gold, *Langmuir*, **19**(10), 4351–4357.

137. Chapman, R. G., *et al.* (2000). Preparation of mixed self-assembled monolayers (SAMs) that resist adsorption of proteins using the reaction of amines with a SAM that presents interchain carboxylic anhydride groups, *Langmuir*, **16**(17), 6927–6936.

138. Fryxell, G. E., *et al.* (1996). Nucleophilic displacements in mixed self-assembled monolayers, *Langmuir*, **12**(21), 5064–5075.

139. Folkers, J. P., *et al.* (1994). Phase-behavior of 2-component self-assembled monolayers of alkanethiolates on gold,. *J. Phys. Chem.*, **98**(2), 563–571.

140. Stranick, S. J., *et al.* (1994). Phase-separation of mixed-composition self-assembled monolayers into nanometer-scale molecular domains, *J. Phys. Chem.*, **98**(31), 7636–7646.

141. Bertilsson, L., and Liedberg, B. (1993). Infrared study of thiol monolayer assemblies on gold—preparation, characterization, and functionalization of mixed monolayers, *Langmuir*, **9**(1), 141–149.

142. Lee, J. W., *et al.* (2005). Characterization of a self-assembled monolayer of thiol on a gold surface and the fabrication of a biosensor chip based on surface plasmon resonance for detecting anti-GAD antibody, *Bio sen. Bio electron.*, **20**(7), 1422–1427.

143. Spinke, J., *et al.* (1993). Molecular recognition at self-assembled monolayers—the construction of multicomponent multilayers, *Langmuir*, **9**(7), 1821–1825.

144. Hobara, D., Uno, Y., and Kakiuchi, T. (2001). Immobilization of horseradish peroxidase on nanometre-scale domains of binary self-assembled monolayers formed from dithiobis-N-succinimidyl propionate and 1-tetradecanethiol on Au(111), *Phys. Chem. Ch. Ph.*, **3**(16), 3437–3441.

145. Berggren, C., and Johansson, G. (1997). Capacitance measurements of antibody-antigen interactions in a flow system, *Anal. Chem.*, **69**(18), 3651–3657.

146. Kang, C. D., *et al.* (2006). Performance enhancement of real-time detection of protozoan parasite, cryptosporidium oocyst by a modified

surface plasmon resonance (SPR). Biosensor, *Enzyme Mi crob Tech.*, **39**(3), 387–390.

147. Wilson, D. L., *et al.* (2001). Surface organization and nanopatterning of collagen by dip-pen nanolithography, *P. Natl. Acad. Sci. USA*, **98**(24), 13660–13664.

148. Lee, K. B., Lim, J. H., and Mirkin, C. A. (2003). Protein nanostructures formed via direct-write dip-pen nanolithography, *J. Am. Chem. Soc.*, **125**(19), 5588–5589.

149. Piletsky, S., *et al.* (2003). Surface functionalization of porous polypropylene membranes with polyaniline for protein immobilization, *Bio tech. Bio eng.*, **82**(1), 86–92.

150. Rich, R. L., and Myszka, D. G. (2000). Advances in surface plasmon resonance biosensor analysis, *Curr. Opin . Bio tech.*, **11**(1), 54–61.

151. Fuentes, M., *et al.* (2004). Reversible and strong immobilization of proteins by ionic exchange on supports coated with sulfate-dextran, *Bio technol. Progr.*, **20**(4), 1134–1139.

152. Chi, Q. J., *et al.* (2001). Ordered assembly and controlled electron transfer of the blue copper protein azurin at gold (111) single-crystal substrates, *J. Phys. Chem. B*, **105**(20), 4669–4679.

153. Lahiri, J., *et al.* (2000). Method for fabricating supported bilayer lipid membranes on gold, *Langmuir* , **16**(20), 7805–7810.

154. Su, L. Y., *et al.* (2005). Characterization of cyanide binding to cytochrome c oxidase immobilized in electrode-supported lipid bilayer membranes, *J. Electroanal. Chem.*, **581**(2), 241–248.

155. Suraniti, E., *et al.* (2007). Construction of hybrid bilayer membrane (HBM) biochips and characterization of the cooperative binding between cytochrome-c and HBM, *Langmuir* , **23**(12), 6835–6842.

156. Zhang, H., *et al.* (2003). Biofunctionalized nanoarrays of inorganic structures prepared by dip-pen nanolithography, *Nanotechnology*, **14**(10), 1113–1117.

157. Wrobel, N., *et al.* (2003). Covalent immobilization of oligonucleotides on electrodes, *Colloid . Surface. B*, **32**(2), 157–162.

158. Ostblom, M., *et al.* (2006). Structure and desorption energetics of ultrathin D2O ice overlayers on serine- and serinephosphate-terminated self-assembled monolayers, *J. Phys. Chem. B*, **110**(4), 1695–1700.

159. Tsai, M. Y., Sun, Y. T., and Lin, J. C. (2007). Surface characterization and platelet compatibility evaluation of the binary mixed self-assembled monolayers, *J. Colloid Interf. Sci.*, **308**(2), 474–484.

160. Gray, J. J. (2004). The interaction of proteins with solid surfaces, *Curr. Opin. Struc. Biol.*, **14**(1), 110–115.

161. Khan, F., He, M. Y., and Taussig, M. J. (2006). Double-hexahistidine tag with high-affinity binding for protein immobilization, purification, and detection on Ni-nitrilotriacetic acid surfaces, *Anal. Chem.*, **78**(9), 3072–3079.

162. Disley, D. M., *et al.* (1998). Covalent coupling of immunoglobulin G to self-assembled monolayers as a method for immobilizing the interfacial-recognition layer of a surface plasmon resonance immuno-sensor, *Biosen. Bioelectron.*, **13**(11), 1213–1225.

163. Caruso, F., Rodda, E., and Furlong, D. N. (1996). Orientational aspects of antibody immobilization and immunological activity on quartz crystal microbalance electrodes, *J. Colloid Interf. Sci.*, **178**(1), 104–115.

164. Rao, S. V., Anderson, K. W., and Bachas, L. G. (1998). Oriented immobilization of proteins, *Mikrochim. Acta*, **128**(3–4), 127–143.

165. D'Souza, S. F., and Godbole, S. S. (2002). Immobilization of invertase on rice husk using polyethylenimine, *J. Biochem. Biophys. Meth.*, **252**(1), 59–62.

166. Choi, H. J., *et al.* (2005). Micropatterning of biomolecules on glass surfaces modified with various functional groups using photoactivatable biotin. *Anal. Biochem.*, **347**(1), 60–66.

167. Mateo, C., *et al.* (2003). Epoxy-amino groups: a new tool for improved immobilization of proteins by the epoxy method, *Biomacromolecules*, **4**(3), 772–777.

168. Fernandez-Lafuente, R., *et al.* (1993). Preparation of activated supports containing low pK amino-groups—a new tool for protein immobilization via the carboxyl coupling method, *Enzyme Microb. Tech.*, **15**(7), 546–550.

169. Villalonga, R., Cao, R., and Fragoso, A. (2007). Supramolecular chemistry of cyclodextrins in enzyme technology, *Chem. Rev.*, **107**(7), 3088–3116.

170. Ludden, M. J. W., Reinhoudt, D. N., and Huskens, J. (2006). Molecular printboards: versatile platforms for the creation and positioning of supramolecular assemblies and materials, *Chem. Soc. Rev.*, **35**(11), 1122–1134.

171. Ludden, M. J. W., *et al.* (2007). Molecular printboards as a general platform for protein immobilization: a supramolecular solution to nonspecific adsorption, *Angew. Chem. Int. Edt.*, **46**(22), 4104–4107.

172. Ludden, M. J., *et al.* (2008). Assembly of bionanostructures onto beta-cyclodextrin molecular printboards for antibody recognition and lymphocyte cell counting, *J. Am. Chem. Soc.*, **130**(22), 6964–6973.

173. Gonzalez, M., *et al.* (1997). Interaction of biotin with streptavidin—thermostability and conformational changes upon binding, *J. Bio l. Chem.*, **272**(17), 11288–11294.

174. Weber, P. C., *et al.* (1992). Crystallographic and thermodynamic comparison of natural and synthetic ligands bound to streptavidin, *J. Am. Chem. Soc.*, **114**(9), 3197–3200.

175. Muller, W., *et al.* (1993). Attempts to mimic docking processes of the immune-systemr— ecognition-induced formation of protein multilayers, *Sci e nce*, **262**(5140), 1706–1708.

176. Pavlickova, P., *et al.* (2002). Antibody detection in human serum using a versatile protein chip platform constructed by applying nanoscale self-assembled architectures on gold, *J. Proteome Res.*, **1**(3), 227–231.

177. Pei, R., *et al.* (2001). Amplification of antigen-antibody interactions based on biotin labeled protein-streptavidin network complex using impedance spectroscopy, *Bio sens. Bio electron.*, **16**(6), 355–361.

178. Lindmark, R., Thoren-Tolling, K., and Sjoquist, J. (1983). Binding of immunoglobulins to protein A and immunoglobulin levels in mammalian sera, *J. Immunol. Meth.*, **62**(1), 1–13.

179. Minunni, M., *et al.* (1996). A quartz crystal microbalance displacement assay for Listeria monocytogenes, *Anal. Chim . Acta*, **325**(3), 169–174.

180. Horisberger, M., and Clerc, M. F. (1985). Labelling of colloidal gold with protein A: a quantitative study, *His tochemis try*, **82**(3), 219–223.

181. Kronvall, G. (1973). Rapid slide-agglutination method for typing pneumococci by means of specific antibody adsorbed to protein A-containing staphylococci, *J. Med. Mic rob l.*, **6**(2), 187–190.

182. Harboe, M., and Folling, I. (1974). Recognition of two distinct groups of human IgM and IgA based on different binding to staphylococci, *Scand. J. Immunol.*, **3**(4), 471–82.

183. Anderson, G. P., *et al.* (1997). Effectiveness of protein A for antibody immobilization for a fiber optic biosensor, *Bio sen. Bio electron.*, **12**(4), 329–336.

184. Kanno, S., *et al.* (2000). Assembling of engineered IgG-binding protein on gold surface for highly oriented antibody immobilization, *J. Bio technology*, **76**(2–3), 207–214.

185. Shen, G., *et al.* (2011). Improvement of antibody immobilization using hyperbranched polymer and protein A. *Anal. Biochem.*, **409**(1), 22–27.

186. Johnson, C. P., *et al.* (2003). Engineered protein A for the orientational control of immobilized proteins. *Bioconjugate Chem.*, **14**(5), 974–978.

187. Lee, W., *et al.* (2004). Nanoscale fabrication of protein A on self-assembled monolayer and its application to surface plasmon resonance immunosensor, *Enzyme Microb. Tech.*, **35**(6–7), 678–682.

188. Bartlett, P. N., and Cooper, J. M. (1993). A review of the immobilization of enzymes in electropolymerized films. *J. Electroanal. Chem.*, **362** (1–2), 1–12.

189. Bidan, G. (1992). Electroconducting conjugated polymers—new sensitive matrices to build up chemical or electrochemical sensors—a review, *Sensor. Actuat. B-Chem.*, **6**(1–3), 45–56.

190. Cosnier, S. (1999). Biomolecule immobilization on electrode surfaces by entrapment or attachment to electrochemically polymerized films—a review, *Biosen. Bioelectron.*, **14**(5), 443–456.

191. Aizawa, M., *et al.* (1985). Electrochemical cation doping of a poly-thienylene film, *J. Chem. Soc.-Chem. Comm.*, (5), 264–265.

192. Gerard, M., Chaubey, A., and Malhotra, B. D. (2002). Application of conducting polymers to biosensors, *Biosen. Bioelectron.*, **17**(5), 345–359.

193. Korri-Youssoufi, H., Richard, C., and Yassar, A. (2001). A new method for the immobilisation of antibodies in conducting polymers, *Mat. Sci. Eng. C-Bio. S.*, **15**(1–2), 307–310.

194. Dupoet, P. D., *et al.* (1990). Direct electron-transfer with glucose-oxidase immobilized in an electropolymerized poly(*n*-methylpyrrole) film on a gold microelectrode, *Anal. Chim. Acta*, **235**(2), 255–263.

195. Si, S. H., *et al.* (1997). Preparation of a piezoelectric immunosensor for the detection of salmonella paratyphi A by immobilization of antibodies on electropolymerized films, *Fresenius J. Anal. Chem.*, **357**(8), 1101–1105.

196. Loyprasert, S., *et al.* (2010). Sub-attomolar detection of cholera toxin using a label-free capacitive immunosensor, *Biosen. Bioelectron.*, **25**(8), 1977–1983.

197. Li, Q., *et al.* (2010). A novel impedance immunosensor based on *O*-phenylenediamine modified gold electrode to analyze abscisic acid, *Chinese Chem. Lett.*, **21**(4), 472–475.

198. Huang, H. H., *et al.* (2008). Impedimetric immunosensor with on-chip integrated electrodes for high-throughput screening of liver fibrosis markers. *J. Anal. Chem.*, **63**(5), 492–498.

199. Grosjean, L., *et al.* (2005). A polypyrrole protein microarray for antibody-antigen interaction studies using a label-free detection process, *Anal. Bio chem.*, **347**(2), 193–200.

200. Livache, T., *et al.* (1998). Electroconducting polymers for the construction of DNA or peptide arrays on silicon chips, *Biosen. Bioelectron.*, **13**(6), 629–634.

201. Livache, T., *et al.* (1994). Preparation of a DNA matrix via an electrochemically directed copolymerization of pyrrole and oligonucleotides bearing a pyrrole group, *Nucleic Acids Res.*, **22**(15), 2915–2921.

202. Livache, T., *et al.* (2001). Polypyrrole electrospotting for the construction of oligonucleotide arrays compatible with a surface plasmon resonance hybridization detection, *Synthetic Met.*, **121**(1–3), 1443–1444.

203. Long, E. C., and Barton, J. K. (1990). On demonstrating DNA intercalation, *Accounts Chem. Res.*, **23**(9), 271–273.

204. Nakamura, F., *et al.* (2000). Immobilization of DNA on self-assembled monolayer, *Mol. Cryst. Liq. Cryst.*, **349**, 219–222.

205. Sato, S., Hirano, A., and Takenaka, S. (2010). Selective immobilization of double stranded DNA on a gold surface through threading intercalation of a naphthalene diimide having dithiolane moieties, *Anal. Chim. Acta*, **665**(1), 91–97.

206. Wang, G. L., Zhang, J., and Murray, R. W. (2002). DNA binding of an ethidium intercalator attached to a monolayer-protected gold cluster, *Anal. Chem.*, **74**(17), 4320–4327.

207. Higashi, N., Takahashi, M., and Niwa, M. (1999). Immobilization of DNA through intercalation at self-assembled monolayers on gold, *Langmuir*, 1**15**(1), 111–115.

208. Li, F., *et al.* (2010). Gold nanoparticles modified electrode via a mercapto-diazoaminobenzene monolayer and its development in DNA electrochemical biosensor, *Biosens. Bioelectron.*, **25**(9), 2084–8.

209. Corgier, B. P., Marquette, C. A., and Blum, L. J. (2005). Diazonium-protein adducts for graphite electrode microarrays modification: direct and addressed electrochemical immobilization, *J. Am. Chem. Soc.*, **127**(51), 18328–18332.

210. Polsky, R., *et al.* (2007). Diazonium-functionalized horseradish peroxidase immobilized via addressable electrodeposition: direct

electron transfer and electrochemical detection, *Langmuir*, **23**(2), 364–366.

211. Civit, L., Fragoso, A., and O'Sullivan, C. K. (2010). Thermal stability of diazonium derived and thiol-derived layers on gold for application in genosensors, *Electrochem. Comm.*, **12**(8), 1045–1048.

212. Kessler, D., Roth, P. J., and Theato, P. (2009). Reactive surface coatings based on polysilsesquioxanes: controlled functionalization for specific protein immobilization, *Langmuir*, **25**(17), 10068–10076.

Chapter 4

Magnetic Nanoparticle Surface Functionalization for Biomedical Applications

F. Geinguenaud, C. de Montferrand, H. Jouni, E. Guenin, I. Milosevic, F. Benyettou, Y. Lalatonne, and L. Motte

University Paris 13, CSPBAT Lab, LPBS, 74 Rue Marcel Cachin, Bobigny, 93009 Cedex, France

laurence.motte-torcheux@univ-paris13.fr

4.1 Introduction

In the field of biomedical applications, magnetic nanoparticles have received considerable attention because they offer unique advantages over other types of nanoparticles as nanoscale noble metals and semiconductors. The most commonly utilized forms of magnetic nanoparticles are iron oxide γ-Fe_2O_3 (maghemite) and Fe_3O_4 (magnetite) and their corresponding ferrites ($MnFe_2O_4$ and $CoFe_2O_4$).

Smaller than approximately 30 nm, these particles exhibit superparamagnetism. That is, the nanoparticles consist of a single crystal domain and thus present a single magnetic moment. With the help of thermal energy, the magnetic moment can overcome

Nanoantenna: Plasmon-Enhanced Spectroscopies for Biotechnological Applications
Edited by Marc Lamy de la Chapelle and Annemarie Pucci

ISBN 978-981-4303-61-3 (Hardcover), 978-981-4303-62-0 (eBook)
www.panstanford.com

the anisotropy barrier and spontaneously flip from one direction of anisotropy to another. The magnetic moment of a particle can rotate through two distinct mechanisms: Néel and Brownian relaxation; thus, over time the net magnetization of one of these particles averages to zero. An ensemble of magnetic nanoparticles consequently displays negligible magnetic remanence. When placed inside an external magnetic field, the magnetic moments align in the direction of the field lines and yield a fixed magnetization direction.

The intrinsic interaction of magnetic nanoparticles with applied magnetic field gradients makes these particles attractive for a large number of biomedical applications, such as magnetic separation systems for biomolecules and cells, magnetic resonance imaging contrast enhancement, therapies like hyperthermia, and drug delivery, as well as for multidetection systems based on biosensors.[1–6] For a specific application, the nanoparticle surface needs to be specifically functionalized. That is to say, appropriate functional groups have to be provided at the nanoparticle surface to enable specific interactions with the molecules needed in different applications.

This chapter summarizes recent advances in the field of nanoparticle surface functionalization. We focus on the surface functionalization of magnetic nanoparticles, but the chemistry we discuss is equally relevant to other nanoparticle constructs. In Section 4.2 we briefly discuss nanoparticle synthesis that induces either hydrophilic or hydrophobic nanoparticles. Section 4.3 discusses the many new and highly specific conjugation methods that have been developed in recent years to functionalize nanoparticle surface. Section 4.4 presents a long-term perspective that highlights emerging research in imaging, drug delivery, hyperthermia, and biodetection.

4.2 Nanoparticle Synthesis

Synthesis methods for metal oxide nanoparticles are typically grouped into two categories: top-down and bottom-up. Top-down methods involve dividing a massive solid into smaller portions. This approach may require milling or attrition, chemical methods, and volatilization of the solid, followed by condensation of the volatilized components. The bottom-up method involves condensing atoms or molecular entities in a gas phase or solution. The latter approach is far more popular in nanoparticle synthesis. Some of the most common synthesis methods are shown in Fig. 4.1.

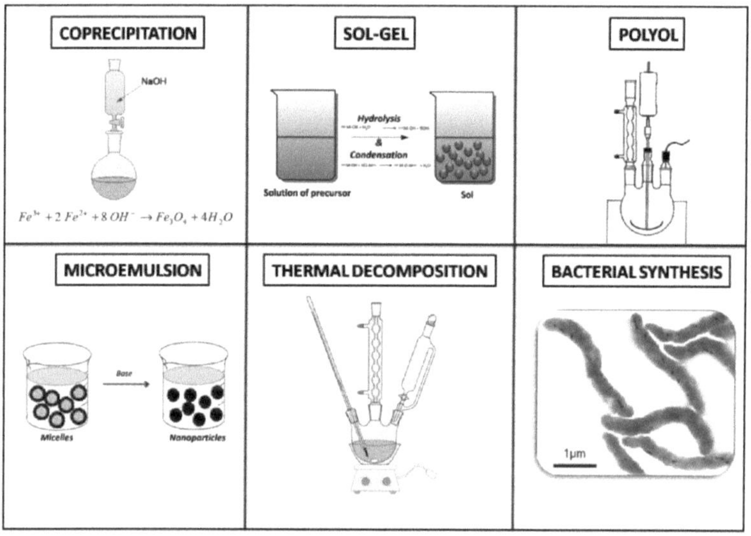

Figure 4.1 Schematic representation of the different chemical synthesis.

4.2.1 Coprecipitation Method

The coprecipitation method is a synthesis that leads to the formation of ferrite nanoparticles. Ferrites $M^{II}Fe_2O_4$ (M = Fe, Co, Ni, Mn, Zn, Cd and so on) are ferric oxides of divalent metal. They crystallize in a spinel cubic structure. The coprecipitation method involves alkalizing an aqueous mixture of Fe^{3+} and M^{2+} ions. Ferrite nanoparticles typically between 3 and 10 nm are obtained, depending on the experimental conditions. The size and shape of the particles obtained are highly dependent on synthesis conditions and in particular

- the ratio of metal ions
- temperature
- pH

- the presence of molecules that can be coordinated to the particle surface.

Controlling the morphology and size polydispersity of the nanoparticles has been a scientific challenge since the early 1980s.[7] These studies highlight various important parameters, such as temperature, heating time, ionic strength, and pH. Ionic strength and the pH of the reaction are the main factors in controlling size polydispersity and average size of iron oxides by the coprecipitation aqueous method. In general, the higher the pH and ionic strength, the smaller the particles.[8] Vayssières *et al.* also observed a critical pH value at fixed ionic strength, above which the particle size no longer changes during aging. If the reaction medium is below this critical pH, the particles tend to grow even after their formation.[9]

4.2.2 Sol–Gel Method

The sol-gel process involves the evolution of inorganic networks through formation of a colloidal suspension[10] and gelation of the sol to form a network in a continuous liquid phase (gel). Three reactions generally describe the sol–gel process: hydrolysis, alcohol condensation, and water condensation. Many parameters can be tuned, including pH, temperature and time of reaction, reagent concentrations, catalyst nature and concentration, H_2O/Si molar ratio, aging temperature and time, and drying. These parameters can affect the rate of hydrolysis and condensation reaction. The precursors used for this synthesis are metal or metalloid elements surrounded by various reactive ligands.[11]

Among other preparative techniques the sol–gel process has proved to be a convenient method for synthesizing nanocomposites. Tadić *et al.*[12] used as starting point a solution prepared by mixing tetraethylorthosilicate (TEOS), water, and ethanol, as well as an aqueous solution of metal nitrate. Silica is convenient because of its nontoxic nature, high biocompatibility, ability to prevent agglomeration, temperature resistance, chemical inertness, controllable shell thickness, and adjustable pore diameter. Considerable attention has been paid to the influence of the processing parameters such as the concentration of components, pH of the solution, and heat treatment temperature and duration on the physical and chemical properties of these materials. A particle size of a few nanometers ($d < 10$ nm) can be obtained with a narrow size distribution and uniform dispersion.[13]

4.2.3 Polyol Synthesis

The principle of polyol synthesis is based on the reduction of metallic salts in a dialcohol solution. The nano-objects are collected from the dialcohol solution. The polyol synthesis process, developed by F. Fiévet *et al.*,[14] involves dispersing metal precursors (hydroxide, acetate, oxide, etc.) in a polyol solution (usually α-diol) and then warming them without exceeding the polyol boiling point. Polyols are used as solvents because they can act as reducing and polar solvents and also because of their high boiling temperature, which allows them to solubilize a large number of metal salts and to reduce the latter in a large temperature range. The main advantage of the polyol process, aside from easily providing several grams of nano-objects, comes from the richness of the parameters which control the reaction. This allows us to obtain a large variety of nanoparticles shapes, dimensions and composition.[15–19]

4.2.4 Synthesizing Nanoparticles in Microemulsion

The use of microemulsions to synthesize nanoparticles is one of the most promising methods. Microemulsion technology has been used to synthesize pure metal nanoparticles as well as to prepare bimetallic nanoparticles.[20–23] In the case of bi- and multimetallic nanoparticles, the atomic composition can be modified as needed.

The principle is based on the decomposition by a weak base, for example dimethylamine,[24] of micelles formed by iron or cobalt dodecyl sulfate molecules. Using iron and/or cobalt dodecylbenzenesulfonates leads to similar results, as shown by Lee *et al.*[25] Micelle formation is also obtained by using cetyltrimethylammonium bromide.[26] For example, under these conditions, cobalt ferrite rods of micrometer size are obtained.

4.2.5 Thermal Decomposition

Thermolysis of the organic precursor leads to the formation of germs. The number of germs formed depends on the stability of the precursors used, so the phases of nucleation and growth can be controlled. Organic ligands present in the medium control the size dispersion of particles by a more or less strong interaction with the particle surface. The key factors of this method are the ratio of reagents, the synthesis temperature and the reaction time. If the precursor is used in the zerovalent state, thermolysis leads to the formation of metallic particles, which will be then oxidized by reflux in air or by the addition of an oxidant. This approach has enabled production of semiconductor quantum dots and nanowires, metallic and metal oxide magnetic nanostructures, and semimetal and noble metal nanoparticles.[27–34] For example, Sun and employed high-temperature (100°C to 300°C) reduction of metal salts in the presence of stabilizing agents to produce monodisperse Co nanoparticles of 2–12 nm in diameter.[35,36]

4.2.6 Bacterial Synthesis

Microbial synthesis of nanoparticles is a green chemistry approach that interconnects nanotechnology and microbial biotechnology. Many biosynthesis processes, such as synthesis of gold, silver, gold-silver alloy, selenium, tellurium, platinum, palladium, silica, titania, zirconia, quantum dots, magnetite and uraninite nanoparticles by bacteria, actinomycetes, fungi, yeasts, and viruses, have been reported.[37–39] Nair and Pradeep show that when *Lactob cl lus* strains (present in buttermilk) are exposed to large concentrations of metal ions they produce microscopic gold, silver, and gold-silver alloy crystals of well-defined morphology. The nanoparticles are produced intracellularly and, remarkably, the cells preserve their viability even after crystal growth.[40]

Magnetosomes are magnetostatic bacteria (3 μm–long cells) occuring widely in the habitats of marine and freshwater sediments. They produce intracellular membrane-bound magnetite. For instance, in *Magnetospr l lum magnetotactc um* the produced magnetite single domain nanoparticles are aligned in a chain parallel to the longitudinal axis of the cell.[41]

4.3 Nanoparticle Surface Functionalization

In its most basic form, a biomedical mNP comprises an inorganic nanoparticle core and a biocompatible surface coating. The biocompatible surface coating provides stability under physiological conditions and may allow further modifications to provide specific properties. As discussed above, such nanohybrid materials are prepared using one of two approaches depending mainly on the nanocrystals synthesis route and surface functionalities.[42] Water-compatible iron oxide nanocrystals are obtained by coprecipitation[43,44] or by using normal micelles.[45–47] On the other hand, precipitation in reverse microemulsions or thermal decomposition of organometalic coumpounds in the presence of hydrophobic ligands[31,35] induces formation of nanocrystals that are soluble in nonpolar organic solvents. Once obtained, both nanocrystals type should be coated to be suitable for biological applications and to be able interact with other systems such as ligands, receptors, or drugs. For nanocrystals soluble in polar solvent, the hydrophobic surfactant coating needs to be replaced by a hydrophilic one. This is achieved using ligand exchange and/or surface silanization and polymer or micelle coating.[48] For water-soluble nanocrystals this coating is allowed by the abundant surface hydronium or hydroxyl groups (depending on pH) which could interact with functional group with high coordination capacity (carboxylic, phosphonate, etc.). This surface coating could be done in either of two ways (Fig. 4.2).

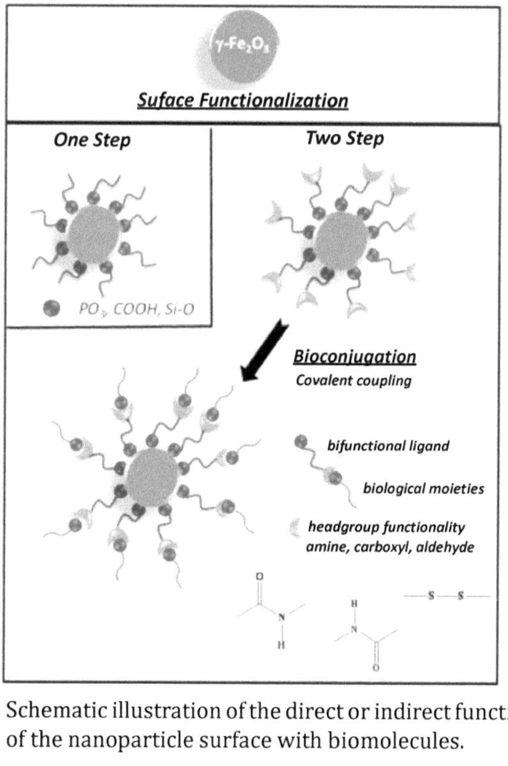

Figure 4.2 Schematic illustration of the direct or indirect functionalization of the nanoparticle surface with biomolecules.

Nanocrystals could be formulated directly with their final properties in a one-step strategy or they could be coated with biocompatible, bifunctional ligands that bind to the nanocrystal surface on one end and that have head-group functionality (amine, carboxyl, aldehyde groups) on the other end. This so-called indirect functionalization is by far the most used strategy because it allows precise molecular control on the surface of the particles and allows the magnetic nanoplatform surface to be decorated with a diverse number of biological properties.[49]

The engineering of the particle surface is a key step in creating bioactive mNP, for which there are several practical methods.

4.3.1 Hydrophobic Nanocrystals

High-temperature organometallic synthetic routes produce nanoparticles with hydrophobic surface ligands such as oleic acid (OA), trioctylphosphine oxide(TOPO), and oleyl amine (OLA). There are four main routes to modifying hydrophobic nanoparticles and render them soluble in an aqueous solution, as illustrated in Fig. 4.3 for a native oleic acid-capped nanoparticle.

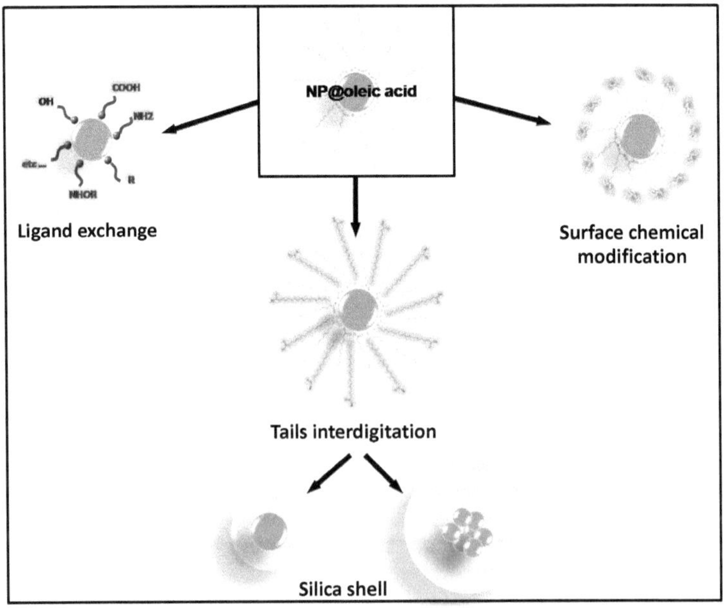

Figure 4.3 From a hydrophobic (oleic acid capped) nanoparticle to a water-soluble nanoprobe.

4.3.2 Ligand Exchange

In the first route, called ligand-exchange, the native monolayer of hydrophobic surface ligands is exchanged with ligands containing head groups that bind on one end of the magnetic nanoparticle surface and hydrophilic tails that interact with the aqueous solvent at the other end (i.e., a bifunctional molecule). Thiols, phosphine oxides, phosphonates, or carboxylates groups are commonly used in the case of chalcogenides, oxide nanoparticles, and noble metal nanoparticles. The complexing strength of the surface anchoring group must be high enough to ensure maximum surface coverage. Many different hydrophilic ligands can be grafted onto the NPs, such as polypeptides,[50] bisphosphonates,[51,52] betaine hydro-chloride,[53] silanes,[54] and polymers.[55,56] They are reported in Table 4.1, with the transfer conditions, duration, and temperature. The transfer duration and temperature vary widely. Some of those ligand exchanges are carried out in very easy conditions, in a few minutes with sonication,[52,57] taking place with or without a transfer agent. The transfer agent acts as a ligand intermediate to help the water solubilization of the nanoparticle. Tetramethylammonium hydroxide (TMAOH) is the most commonly used transfer agent (see Table 4.1): once around the particle, this quaternary ammonium salt brings a hydrophilic surface charge, which is then removed by the final ligand.

Table 4.1 Ligand exchanges from oleic acid and transfer conditions

Initial ligand	Solvants	Transfer agent	Final ligand	Transfer duration	Transfer temp.	Ref.
Oleic acid	Ethanol: water 4:20	TMAOH	APAS (Alendronate)	12 h	Ambiant	51
Oleic acid	Hexane: water 5:5	TMAOH	Citrate	12 h	Ambiant	58
Oleic acid	Hexane: ethanol	TMAOH	TMAOH, GSH	5 min (US)		57, 59, 60
Oleic acid	Water	TMAOH	Citrate and polylysine-block-polyaspartic acid			50
Oleic acid	Hexane: water		Silane	72 h	Ambiant	54

(*Cont'd*)

Table 4.1 (*Continued*)

Initial ligand	Solvants	Transfer agent	Final ligand	Transfer duration	Transfer temp.	Ref.
Oleic acid	Ethanol	Betaine-HCl	Betaine-alkyl	40 h	Ambiant	53
Oleic acid, TOPO	Toluene		Butylamine			61
Oleic acid	Chloroform: water 2:2		Cyclodextrin (pH12)	5 h	Ambiante	62
TOPO	Chloroform: water 1:1	Ferrocenyl-thiols + cyclodextrin	Ferrocene@ cyclodextrin	6 d	40°C	63
Oleic acid	Toluene: DMSO 1:1		DMSA	14 h	Ambiant	64
Oleic acid	Chloroform: DMSO 1:1		G5–Ac(102)–FA(5)–6T(3) folic acid–dendrimer	12 h	60°C	65
Oleic acid	Dichloro-benzene: diméthyl-formamide 7.5:7.5		Citric acid	24 h	100°C	66
Oleic acid, TOPO	Toluene: glycol 1:8		PAA, PAAm	1 h under N_2	240°C	55
Oleic acid	1,4 dioxane		Thermo-responsive polymers	48 h	60°C	56
Oleic acid	Hexane: water 5:5		DA-BP (bisphos-phonate)	30 min (US)		52
Tetraalkyl ammonium salts	Toluene: water 1:1		DiMethylAmino Pyridine	1 h	Ambiant	67
Alkane-thiolate	THF		Thiocarboxylic acids	2 d under Ar	Ambiant	68

4.3.3 Surface Chemistry Modification

Another approach is the chemical modification of the oleic acid tail so that it has a hydrophilic terminal function on the surface of the NPs. The oleic acid molecule has a double-link carbon-carbon molecule in the middle of its hydrocarbon chain; the ozonolyse of this double bond via potassium permanganate ($KMnO_4$) transforms oleic acid into azaleic acid and yields second carboxylic acid functionality.[53,69] The azaleic acid coating is water-soluble and the outer carboxylic acid function allows the direct biofunctionnalization of the NPs (Fig. 4.4).

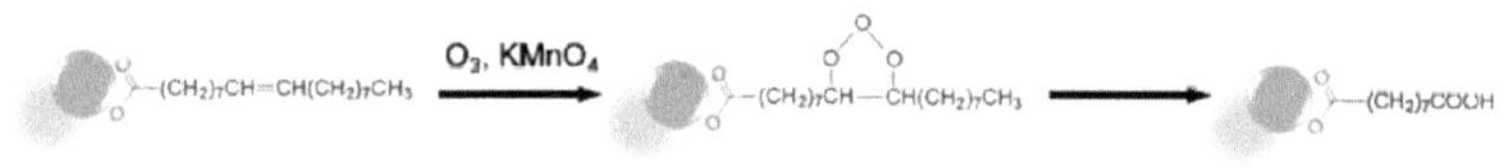

Figure 4.4 Transformation of oleic acid into azaleic acid via ozonolyse.

4.3.4 Tails Interdigitation

A third route for transferring NPs in aqueous media consists of keeping the first ligand coating and adding a second one composed of equivalent-structured molecules. Interdigitation between hydrocarbon tails of acid oleic and hydrophilic molecules takes place.[70,71] This "swift" transfer is used as intermediate step for silica[72] or polymere[73] shell formation. The tails interdigitation is softer than ligand place-exchange. The surface of the NPs is not modified, but the bilayer coating can be sensitive to pH variation or salts addition, and is thus less stable than grafted hydrophilic coating.

4.3.5 Silica or Polymer Shell

The last way for obtaining water-soluble NPs is the encapsulation inside a silica shell. An intermediate step is the bilayer formation, followed by the addition of tetraethyl orthosilicate (TEOS). In that way, single NP[74] or several NPs[75] can be embedded in a mesoporous silica shell. The hydrophilic shell can also be a polymer: the intermediate formation of an oleic acid-SDS bilayer around NPs aggregates is followed by the miniemulsification of styrene droplets. The polymerization of a styrene monomer inside the droplets drives to the NPs polystyrene bead formation (miniemulsion process).[76]

These phase transfer strategies allow to take advantage of nonaqueous-based synthesis, which produces nanoparticles with narrow size distributions, fewer crystalline defects, and more controllable morphologies, for application in biological and medical fields. Once particles are hydrophilic, they can be functionalized with biomolecules via direct, covalent functionalization or click chemistry for vectorization, *in vitro* selection to screen libraries of nucleic acid sequences, and point-of-care clinical diagnostics (magnetic immunoassays). (See Sections 4.3.2, 4.3.3 and 4.4.)

4.3.6 Direct Surface Functionalization

The direct functionalization of a nanoparticle is a reaction that allows the binding of a molecule, such as a bioactive molecule, onto the surface of a particle without any intermediate or additional reactive. Once the molecule is bound, the particle can be used for its own therapeutical properties or as an intermediate for further reactions (see Section 4.3.3). The general technique for direct functionalization involves adsorption mechanisms between the surface of a nanoparticle and a molecule of interest, such as between an adsorbent and an adsorbate. Adsorption occurs at the surface of the nanocrystal, where many lattice defects are observed and where adsorbates can be attracted to the surface via electrostatic forces between atoms. Three types of adsorption mechanisms are generally described: physisorption, chemisorption, and electrostatic attraction.

In physical adsorption, the binding of the adsorbate is mainly guided by van der Waals dipole-dipole and dipole-quadrupole interactions. Owing to the nature of the forces applied, which do not involve significant changes in the electronic structure of the adsorbed species, physisorption involves generally low binding energies and therefore is a fully reversible phenomenon that can lead to desorption of the molecules. Concerning the small size of the molecules, this last characteristic contraindicates their application in the biomedical domain; instead, physisorption is better suited to industrial applications such as gas separation or purification. However, the binding of a polymer could lead to a relative strong affinity due to a cooperative effect of the forces engaged.

The forces involved in chemisorptions, on the other hand, lead to drastic changes in the electronic state of the adsorbate, with binding energies characteristic of a covalent bond. As a consequence of this process, which can be irreversible, the modifications can be easily detected by physical methods such as infrared or UV-visible absorption spectroscopy (Fig. 4.5). In order to avoid a strong desorption phenomenon, chemisorption via carboxylate, sulfate, thiol, phosphate groups, or derivatives is preferred for anchoring a small molecule on a particle. For this purpose, the design of bifunctional molecules with an anchoring head and a bioactive tail is of great interest.

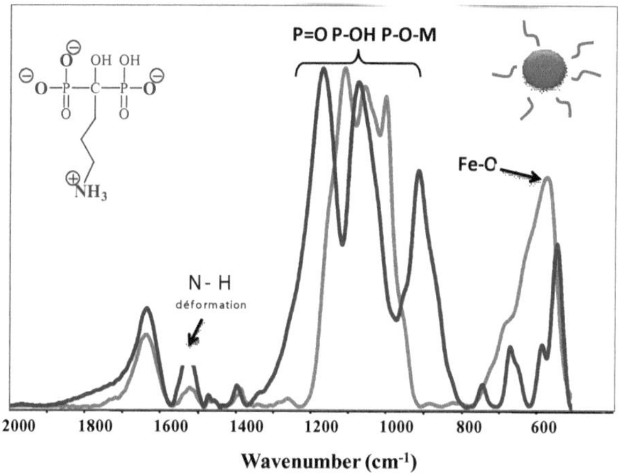

Figure 4.5 Infrared spectra of free alendronate (blue) and γFe_2O_3 alendronate particles (red). Reproduced from Benyettou *et al.*[87] See also Color Insert.

The direct functionalization approach is smart, of course, in terms of simplicity—most of the syntheses are carried out in water at ambient temperature—but also in terms of metabolization of the nanodrug. Indeed, the addition of a linker between the drug and the nanoparticle could favor the interaction of the drug with its target by reducing steric hindrance with the particle, but it could also lead to an additional toxicity due to the nature of the linker itself. This is the case, for instance, with a type of nanoparticle that has been developed for the vectorization of DNA into the cells. McIntosh *et al.* and then Sandhu *et al.* Rotello *et al.* have elaborated cationically functionalized gold nanoparticles as DNA-binding agents.[77,78] They have functionalized nanoparticles with a positively charged 11-trimethylammonium-undecanethiol that can attract the polyanionic DNA through electrostatic interactions. They report high levels of transfection for a pET-16b plasmid into 293T mammalian cells, even eight times higher than with polyethylenimine (PEI), a widely used transfection agent. Nevertheless, they observed an important toxicity of the particles at a concentration only two-fold in excess of that needed to provide maximal transfection activity. In a subsequent study, Goodman *et al.* demonstrated that the toxicity of their system is related to the cationic amphiphilic 11-trimethylammonium-undecanethiol linkers that interact with the cell membrane and provoke the lysis of the lipidic bilayer.[79]

An alternative method for DNA delivery, using gold nanoparticles, is to tether an oligonucleotide directly onto the gold surface via sulfur chemistry.[80,81] This technique was first developed for the formation of self-assembled monolayers (SAMs) of thiol on gold planar surfaces. Typically, molecules used for this purpose are alkanethiols comporting a –SH head group that can bind the gold surface through chemisorption and formation of a thiolate S–Au bond. Thanks to progress in nucleic acids synthesis, it is now easy to find commercial oligonucleotides containing a thiol or a disulfide modification tethered by a short linker to the 5′- or 3′-end of the DNA strand. Of course, this process of adsorption can be extended to every bioactive molecule comporting a sulfur atom. This sulfur atom could be added during the synthesis at the end of the molecule, but it could also come from a native cysteine residue in the case of peptides or proteins.

Mandal *et al.*[82] have developed gold nanoparticle-conjugated peptides in an attempt to specifically target subcellular organelles. The peptide they used was created by combining two specific sequences: a cell-penetrating peptide, tethered to a nuclear localization signal (NLS). This peptide was synthesized with a terminal cysteine residue to enable its conjugation to the gold surface. TEM micrographs of the TE85 cell line incubated with the bioconjugate particles reveal the presence, in the cytoplasm or close to the nuclear membrane, of numerous particles probably still enclosed in endosomal vesicles. Curiously, the synthesis protocol of gold particles, reduction of tetrachloroauric acid by sodium borohydride followed by the filtration of the particle, and then the adsorption of the peptide, seems to be efficient without the need for additional coating to stabilize the gold particles before adsorption of the peptide.[82]

Unfortunately, nonpassivated gold nanoparticles are often subject to a rapid and nonreversible aggregation. To avoid this phenomenon, particles are surface-passivated with citrate or alkanethiols. As a consequence, adsorption of thiolated bioactive molecules on these particles is not, strictly speaking, a direct functionalization mechanism and could be considered at least as a partial ligand exchange mechanism. Depending on the quantity of thiol added, the replacement of the initial coating could be partial or total.

Another method for real, direct functionalization of nanoparticles is to use metallic oxide particles. These nanoparticles can be stabilized in a colloidal solution simply by adjusting the pH of the solution, eliminating the need for additional coating of the molecules. At low salinity, the stability of the colloidal solution is controlled by electrostatic repulsions between the particles. This phenomenon is generally related to the presence of defects in the crystal lattice at the surface of the particle. Indeed, at the solid-liquid interface, the coordination of the metallic cation is conserved via chemisorption of a water molecule, leading to the hydroxylation of the metal. This confers to the nanoparticle an acid-base behavior in Brönstedt terms.[83]

When the pH is below the isoelectric point (IEP), particles are positively charged ($M{-}OH_2^+$, where M stands for metal atom); beyond this point, as pH increases, the surface charge becomes negative ($M{-}O^-$), as confirmed by zeta potential values. We can take advantage of this surface property to drive the adsorption of molecules having an opposite charge onto a nanoparticle.

Oxide ceramics such as ZrO_2 are used in high-tech ceramics in the medical sector (hip joints, dental restorations and fillings, etc.) as well as in the electronic industry. Wang *et al.* studied ZrO_2 particles coated with positively charged polyethylenimine (PEI).[52] The intent of their investigation was to determine, by spectroscopic methods, how the adsorbed PEI affects the polymer/solid interfacial properties of nanosized ZrO_2 suspensions. They conclude by attributing the adsorption process of PEI on ZrO_2 particles to a mechanism implying mainly hydrogen bonding. Nevertheless, they noticed an increased amount of adsorbed PEI when the reaction occurs at a pH value higher than the isoelectric point of the zirconium oxide; almost twice as many molecules adsorbed at pH 8 than at pH 5. In these experimental conditions, that is, pH 8, particles are negatively charged, while the PEI molecules are still positively charged. All these results lead Wang *et al.* to propose that when the reaction occurs at a pH value higher than the IEP of the particles, another mechanism, called electrostatic interaction, takes effect alongside hydrogen bonding. This mechanism is general and can be extended to other metallic oxide particles with various adsorbates.

Bisphosphonates (BPs) are widely used as inhibitors of bone resorption in the treatment of bone disorders like Paget's disease, osteoporosis, and solid tumor bone metastases due to their high affinity for calcium ions.[84] BPs can also be anchored to the surface of a metal oxide through the formation of strong M–O–P bonds via heterocondensation reaction and coordination of the phosphoryl oxygen. Coordination of one phosphonic acid can lead to the formation of three M–O–P bonds with the metal oxide surface thus conferring to the BPs a great binding affinity to the γFe_2O_3 nanoparticles.[85] Motte *et al.* successfully applied this technique to the surface passivation of maghemite nanocrystals (γFe_2O_3) using different bisphosphonates.[84,85] They anchored onto their particles a bisphosphonate antitumor agent called alendronate (Fosamax®) and they obtained significant results concerning the growth inhibition of an induced tumor in a nude mouse model.[86] The FTIR spectroscopic data presented in Fig. 4.5 are indicative of large modifications in the P–O stretching region, confirming the chemical adsorption of phosphonic acids onto the maghemite surface.

Given the number of biomolecules and/or bioactive molecules containing carboxylic or phosphate functions, applications of direct functionalization on various metallic particles will certainly increase during the next years. To conclude, direct functionalization is an easy way to obtain bioactive nanoparticles for a further utilization in many biomedical applications, such as drug delivery, cell imaging, sensors, or other molecular biology tools. Moreover, direct functionalization, by decreasing synthesis intermediates and derivatives, by reducing the utilization of organic solvent, and by reducing the temperature of reaction, could be considered as an alternative technique for achieving green chemistry goals in nanotechnology.

4.3.7 Two-Step Surface Functionalization

To insert biomolecules on the particles surface with control of their architecture and surface density, several chemical methodologies are described for the inorganic nanoparticles surface functionalization. The two more widespread are the classical technique of formation of amide, ester, carbamate, and other bonds and the techniques called click chemistry. A third methodology, which does not use covalent linkage but strong biological interaction is also described, biotin-streptavidin linkage being the best-known example.

4.3.7.1 Covalent coupling

Covalent linkages are strong and stable bonds, which can be specifically formed between functional groups usually found in biomolecules, typically amino, carboxylic acid, and thiol groups and other corresponding groups on the NP surface. It permit to introduce on the nanoparticle surface many type of biomolecules such as fluorescent molecules, drugs, targeting molecules,[4] ... (Fig. 4.6).

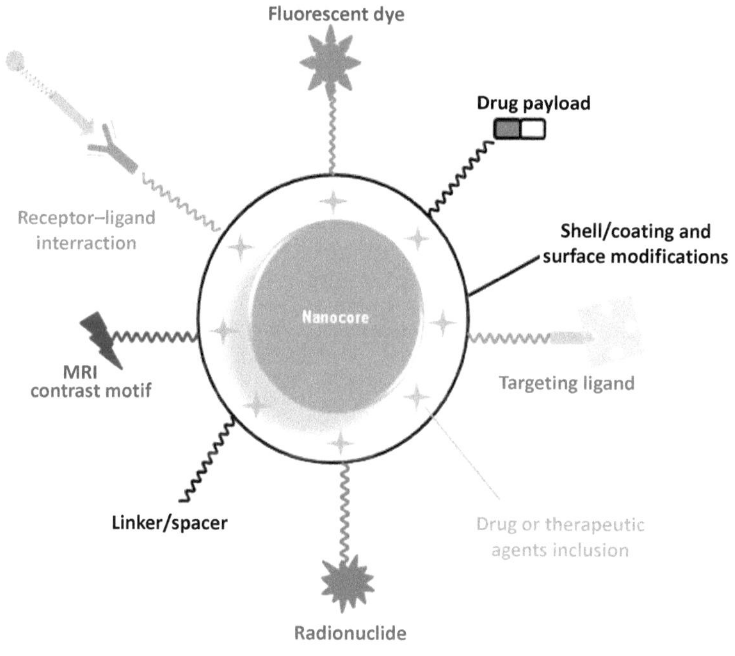

Figure 4.6 Optionally functionalized and devised nanoparticles could be achieved for individualized diagnosis and treatments.

Attaching ligands to nanoparticles requires diverse chemistries. The direct approach is to use a conventional bioconjugation strategy. As reported in Veiseh,[88] NP surfaces functionalized with amine, sulfhydryl, aldehyde, carboxylic acid, and active hydrogen functional groups can be targeted. These strategies are particularly suitable for small molecule conjugation but were extended to peptide proteins and antibodies. Most commonly, amines or carboxylic acids are present on the nanoparticle surface and carbodiimide coupling chemistry is used for biomolecule conjugation. Usually, these functional groups are added to the NP surface via its first surface functionalization step (see Fig. 4.2), which can dictate both the type and number of functional groups on each NP. For example, PEG-coated SPION was reported to have 26 reactive amino groups per NP.[89] Lalatonne *et al.* used molecules of the bisphosphonate family: the bisphosphonate moiety binds to the surface and exposes a functional group (amino, carboxylic acid) at the surface.[5,84] With an iron oxide nanoparticle coated with alendronate, and an antitumoral bisphophonate molecule with amine as an end-terminal functional group, it was shown that the nanoparticle surface presented 1200 reactive amino groups per NP. In other work,[86] such a nanoplatform was used for the covalent coupling of fluorescein as fluorophore, leading to the elaboration of magneto-fluorescent iron- oxide nanoparticle with both therapeutic and dual imageries (fluorescence and MRI) properties.

4.3.7.2 Biofunctionalization carbodiimide coupling

Amine and carboxylic functions are the more commonly used functions for biofunctionalization. This is because amine and carboxylic nanoparticles can be synthesized with relative ease and because of the importance of these two functionalities in biological macromolecules. Moreover, amine-coated nanoparticles are easily modified by linkers to create new functionalities such as iodoacetyl or maleimide.[90]

Amine-functionalized nanoparticles can be made to react with reactive ligands such anhydride, epoxyde, and isothiocyanate.[91] But the usual coupling strategy is to form a stable peptidic ligand with a carboxylic ligand using carbodiimide chemistry. The most common carbodiimide coupling strategy uses 1-ethyl-3-(dimethylaminopropyl) carbodiimide hydrochloride (EDC or EDAC) (Fig. 4.7) as the coupling agent because this derivative is completely soluble in water (unlike N,N′-dicyclohexylcarbodiimide DCC or N,N′-diisopropylcarbodiimide, the common carbodiimides with organic solubility). 1-Ethyl-3-(3-dimethylaminopropyl)-carbodiimide (EDC) will allow the formation of an amide bond between a carboxylic acid and an amine by activating carboxyl to form a derivative of the

O-urea. N-hydroxysuccinimide (NHS) or sulfo-NHS is usually used to enhance the coupling with EDC (Fig. 4.7). The reaction involves the formation of an active intermediate NHS ester, which then reacts with the amine to produce the amide bond. The coupling mixture EDC/NHS is described as more reactive, giving higher yields, and has been widely applied to covalent coupling of protein and enzymes.[92,93] The coupling rate of this reaction could also be greatly enhanced by the used of microwave energy.[86]

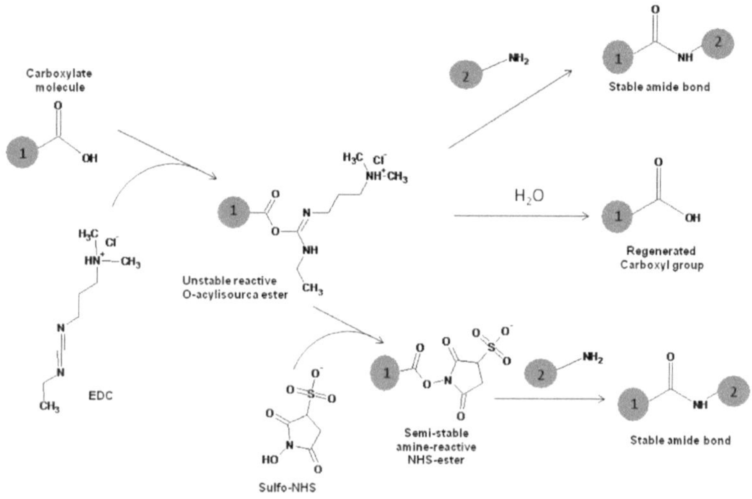

Figure 4.7 Different activation mechanisms for covalent coupling on carboxylic surface functionalized nanoparticles.

These functional groups are readily interchangeable. Reaction of amine-functionalized particles with succinic anhydride or glutaric anhydride converts the surface to carboxylates, whereas the activation of carboxylic acid–functionalized particles with 1-ethyl-3-(di methylaminopropyl)carbodiimide (EDC) and N-hydroxysuccinimide (NHS) or N-hydroxysulfosuccinimide (sulfo-NHS) followed by a reaction with a diamine results in amine-functionalized nanoparticles.

The reverse methodology—using carboxylic acid–functionalized nanoparticles on which amine substrates are conjugated—is also widely used.[94] In this case, carboxylic acid–coated particles require activation prior to conjugation to the amine functionality. Once again the EDC/NHS mixture is the prominent reactant used.[84] Although fairly tedious, the reaction of ligands containing hydroxyl groups can also be accomplished with carboxylic acid–coated particles.[90]

4.3.7.3 Maleimide and iodoacetyl coupling

Maleimide and iodoacetyl are usual reactive ligands for thiol-containing substrates (Fig. 4.8). They are also often utilized to functionalize sulfhydril-coated nanoparticles. Maleimide coupling has been used to conjugate biomolecules such as DNA,[95] herceptin,[96] and proteins[97] to NPs.[98] This reaction leads to stable linkage, but interparticle disulfide crosslinking is an inconvenient side effect of the method. The reverse methodology consists of functionalizing the nanoparticle surface with maleimide or iodoacetyl in order to get these functions to react with the thiol contained in biomolecules. For example, Brinley *et al.*[99] used sulfosuccinimidyl-4-(maleimido methyl)cyclohexane-1-carboxylate (sulfo-SMCC) to functionalize amine-coated nanoparticles with maleimide.

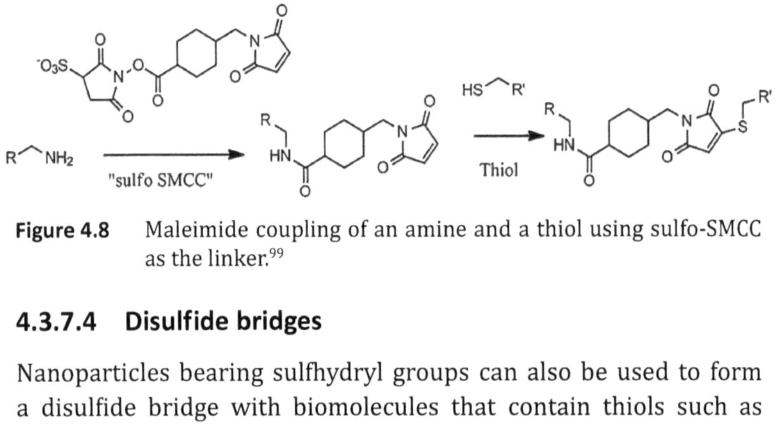

Figure 4.8 Maleimide coupling of an amine and a thiol using sulfo-SMCC as the linker.[99]

4.3.7.4 Disulfide bridges

Nanoparticles bearing sulfhydryl groups can also be used to form a disulfide bridge with biomolecules that contain thiols such as peptides, proteins, or antibodies. Moreover, disulfide bridges have been used for the reversible chemical coupling of NPs. Tian *et al.*[100] showed that the oxidation and reduction of disulfide bridges between silica NPs and Fe_3O_4 NPs was facilitated by glutathione disulfide (oxidative bond formation) and dithiothreitol (DTT) (reductive bond cleavage). While current work appears to be focused on the formation of hybrid nanostructures, this approach could also be adapted for drug delivery because cleaving agents such as glutathione[101] are present at appreciable concentrations *ṅ vv̇ o*.

4.3.8 Click Chemistry

Click chemistry, a synthetic approach to covalently linking two molecules, has had an impact on modern chemistry and biological chemistry this past decade. Click chemistry has been used in various domains such as medicinal chemistry, material and polymer chemistry, imaging, and so on.[102–105] It has therefore been broadly used for nanoparticle functionalization recently. The concept of click chemistry was introduced by K. Barry Sharpless in 2001.[106] The term describes chemical reactions that can covalently bond two different species rapidly and efficiently (ideally quantitative reactions without byproduct formation). These reactions must be specific, chemoselectives and should be realized at ambient temperatures, in aqueous or hydro-alcoholic media, and in the presence of air or moisture. Few chemical reactions have been described that have been accepted as click chemistry reactions: dipolar [4 + 2] cycloaddition (Diels-Alder reaction), oxime or hydrazone formation, thiol-ene and thiol-yne reactions, and the 1,3 dipolar cycloaddition between azide and alkyne catalyzed by copper (I). This later reaction, originally described by Huisgen[107] and re-introduced by Sharpless, is by far the most common example of a click chemistry reaction (Fig. 4.9).[108,109]

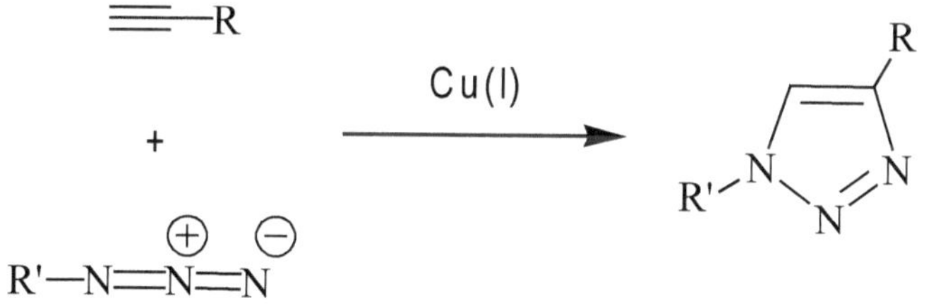

Figure 4.9 1,3 dipolar cycloaddition between azide and alkyne catalyzed by copper (I).

Due to its specificities, click chemistry provides very attractive possibilities for bioconjugation reactions. The speed, efficiency, mild conditions, and possibility of use in aqueous media of these reactions render possible the coupling of biomolecules sensible to classical, harsh conditions. The chemospecificity of the reaction also ensures reaction in a precise location without the need of complex protection and deprotection steps. Functionality used in click chemistry (such as azide and alkyne) are not usually found on biomolecules, so click chemistry is said to be "bioorthogonal." These features are of great interest when a precise active site (such as a targeting site) on the biomolecule is needed. Finally, the chemical bond created using click chemistry (for example the triazole ring formed in Huisgen reaction) is rigid and stable with regard to biological degradation.

4.3.8.1 Nanoparticle targeting and labeling

Up to now several groups have used click chemistry to tailor a nanoparticle surface for biomedical applications. Most of them used a 1,3 cycloaddition reaction to surface functionalized inorganic nanoparticles with diverse molecules: targeting, labeling, or bio-active molecules (Fig. 4.10). The main goal of these functionalizations is to allow inorganic nanoparticles targeting, the major application being imaging, hyperthermia, or biosensor development. In terms of targeting, molecules added at the surface of inorganic NP are diverse in properties.

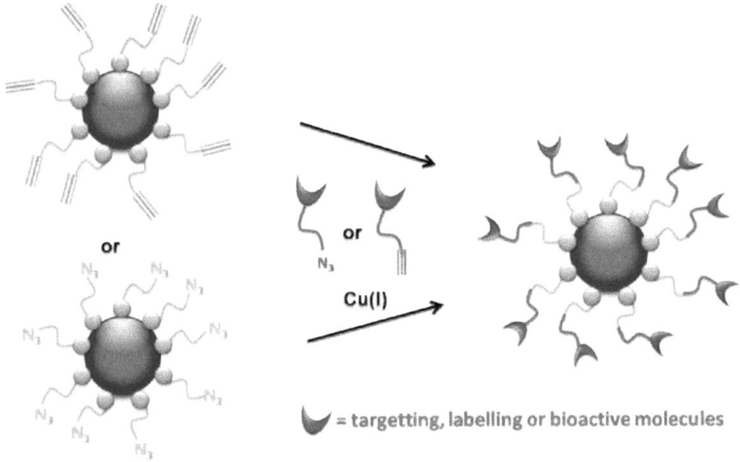

Figure 4.10 1,3 cycloaddition reaction used to surface functionalized inorganic nanoparticles with diverse molecules.

Chikae *et al.*[110] for example, covalently attached sialic acid to the surface of gold nanoparticles immobilized on a carbon electrode. The resulting nanobioelectronic detection system of saccharide–protein interactions was used to detect Alzheimer's amyloid-beta ($A\beta$).

4.3.8.2 Folic acid

Folic acid is also a common biomolecule used for targeting.[111,112] Hayashi *et al.*[113,114] developed folic acid-functionalized Fe_3O_4 nanoparticles from Huisgen 1,3 dipolar cycloaddition between azide-functionalized magnetite NPs and alkyne-bound folic acid. The azide-functionalized magnetite NPs were further functionalized with cyclodextrine as a drug container. The authors showed that this functionalization enhanced the cellular affinity of SPIONs to glioma cells and allowed a better entry into the cells. Folic acid was also used by Santra *et al.*[115] to build a biocompatible system for theranostics. It consists of iron oxide nanoparticles coated with a polymeric matrix of poly- (acrylic acid) bearing alkyne functions on its surface. This system was able to react with folate azide and co-encapsulate a lipophilic near-infrared (NIR) dye and an anticancer drug within the hydrophobic pockets present in the polymeric matrix. Enhanced internalization in lung carcinoma A549 cells was observed upon incubation with the folate-functionalized nanoparticles, and taxol-loaded particles induced cell death.

4.3.8.3 Peptide targeting

Peptide targeting is also a current methodology for determining precise biological location.[116] Von Maltzahn *et al.*[117] developed superparamagnetic iron oxide nanoparticles modified to display azido-PEG groups on their surfaces. The authors covalently clicked alkyne modified cyclic LyP-1 targeting peptides to these nanoparticles in order to direct their binding to p32-expressing tumor cells. This targeting was assessed *in vitro* and *in vivo* on mice bearing human MDA-MB-435 cancer xenografts and injected intravenously via the tail vein. More recently, Martin *et al.*[118] synthesized bombesin-functionalized iron oxide nanoparticles for their specific uptake in prostate cancer cells. Dextran-coated superparamagnetic iron oxide nanoparticles functionalized with azides and as fluorescent rhodamine derivative were reacted with a pan-bombesin peptide bearing an alkyne function. The peptide-functionalized nanoparticles were then demonstrated to be selectively taken up by PC-3 prostate cancer cells relative to unfunctionalized nanoparticles; this uptake was inhibited by the presence of free peptide, confirming the specificity of the interaction.

4.3.8.4 ^{18}F-labeled nanoparticles for PET-CT imaging

Another example of click chemistry is the work done by Devaraj *et al.*,[119] who used ^{18}F-labeled nanoparticles for *in vivo* PET-CT imaging. Monocrystalline iron oxide nanocompounds coated with dextran were crosslinked with epichlorin hydrin, aminated, and labeled with near-infrared fluorochrome Vivotag-680. The remaining amine functions at the surface of the nanoparticles were linked to an azido PEG, which was reacted with a ^{18}F-labeled PEG alkyne. The obtained nanosystem showed very high signal-to-noise in *in vivo* dynamic PET imaging and pharmacokinetic analysis were conducted.

4.3.8.5 Biomacromolecule coupling

The most challenging task in nanoparticle functionalization by click chemistry is the coupling of biomacromolecules such as antibodies, enzymes, or DNA. One of the first examples of such bionanoconjugation was given by Brennan *et al.*,[120] who created functional hybrids of lipases and gold nanoparticles. They functionalized gold nanoparticles with azide functions and then clicked to acetylene-functionalized lipase molecules. This resulted in linkage of approximately seven fully active lipase molecules per nanoparticle.

More recently Zhang *et al.*[121] proposed the synthesis of a nanobioprobe by conjugating horse radish peroxydase (HRP) to gold nanoparticles. The nanoparticles were coated with lipophile polymers bearing azide functions and coupled to alkyne modified HRP. The number of HRPs bound on the surfaces of AuNPs could be controlled by adjusting the molar ratio of reactants. This HRP-AuNP conjugate retained the catalytic activity of HRP molecules, as confirmed by a colorimetric reaction.

Cutler *et al.*[122] used the copper-catalyzed azide-alkyne reaction to form a dense monolayer of oligonucleotides on a superparamagnetic nanoparticle core. They coupled oligonucleotide modified with a terminal alkyne to azide-functionalized iron oxide particles. This method afforded particles that exhibit properties such as sharp melting transitions and high cellular uptake, indicative of their dense functionalization.

A last example of bioconjugation assisted by click chemistry is the work done by Thorek *et al.*[123] They carried out a comparative analysis of nanoparticle-antibody conjugations between carbodiimide methodology and click chemistry. They evaluated the efficiency of conjugations between antibodies and superparamagnetic iron oxide nanoparticles coated with dextran and bearing azide or carboxylic functions. The conjugation was done using click chemistry with alkynated antibodies or by formation of carbodiimide crosslinking to native antibodies. The click reaction allowed for a higher extent and efficiency of labeling compared with carbodiimide, thus requiring fewer antibodies. Further, conjugates prepared via the click reaction exhibited improved binding to target receptors.

4.3.8.6 Limitations

Though click chemistry coupling via 1,3 cycloaddition on nanoparticles has many unique advantages, there is a limitation to its use. The copper catalyst needed to complete the reaction must be removed before use in several applications. First, use of copper *in vivo* may be linked to several disorders.[102] Second, copper ions can strongly inhibit quantum dot luminescence. Fortunately, metal-free catalysis solutions have been developed as an alternative.[124] Moreover, click chemistry reactions that do not involve metal catalysis have been described that may certainly be used in the future for inorganic nanoparticle biofunctionalization.

4.3.8.7 Noncovalent interactions

Another way to surface functionalized inorganic nanoparticles is to create a noncovalent but highly strong and specific interaction such as the one found in nature, for example in protein-ligand interactions. Functionalization of nanoparticles with biomolecules by specific affinity interactions uses this long known methodology developed initially for the functionalization of macrosize supports. Among all the techniques described, the leading one is still the streptavidin–biotin conjugation method. Several examples of utilization of the biotine-streptavidine interaction to functionalized nanoparticles with biomolecules can be found in the literature.[2,125–127] Two recent examples are particularly representative of this methodology.

The first one is the work presented by Charalambous *et al.*,[128] which proposed a functionalization of streptavidin-coated quantum dots by a *biotinylated* peptide corresponding to half of the intein-mediated ligation system (IC-QDs). Inteins are polypeptide sequences that are able to self-excise, rejoining the two flanking extein sequences by a native peptide bond. The RNA encoding for the target protein (EGFP° and the other half of split intein were delivered into Xenopus embryos via microinjection together with the IC-QDs). *In vivo* association of the intein halves in the cytosol triggered protein trans-splicing, resulting in the ligation of the QD to the target protein through a peptide bond. Use of NIR-emitting QDs allowed monitoring of QD-conjugates within the embryo at depths where EGFP is undetectable, demonstrating the advantages of QDs for this type of experiment.

The second example is a straightforward biot-streptavidin interaction presented by Brennan *et al.*[120] In this work the authors coated gold nanoparticles with thioalkylated PEG-biotin. They then used streptavidin as a "molecular glue" between their gold biotinylated particle and biotinylated lipase molecules. Moreover, lipase variants with biotin labels incorporated at different sites on the protein's outer surface have been used to control the orientation of the enzyme. The authors then proved that enzymatic activity appears to be orientation dependent, with a significant decrease if the active site is facing the nanoparticle.

4.4 Some Biomedical Applications

Since 2004, the National Institutes of Health has referred to applications of nanotechnology for treatment, diagnosis, monitoring, and control of biological systems as *nanomedicine*.[129] Because nanomaterials are similar in scale to biologic molecules and systems yet can be engineered to have various functions, nanotechnology is potentially useful for medical applications. Hence, nanomedicine is a large subject area that includes biomedical applications such as imaging, biological labeling and separation, drug delivery and transfection, hyperthermia, and immunoassays (Fig. 4.11).

Nanomaterials are now being designed to aid the transport of diagnostic or therapeutic agents through biologic barriers and to mediate molecular interactions. Unlike macroscopic materials, nanomaterials have a high ratio of surface area to volume as well as tunable optical, electronic, magnetic, and biologic properties, and they can be engineered to have different sizes, shapes, chemical compositions, surface chemical characteristics, and hollow or solid structures.[130,131] These properties are being incorporated into new generations of drug-delivery vehicles, contrast agents, and diagnostic devices, some of which are currently undergoing clinical investigation or have been approved by the Food and Drug Administration for use in humans. Examples of the nanomaterials used or developed in medicine are provided Table 4.2.[75]

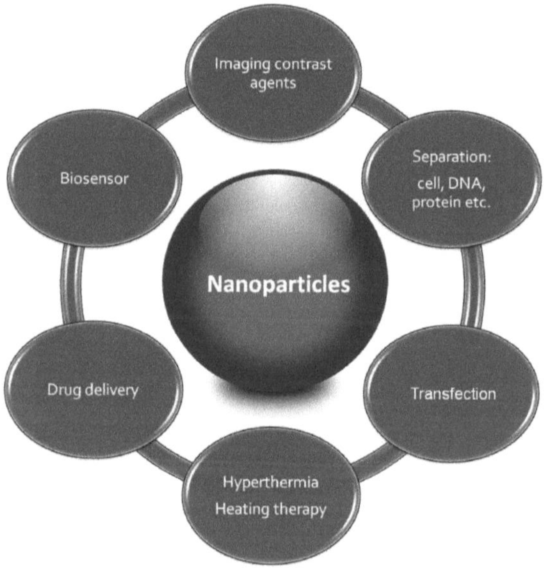

Figure 4.11 Examples of nanotechnology biomedical applications.

Table 4.2 Examples of nanomaterials for biomedical applications

Application	Nanomaterial	Trade name	Manufacturer	Target
Imagin g				
MRI contrast agent	Iron oxide	Feridex	Guerbet	Liver
MRI contrast agent	Iron oxide	Resovist	Bayer Schering	Liver
Fluorescent contrast,	Quantum Dots	QD's	Life Technologies	Cell tissues
In-vt ro dia gnostc				

(*Cont'd*)

Table 4.2 (*Continued*)

Application	Nanomaterial	Trade name	Manufacturer	Target
	Gold	Verigene	Nanosphere	Genetic
	Iron Oxide	MIAplex	Magnisense	Immunoassays
Therapy				
Hyperthermia	?	NanoXray	Nanobiotix	Tumors
Drug vehicles	Liposome	Doxil/Caelyx	Ortho Biotech	Tumors

Nanoparticle surface functionalization is a key factor in translating the intrinsic properties of nanoparticles to biomedical applications. First there is a need for a robust way to stabilize nanoparticle dispersions in aqueous media in a variety of processing conditions. Furthermore, molecules such as antibodies, peptides, or polymer allow specifically biological recognition.

In this section we focus on the use of nanomaterials for some medical applications: immunoassays; imaging; nanovehicles and drug carriers; and hyperthermia.

4.4.1 Immunoassays

The development of reliable monitoring methods is essential for the diagnosis of various diseases and the accurate evaluation of environmental problems. Immunoassay, one of the most important of these methods, is based on the exceptional specificity that an antibody has for its own target antigen. This highly specific affinity coupled with a very low limit of detection has made immunoassay a widely accepted analytical method in various fields. Immunoassays using a radio isotopic technique have been heavily utilized since the 1960s[10] because radio isotopic labels have shown superior performance in both high selectivity and low detection limits. The use of radioisotopes, however, has many problems related to handling and disposal; therefore, immunoassays using a nonisotopic technique have increased dramatically and have gained a certain status as a common analytical technique in recent years. The use of magnetic particles in immunological assays has grown considerably, because the particles' magnetic properties permit their easy separation and/or concentration in large volumes, allowing for faster assays and in some cases improved sensitivity over currently available commercial methods. A typical microtiter plate sandwich immunoassay is schematically presented in Fig. 4.12.

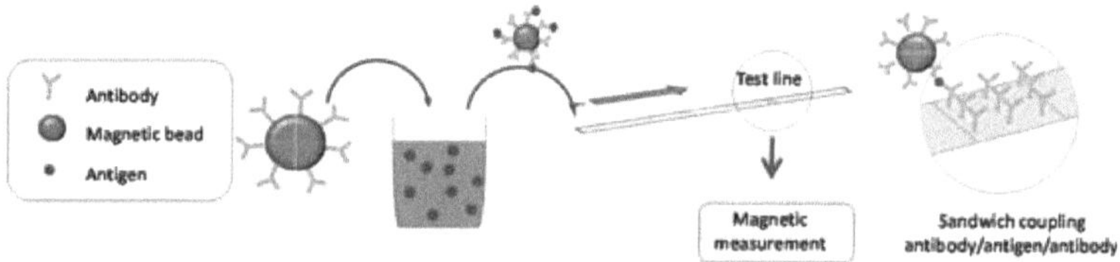

Figure 4.12 Schematic principle for a sandwich immunoassay based on magnetic detection.

One antibody is attached to a superparamagnetic surface (usually a microsphere), and the other antibody is adsorbed onto a strip (test line). The antibody-modified magnetic nanobeads are incubated with the sample. The antibody on the beads will bind the target antigen from the sample. After magnetic separation, the antigen-antibody bead is deposited on the strip and migrates by capillarity. The antigen-antibody beads will bind the antigen on the test line and retain it in the plate, thereby forming an antigen "sandwich" between the two different antibodies.

This binding reaction is measured by detecting a magnetic signal on the test line. An enormous effort has been made to develop new magnetic signal detection devices such as a giant magneto-resistant (GMR),[132] spinvalve,[133] and magnetoelastic sensors.[134] Recently, a new detection and characterization method was developed to take advantage of the nonlinearities of superparamagnetic or ferromagnetic materials for magnetic imunoassays (MIA). The Magnisense company developed two room temperature sensors, called MIAtek® and MIAplex® (magnetic imunoassay and multiplex technology, respectively). The MIAtek® sensor measures a signal proportional to the third derivative of the magnetization at zero magnetic field.[135] Hence, the MIAtek signal is very sensitive and is proportional to the amount of magnetic particles. This allows the detection of nanograms of superparamagnetic materials. MIAtek technology is used for very sensitive monoparametric immunoassays. This technology presents also the advantage of being easily transportable (Fig. 4.13) and can be used for emergency diagnostics.

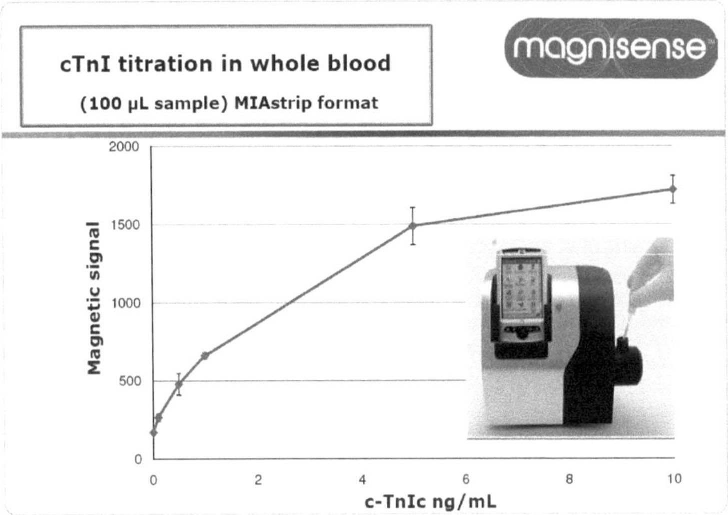

Figure 4.13 MIAtek detector and titration of troponin (c-TnIc) within whole blood. (*Source:* www.magnisense.com.)

For emergency facilities, Magnisense has developed the MIAtroponin® test, based on its MIAtek® technology, which uses magnetic beads as labels. The chosen cardiac marker is troponin I, the most common cardiac marker used in the evaluation of chest pain and acute myocardial infarction. It is a cardiac necrosis diagnostic marker, a specific and sensitive marker of myocardial cell injury, the ideal biomarker in emergency facilities. Indeed, acute coronary syndrome and heart failure means rapid decisions, and appropriate interventions are needed because timely action reduces mortality and complications. Nevertheless, diagnostic equipment, such as echocardiographs, is not always available in an acute setting. Moreover, ECG has been found to be only 50% sensitive for diagnosis of acute myocardial infarction, meaning that half the time the test will be inconclusive. So there is a need for reliable, sensitive, and quick medical tests for accurate and efficient emergency department triage to sort out the cardiac dysfunctions and pulmonary emboli causes and other dyspnoea etiologies.

The major benefits to emergency medicine of this technology are

- quantitative results (Fig. 4.13)

- connectability to the laboratory information management system
- achievable results, which is very interesting within the framework of personal health records
- no influence of the blood hemolysis on the result (the result reading can be checked at any time by doctors or biologists).

Because the device is transportable, there is no need to send samples through a laboratory analysis; a quick scan can be performed directly at the physician's office, emergency department, or clinic or humanitarian mission.

4.4.2 Imagining

Superparamagnetic nanoparticles such as maghemite (γ-Fe_2O_3) are currently used as contrast agents for magnetic resonance imaging (MRI). In nanoparticle-enhanced MRI, a contrast can be observed between tissues loading superparamagnetic iron oxide nanoparticles (SPION) and tissues without SPION, owing to a difference in the precession frequency of the protons. Indeed, the resonance of SPION enhances the static and applied magnetic fields. Then the stronger local magnetic field shortens the relaxation times of protons, resulting in a stronger local MR signal. This property is currently used for MRI contrast enhancement, especially with T2 relaxation time alteration.[136] To investigate MRI signal enhancement effects, aqueous as-prepared nanoparticles were measured on a 4.7T MRI scanner.[137] The measurements were taken at room temperature for various iron concentrations between 0.031 and 1 mM. For T2* relaxation times, axial T2*-weighted SE images were obtained with a TR of 800 ms and TE of 8.6 ms at 4.7T. As shown in Fig. 4.14 with γ-Fe_2O_3@zoledronate nanoparticles, T2* weighted images change drastically in signal intensity with an increasing number of nanoparticles, indicating that the as-synthesized nanoparticles generated MR contrast on transverse (T2*) proton relaxation time weighted sequences. Fig. 14b shows the relaxation rates 1/T2* as a function of the iron concentration. The relaxation rate varied linearly with the iron concentration, as expected. Hence, these hybrid nanoparticles are potential candidates for MRI contrast agents.

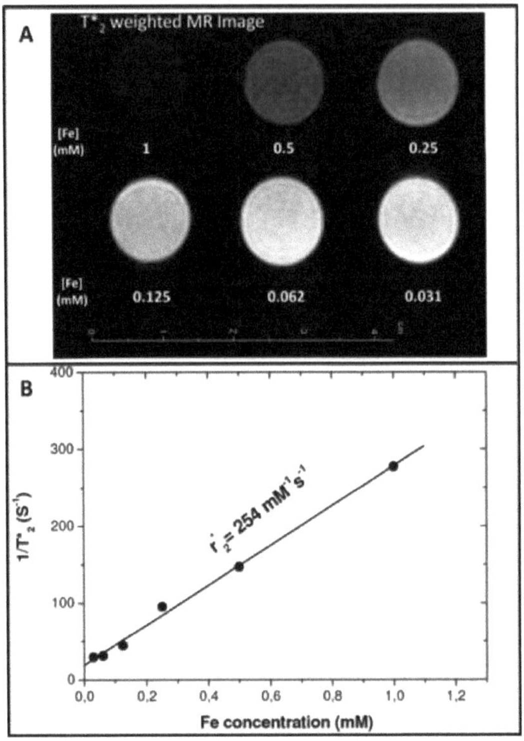

Figure 4.14 (a) T2* weight MR images of aqueous solutions of as-synthesized of γFe_2O_3 nanoparticles at different Fe concentrations. (b) T2* relaxation rates (1/T2*) plotted against the Fe concentration for the various aqueous solutions.

SPION are currently used to detect liver lesions since these particles are taken up by macrophages.[138] SPION are also attractive as sensitive contrast agents for cancer imaging. The nanoparticles remained in the tumors 24 hours after the injection, as compared with 1 hour for gadolinium-chelate contrast agents.[139] The reason for this difference is that the smaller nanoparticles are more easily taken up by tumor cells and diffuse out of the tumor more slowly.[140]

4.4.3 Nanovehicles and Drug Carriers

Most of nanoparticles used for drug targeting are hollow structures such as liposomes (Table 4.2). Nevertheless, inorganic structures are of special interest for theranostic—diagnsotic and therapeutic—applications. For example, antitumor magnetic nanoparticles have been designed by coating iron oxide nanoparticles with a clinically relevant antitumor agent: alendronate.[87] The antitumor agent anchored onto the nanoparticle surface through phosphonate groups. These nanoparticles have both MRI contrast agent and anticancer properties. This nanoparticle functionalization spectacularly improved cell penetration by alendronate and its antitumor effect. In particular, this behavior was enhanced in the presence of an appropriate magnetic field.

In vivo experiments confirmed the therapeutic efficacy of γFe_2O_3@alendronate nanocrystals in the presence of a magnetic field. The particles, loaded with the drug, were concentrated at the target site by an external magnet. Alendronate molecules were then released in the desired area over a long period of time. Therefore, these nanohybrid particles can be considered to be a drug delivery system. The tumor targeting efficiency of γFe_2O_3@alendronate allowed detection of tumors by MRI and at the same time delivery of sufficient amounts of anticancer drug, released from the nanoparticles, to exhibit anticancer activity.

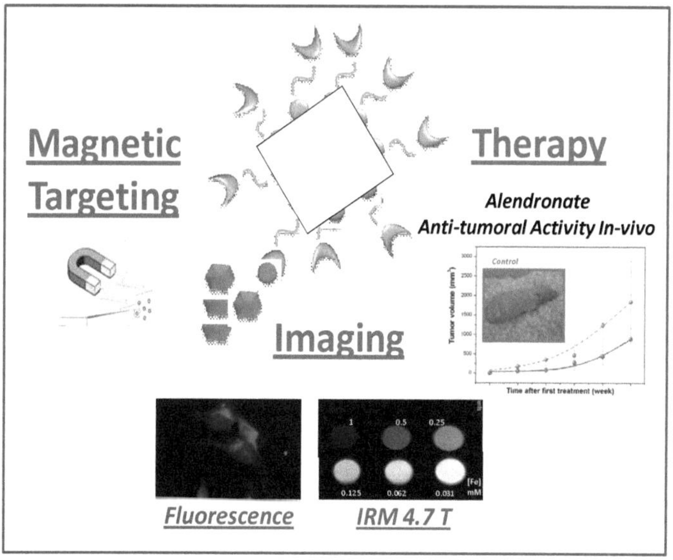

Figure 4.15 An innovative multimodal system that combines magnetic targeting of therapeutic agents with both magnetic resonance and fluorescence imaging into one system.

The amine functionalities as the terminal group of alendronate (coating agents) can be used as precursor groups for the covalent coupling of fluorescence markers. The feasibility of this process was demonstrated by coupling Rhodamine B to the γFe_2O_3@ alendronate hybrid nanomaterial. The system showed fluorescent properties and a high affinity for cells. The composite magnetic and fluorescent nanoparticles are potential candidates as a smart drug delivery system. Indeed, magneto-fluorescent nanocrystals show great potential for biosensing and medical diagnostics, and in particular for *in vitro* and *in vivo* imaging (Fig. 4.15).

4.4.4 Hyperthermia for Cancer Therapy

Hyperthermia, also called *thermal therapy* or *thermotherapy*, is a type of cancer treatment in which body tissue is exposed to high temperatures (above 45°C). Since nanoparticles are excitable within light emission (gold), magnetic field (iron oxide), and x-ray (iron oxide; gold), these particles are of specific interest to delivering local heat (hyperthermia). Several methods of hyperthermia are currently under study, including local, regional, and whole-body hyperthermia.[141–144]

A strategy that slightly derivates from hyperthermia and is based on the fact that metallic nanoparticles boosts x-rays is radiotherapy efficacy. In the presence of low-energy x-rays of around 100 keV, the metallic nanoparticles emit photoelectrons and/or free radicals that travel a very short distance before depositing their energy in the nearest endothelial cell.[145] The nanoXray™ technology was developed by Nanobiotix to dramatically improve the efficiency of radiation therapy in the treatment of cancer (Fig. 4.16). The nanoXray particle is made of a nanocrystal (dense, highly organized core) surrounded by a thin amorphous coating. Particles are injected directly into the tumor site, where they accumulate within cancer cells. The particles function only when "turned on" by outside x-rays during radiation therapy. X-ray energy is highly absorbed by the particle, due to its heavy oxide core. Electrons and free radicals produced by the physical reaction damage the cell's DNA and structures, leading to the death of the cell. *In vivo* experiments on Swiss nude mice xenografted with a colorectal model exhibit the clearance of the tumors after 18 days of treatment (*Source*: Nanobiotix and Oncodesign).

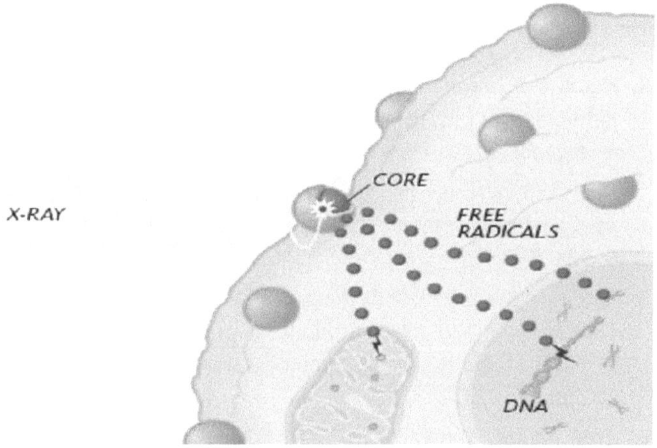

Figure 4.16 Schematic representation of nanoXray mechanism.

4.5 Conclusion

To quickly summarize, today new multifunctional nanoplatforms are designed for bioassays, diagnosis, and therapy applications. The nanoparticle has two essential roles: to act as a probe owing to its specific physical properties (magnetic, optical, or fluorescent for imaging or biosensing) and to carry on its surface specific molecules for therapeutic applications or precursor groups for the covalent coupling of biological recognition molecules, such as antibodies, nucleic acids (*ṅ vt ro* and *ṅ vi̇ o* diagnosis). Hence, nanomedicine is a large subject area and includes biomedical applications such as imaging, biological labeling, drug delivery and transfection, hyperthermia, immunoassays, and so on. Whatever the intended application, the nanoparticle surface functionalization is a key factor. A challenge is the elaboration of a nanoplatform that can be used for the various biomedical applications.

References

1. Chou, L. Y. T., Ming, K., and Chan, W. C. W. (2011). Strategies for the intracellular delivery of nanoparticles, *Chem. Soc. Rev.*, **40**, 233–245.
2. Dave, S. R., and Gao, X. (2009). Monodisperse magnetic nanoparticles for biodetection, imaging, and drug delivery: a versatile and evolving

technology, *Wl ey Interdś cṗ . Rev. Nanomed. Nanob technol.*, **1**, 583–609.

3. Haun, J. B., Yoon, T.-J., Lee, H., and Weissleder, R. (2010). Magnetic nanoparticle biosensors, *Wl ey Interdś cṗ . Rev. Nanomed. Nano-b technol.*, **2**, 291–304.
4. Liu, G., Swierczewska, M., Lee, S., and Chen, X. (2010). Functional nanoparticles for molecular imaging guided gene delivery, *Nano Today*, **5**, 524–539.
5. Motte, L., Benyettou, F., de Beaucorps, C., Lecouvey, M., Milesovic, I., and Lalatonne, Y. (2010). Multimodal superparamagnetic nanoplatform for clinical applications: immunoassays, imaging & therapy, *Faraday Dś cuss.*, **149**, 211–225.
6. Tran, N., and Webster, T. J. (2010). Magnetic nanoparticles: biomedical applications and challenges, *J. Mat. Chem.*, **20**, 8760–8767.
7. Massart, R. (1981). Preparation of aqueous magnetic liquids in alkaline and acidic media, *IEEE Trans. Magn.*, **17**, 1247–1248.
8. Vayssieres, L. (2009). On the effect of nanoparticle size on water-oxide interfacial chemistry, *J. Phys. Chem. C*, **113**, 4733–4736.
9. Vayssières, L., Chanéac, C., Tronc, E., and Jolivet, J. P. (1998). Size tailoring of magnetite particles formed by aqueous precipitation: an example of thermodynamic stability of nanometric oxide particles, *J. Coll. Interf. Sci.*, **205**, 205–212.
10. Yalow, R. S., and Berson, S. A. (1960). Immunoassay of endogenous plasma insulin in man, *J. Clǹ . Invest.*, **39**, 1157–1175.
11. Niederberger, M., and Pinna, N. (2009). *Metal Oxü e Nanopartċ les ǹ Organic Solvents: Synthesś , Formatȯ n, Assemb y and Applċ atȯ n*, Springer: London.
12. Tadic, M., Markovic, D., Spasojevic, V., Kusigerski, V., Remskar, M., Pirnat, J., *et al.* (2007). Synthesis and magnetic properties of concentrated [alpha]-Fe_2O_3 nanoparticles in a silica matrix, *J. Alloys Compd.*, **441**, 291–296.
13. Tadic, M., Kusigerski, V., Markovic, D., Milosevic, I., and Spasojevic, V. (2009). Anomalous behavior of the magnetic hysteresis loop in the [alpha]-Fe_2O_3/SiO_2 nanocomposite, *Mat. Lett.*, **63**, 1054–1056.
14. Fiévet, F., Lagier, J. P., Blin, B., Beaudoin, B., and Figlarz, M. (1989). Homogeneous and heterogeneous nucleations in the polyol process for the preparation of micron and submicron size metal particles, *Solid State Ionċ s*, **323 3**, 198–205.

15. Soumare, Y., Dakhlaoui-Omrani, A., Schoenstein, F., Mercone, S., Viau, G., and Jouini, N. (2010). Nickel nanofibers and nanowires: elaboration by reduction in polyol medium assisted by external magnetic field, *Solid State Comm.*, **151**, 284–288.

16. Piquemal, J.-Y., Viau, G., Beaunier, P., Bozon-Verduraz, F., and Fiévet, F. (2003). One-step construction of silver nanowires in hexagonal mesoporous silica using the polyol process, *Mat. Res. Bull.*, **38**, 389–394.

17. Ung, D., Viau, G., Fiévet-Vincent, F., Herbst, F., Richard, V., and Fiévet, F. (2005). Magnetic nanoparticles with hybrid shape, *Prog. Solid State Chem.*, **33**, 137–145.

18. Souad, A., *et al.* (2006). Magnetic properties of zinc ferrite nanoparticles synthesized by hydrolysis in a polyol medium, *J. Phys.: Condens. Matter*, **18**(39), 9055.

19. Ammar, S., Helfen, A., Jouini, N., Fievet, F., Rosenman, I., Villain, F., *et al.* (2001). Magnetic properties of ultrafine cobalt ferrite particles synthesized by hydrolysis in a polyol medium, *J. Mat. Chem.*, **11**, 186–192.

20. Ahmed, J., Sharma, S., Ramanujachary, K. V., Lofland, S. E., and Ganguli, A. K. (2009). Microemulsion-mediated synthesis of cobalt (pure fcc and hexagonal phases) and cobalt-nickel alloy nanoparticles, *J. Coll. Interf. Sci.*, **336**, 814–819.

21. Zhang, W., Qiao, X., and Chen, J. (2007). Synthesis of silver nanoparticles—effects of concerned parameters in water/oil microemulsion, *Mat. Sci. Eng. B*, **142**, 1–15.

22. Ahmed, J., Ahmad, T., Ramanujachary, K. V., Lofland, S. E., and Ganguli, A. K. (2008). Development of a microemulsion-based process for synthesis of cobalt (Co) and cobalt oxide (Co_3O_4) nanoparticles from submicrometer rods of cobalt oxalate, *J. Coll. Interf. Sci.*, **321**, 434–441.

23. Pal, A., Shah, S., Belochapkine, S., Tanner, D., Magner, E., and Devi, S. (2009). Room temperature synthesis of platinum nanoparticles in water-in-oil microemulsion, *Coll. Surf. A*, **337**, 205–207.

24. Lalatonne, Y., Benyettou, F., Bonnin, D., Lièvre, N., Monod, P., Lecouvey, M., *et al.* (2009). Characterization of magnetic labels for bioassays, *J. Magn. Magn. Mat.*, **321**, 1653–1657.

25. Lee, Y., Lee, J., Bae, C. J., Park, J. G., Noh, H. J., Park, J. H., *et al.* (2005). Large-scale synthesis of uniform and crystalline magnetite nanoparticles using reverse micelles as nanoreactors under reflux conditions, *Adv. Funct. Mat.*, **15**, 2036–2036.

26. Zhang, Z., Rondinone, A. J., Ma, J. X., Shen, J., and Dai, S. (2005). Morphologically templated growth of aligned spinel $CoFe_2O_4$ nanorods, *Adv. Mat.*, **17**, 1415–1419.

27. Bhattacharyya, S., Salvetat, J.-P., Fleurier, R., Husmann, A., Cacciaguerra, T., and Saboungi, M.-L. (2005). One-step synthesis of highly crystalline and high coercive cobalt-ferrite nanocrystals, *Chem. Comm.*, 4818–4820.

28. Sun, S., Murray, C. B., Weller, D., Folks, L., and Moser, A. (2000). Monodisperse FePt nanoparticles and ferromagnetic FePt nanocrystal superlattices, *Scė nce*, **287**, 1989–1992.

29. Murray, C. B., Kagan, C. R., and Bawendi, M. G. (2000). Synthesis and characterization of monodisperse nanocrystals and close packed nanocrystal assemblies, *Ann. Rev. Mat. Sci.*, **30**, 545–610.

30. Murray, C. B., Norris, D. J., and Bawendi, M. G. (1993). Synthesis and characterization of nearly monodisperse CdE (E = sulfur, selenium, tellurium) semiconductor nanocrystallites, *J. Am. Chem. Soc.*, **115**, 8706–8715.

31. Kang, E., Park, J., Hwang, Y., Kang, M., Park, J.-G., and Hyeon, T. (2004). Direct synthesis of highly crystalline and monodisperse manganese ferrite nanocrystals, *J. Phys. Chem. B*, **108**, 13932–13935.

32. Chaudret, B., and Philippot, K. (2007). Revue sur la synthèse de nanoparticules de métaux et d'oxydes métalliques, *Ol Gas Sci. Technol.*, **62**, 799–817.

33. Carenco, S., Boissière, C., Nicole, L., Sanchez, C., Le Floch, P., and Mézailles, N. (2010). Controlled design of size-tunable monodisperse nickel nanoparticles, *Chem. Mat.*, **22**, 1340–1349.

34. Yu, W. W., and Peng, X. (2002). Formation of high-quality CdS and other II–VI semiconductor nanocrystals in noncoordinating solvents: tunable reactivity of monomers, *Angew. Chem. Inter. Edt .*, **41**, 2368–2371.

35. Hyeon, T. (2003). Chemical synthesis of magnetic nanoparticles, *Chem. Comm.*, 927–934.

36. Sun, S., and Murray, C. B. (1999). Synthesis of monodisperse cobalt nanocrystals and their assembly into magnetic superlattices (invited), *J. App. Phys.*, **85**, 4325–4330.

37. Konishi, Y., Ohno, K., Saitoh, N., Nomura, T., Nagamine, S., Hishida, H., *et al.* (2007). Bioreductive deposition of platinum nanoparticles on the bacterium Shewanella algae, *J. Bȯ tech.*, **128**, 648–653.

38. Thakkar, K. N., Mhatre, S. S., and Parikh, R. Y. (2009). Biological synthesis of metallic nanoparticles, *Nanomed. Nanotech. Bȯ l. Med.*, **6**, 257–262.

39. Narayanan, K. B., and Sakthivel, N. (2010). Biological synthesis of metal nanoparticles by microbes, *Adv. Coll. Interf. Sci.*, **156**, 1–13.

40. Nair, B., and Pradeep, T. (2002). Coalescence of nanoclusters and formation of submicron crystallites assisted by lactobacillus strains, *Cryst. Growth Des.*, **2**, 293–298.

41. Alphandéry, E., Lijeour, L., Lalatonne, Y., and Motte, L. (2010). Different signatures between chemically and biologically synthesized nanoparticles in a magnetic sensor: a new technology for multiparametric detection, *Sensor. Actuator. B*, **147**, 786–790.

42. Krishnan, K. M. (2010). Biomedical nanomagnetics: a spin through possibilities in imaging, diagnostics, and therapy, *IEEE Trans. Magn.*, **46**, 2523–2558.

43. Kang, Y. S., Risbud, S., Rabolt, J. F., and Stroeve, P. (1996). Synthesis and characterization of nanometer-size Fe_3O_4 and γ-Fe_2O_3 particles, *Chem. Mat.*, **8**, 2209–2211.

44. Massart, R. (1980). Preparation of aqueous ferrofluids without surfactant — behavior as a function of the pH and the couterions, *C. R. Séances Acad. Sci C*, **291**, 1, 1–3.

45. Lalatonne, Y., Motte, L., Richardi, J., and Pileni, M. P. (2005). Influence of short-range interactions on the mesoscopic organization of magnetic nanocrystals, *Phys. Rev. E*, **71**, 011404.

46. Lalatonne, Y., Motte, L., Russier, V., Ngo, A. T., Bonville, P., and Pileni, M. P. (2004). Mesoscopic structures of nanocrystals: collective magnetic properties due to the alignment of nanocrystals, *J. Phys. Chem. B*, **108**, 1848–1854.

47. Ngo, A. T., and Pileni, M. P. (2000). Assemblies of ferrite nanocrystals: partial orientation of the easy magnetic axes, *J. Phys. Chem. B*, **105**, 53–58.

48. Hao, R., Xing, R., Xu, Z., Hou, Y., Gao, S., and Sun, S. (2010). Synthesis, functionalization, and biomedical applications of multifunctional magnetic nanoparticles, *Adv. Mat.*, **22**, 2729–2742.

49. Weissleder, R., Kelly, K., Sun, E. Y., Shtatland, T., and Josephson, L. (2005). Cell-specific targeting of nanoparticles by multivalent attachment of small molecules, *Nat. Bȯ tech.*, **23**, 1418–1423.

50. Euliss, L. E., Grancharov, S. G., O'Brien, S., Deming, T. J., Stucky, G. D., Murray, C. B., *et al.* (2003). Cooperative assembly of magnetic nanoparticles and block copolypeptides in aqueous media, *Nano Lett.*, **3**, 1489–1493.

51. Yang, H., Li, X., Zhou, H., Zhuang, Y., Hu, H., Wu, H., *et al.* (2010). Monodisperse water-soluble Fe-Ni nanoparticles for magnetic resonance imaging, *J. Alloy. Compd.*, **509**, 1217–1221.

52. Wang, L., Yang, Z., Gao, J., Xu, K., Gu, H., Zhang, B., *et al.* (2006). A biocompatible method of decorporation: bisphosphonate-modified magnetite nanoparticles to remove uranyl ions from blood, *J. Am. Chem. Soc.*, **128**, 13358–13359.

53. Lee, S.-Y., and Harris, M. T. (2006). Surface modification of magnetic nanoparticles capped by oleic acids: characterization and colloidal stability in polar solvents, *J. Coll. Interf. Sci.*, **293**, 401–408.

54. De Palma, R., Peeters, S., Van Bael, M. J., Van den Rul, H., Bonroy, K., Laureyn, W., *et al.* (2007). Silane ligand exchange to make hydrophobic superparamagnetic nanoparticles water-dispersible, *Chem. Mat.*, **19**, 1821–1831.

55. Zhang, T., Ge, J., Hu, Y., and Yin, Y. (2007). A general approach for transferring hydrophobic nanocrystals into water, *Nano Lett.*, **7**, 3203–3207.

56. Robinson, I., Alexander, C., Tung, L. D., Fernig, D. G., and Thanh, N. T. K. (2009). Fabrication of water-soluble magnetic nanoparticles by ligand-exchange with thermo-responsive polymers, *J. Magn. Magn. Mat.*, **321**, 1421–1423.

57. Salgueiri ño-Maceira, V., Liz-Marzán, L. M., and Farle, M. (2004). Water-based ferrofluids from Fe_xPt_{1-x} nanoparticles synthesized in organic media, *Langmuir*, **20**, 6946–6950.

58. Taboada, E., Rodríguez, E., Roig, A., Oró, J., Roch, A., and Muller, R. N. (2007). Relaxometric and magnetic characterization of ultrasmall iron oxide nanoparticles with high magnetization. Evaluation as potential T1 magnetic resonance imaging contrast agents for molecular imaging, *Langmuir*, **23**, 4583–4588.

59. Saita, S., and Maenosono, S. (2006). Equiatomic FePt nanoparticles synthesized via pyrolysis of iron(III) ethoxide and platinum(II) acetylacetonate, *Phys. Status Solidi A*, **203**, 1206–1210.

60. Jiang, J., Gu, H., Shao, H., Devlin, E., Papaefthymiou, G. C., and Ying, J. Y. (2008). Bifunctional Fe_3O_4–Ag heterodimer nanoparticles for two-photon fluorescence imaging and magnetic manipulation, *Adv. Mater.*, **20**, 4403–4407.

61. Sharma, S. N., Kumar, U., Singh, V. N., Mehta, B. R., and Kakkar, R. (2010). Surface modification of CdSe quantum dots for biosensing applications: role of ligands, *Thin Solid Films*, **519**, 1202–1212.

62. Schärtl, W., Xie, R., Ren, T., Sejfic, M., Wenz, G., Heisel, R., *et al.* (2009). Water-soluble, cyclodextrin-functionalized semiconductor nanocrystals: preparation and pH-dependent aggregation and emission properties, *J. Lumin.*, **129**, 1428–1434.

63. Dorokhin, D., Tomczak, N., Han, M., Reinhoudt, D. N., Velders, A. H., and Vancso, G. J. (2009). Reversible phase transfer of (CdSe/ZnS) Quantum Dots between organic and aqueous solutions, *ACS Nano*, **3**, 661–667.

64. Haddad, P. S., Martins, T. M., D'Souza-Li, L., Li, L. M., Metze, K., Adam, R. L., *et al.* (2008). Structural and morphological investigation of magnetic nanoparticles based on iron oxides for biomedical applications, *Mat. Sci. Eng: C*, **28**, 489–494.

65. Landmark, K. J., DiMaggio, S., Ward, J., Kelly, C., Vogt, S., Hong, S., *et al.* (2008). Synthesis, characterization, and *in vitro* testing of superparamagnetic iron oxide nanoparticles targeted using folic acid-conjugated dendrimers, *ACS Nano*, **2**, 773–783.

66. Lattuada, M., and Hatton, T. A. (2006). Functionalization of monodisperse magnetic nanoparticles, *Langmuir*, **23**, 2158–2168.

67. Gittins, D. I., and Caruso, F. (2001). Spontaneous phase transfer of nanoparticulate metals from organic to aqueous media, *Angew. Chem. Inter. Edt.*, **40**, 3001–3004.

68. Simard, J., Briggs, C., Boal, A. K., and Rotello, V. M. (2000). Formation and pH-controlled assembly of amphiphilic gold nanoparticles, *Chem. Comm.*, 1943–1944.

69. Herranz, F., Morales, M. P., Roca, A. G., Vilar, R., and Ruiz-Cabello, J. (2008). A new method for the aqueous functionalization of superparamagnetic Fe_2O_3 nanoparticles, *Contrast Media Mol. I.*, **3**, 215–222.

70. Maity, D., and Agrawal, D. C. (2007). Synthesis of iron oxide nanoparticles under oxidizing environment and their stabilization in aqueous and non-aqueous media, *J. Magn. Magn. Mater.*, **308**, 46–55.

71. Swami, A., Kumar, A., and Sastry, M. (2003). Formation of water-dispersible gold nanoparticles using a technique based on surface-bound interdigitated bilayers, *Langmuir*, **19**, 1168–1172.

72. Liong, M., Lu, J., Kovochich, M., Xia, T., Ruehm, S. G., Nel, A. E., *et al.* (2008). Multifunctional Inorganic Nanoparticles for Imaging, Targeting, and Drug Delivery, *ACS Nano*, **2**, 889–896.

73. Pellegrino, T., Manna, L., Kudera, S., Liedl, T., Koktysh, D., Rogach, A. L., *et al.* (2004). Hydrophobic nanocrystals coated with an amphiphilic polymer shell: a general route to water soluble nanocrystals, *Nano Letters*, **4**, 703–707.

74. Lee, J., Lee, Y., Youn, J. K., Na, H. B., Yu, T., Kim, H., *et al.* (2008). Simple synthesis of functionalized superparamagnetic magnetite/silica core/shell nanoparticles and their application as magnetically separable high-performance biocatalysts, *Small*, **4**, 143–152.

75. Kim, J., Lee, J. E., Lee, J., Yu, J. H., Kim, B. C., An, K., *et al.* (2005). Magnetic Fluorescent Delivery Vehicle Using Uniform Mesoporous Silica Spheres Embedded with Monodisperse Magnetic and Semiconductor Nanocrystals, *J. Am. Chem. Soc.*, **128**, 688–689.

76. Landfester, K. (2006). Synthesis of colloidal particles in miniemulsions, *Annual Revė w Materi̇a ls Res.*, **36**, 231.

77. McIntosh, C. M., Esposito, E. A., Boal, A. K., Simard, J. M., Martin, C. T., and Rotello, V. M. (2001). Inhibition of DNA transcription using cationic mixed monolayer protected gold clusters, *J. Am. Chem. Soc.*, **123**, 7626–7629.

78. Sandhu, K. K., McIntosh, C. M., Simard, J. M., Smith, S. W., and Rotello, V. M. (2001). Gold nanoparticle-mediated transfection of mammalian cells, *Bi̇o conjugate Chem.*, **13**, 3–6.

79. Goodman, C. M., McCusker, C. D., Yilmaz, T., and Rotello, V. M. (2004). Toxicity of gold nanoparticles functionalized with cationic and anionic side chains, *Bi̇o conjugate Chem.*, **15**, 897–900.

80. Sato, K., Hosokawa, K., and Maeda, M. (2003). Rapid aggregation of gold nanoparticles induced by non-cross-linking DNA hybridization, *J. Am. Chem. Soc.*, **125**, 8102–8103.

81. Storhoff, J. J., Elghanian, R., Mucic, R. C., Mirkin, C. A., and Letsinger, R. L. (1998). One-pot colorimetric differentiation of polynucleotides with single base imperfections using gold nanoparticle probes, *J. Am. Chem. Soc.*, **120**, 1959–1964.

82. Mandal, D., Maran, A., Yaszemski, M., Bolander, M., and Sarkar, G. (2009). Cellular uptake of gold nanoparticles directly cross-linked with carrier peptides by osteosarcoma cells, *J. Mater. Sci.: Mater. Med.*, **20**, 347–350.

83. Neirinck, B., Soccol, D., Fransaer, J., Biest, O. V. D., and Vleugels, J. (2010). Influence of short chain organic acids and bases on the wetting properties and surface energy of submicrometer ceramic powders, *J. Coll. Interf. Sci.*, **348**, 654–660.

84. Lalatonne, Y., Paris, C., Serfaty, J. M., Weinmann, P., Lecouvey, M., and Motte, L. (2008). Bis-phosphonates-ultra small superparamagnetic iron oxide nanoparticles: a platform towards diagnosis and therapy, *Chem. Comm.*, 2553–2555.

85. Mutin, P. H., Guerrero, G., and Vioux, A. (2005). Hybrid materials from organophosphorus coupling molecules, *J. Mat. Chem.*, **15**, 3761–3768.

86. Benyettou, F., *et al.* (2011). Microwave assisted nanoparticle surface functionalization, *Nanotechnology*, **22**, 055102.

87. Benyettou, F., Lalatonne, Y., Chebbi, I., Di Benedetto, M., Serfaty, J. M., Lecouvey, M., and Motte, L. (2011). A multimodal magnetic resonance imaging nanoplatform for cancer theranostics, *Phys. Chem. Chem. Phys.*, **13**, 10020–10027.

88. Veiseh, O., Gunn, J. W., and Zhang, M. (2010). Design and fabrication of magnetic nanoparticles for targeted drug delivery and imaging, *Adv. Drug Deliv er. Rev.*, **62**, 284–304.

89. Gunn, J., Wallen, H., Veiseh, O., Sun, C., Fang, C., Cao, J., *et al.* (2008). A multimodal targeting nanoparticle for selectively labeling T cells, *Small*, **4**, 712–715.

90. McCarthy, J. R., and Weissleder, R. (2008). Multifunctional magnetic nanoparticles for targeted imaging and therapy, *Adv. Drug Deliv er. Rev.*, **60**, 1241–1251.

91. Sun, E. Y., Josephson, L., Kelly, K. A., and Weissleder, R. (2005). Development of nanoparticle libraries for biosensing, *Bio conjugate Chem.*, **17**, 109–113.

92. Caruso, F., Rodda, E., Furlong, D. N., Niikura, K., and Okahata, Y. (1997). Quartz crystal microbalance study of DNA immobilization and hybridization for nucleic acid sensor development, *Anal. Chem.*, **69**, 2043–2049.

93. Frey, B. L., and Corn, R. M. (1996). Covalent attachment and derivatization of poly(*l*-lysine) monolayers on gold surfaces as characterized by polarization-modulation ft-ir spectroscopy, *Anal. Chem.*, **68**, 3187–3193.

94. Chan, W. C. W., and Nie, S. (1998). Quantum Dot bioconjugates for ultrasensitive nonisotopic detection, *Scie nce*, **281**, 2016–2018.

95. Dubertret, B., Skourides, P., Norris, D. J., Noireaux, V., Brivanlou, A. H., and Libchaber, A. (2002). *In viv o* imaging of quantum dots encapsulated in phospholipid micelles, *Scie nce*, **298**, 1759–1762.

96. Huh, Y.-M., Jun, Y.-W., Song, H.-T., Kim, S., Choi, J.-S., Lee, J.-H., *et al.* (2005). *In viv o* magnetic resonance detection of cancer by using multifunctional magnetic nanocrystals, *J. Am. Chem. Soc.*, **127**, 12387–12391.

97. Bonasio, R., Carman, C. V., Kim, E., Sage, P. T., Love, K. R., Mempel, T. R., *et al.* (2007). Specific and covalent labeling of a membrane

protein with organic fluorochromes and quantum dots, *P. Natl. Acad. Sci.*, **104**, 14753–14758.

98. Sun, X.-L., Stabler, C. L., Cazalis, C. S., and Chaikof, E. L. (2005). Carbohydrate and protein immobilization onto solid surfaces by sequential diels-alder and azide-alkyne cycloadditions, *Bb conjugate Chem.*, **17**, 52–57.

99. Brinkley, M. (1992). A brief survey of methods for preparing protein conjugates with dyes, haptens and crosslinking reagents, *Bb conjugate Chem.*, **3**, 2–13.

100. Tian, L., Shi, C., and Zhu, J. (2007). Reversible formation of hybrid nanostructures via an organic linkage, *Chem. Comm.*, 3850–3852.

101. El-Sayed, M. E. H., Hoffman, A. S., and Stayton, P. S. (2005). Rational design of composition and activity correlations for pH-sensitive and glutathione-reactive polymer therapeutics, *J. Control. Release*, **101**, 47–58.

102. Hein, C., and Wang, L. X. (2008). Click chemistry, a powerful tool for pharmaceutical sciences, *Pharm. Res.*, **25**, 2216–2230.

103. Lutz, J.-F. (2007). 1,3-dipolar cycloadditions of azides and alkynes: a universal ligation tool in polymer and materials science, *Angew. Chem. Inter. Edt .*, **46**, 1018–1025.

104. Nandivada, H., Jiang, X., and Lahann, J. (2007). Click chemistry: versatility and control in the hands of materials scientists, *Adv. Mat.*, **19**, 2197–2208.

105. Pieters, R. J., Rijkers, D. T. S., and Liskamp, R. M. J. (2007). Application of the 1,3-dipolar cycloaddition reaction in chemical biology: approaches toward multivalent carbohydrates and peptides and peptide-based polymers, *QSAR Comb. Sci.*, **26**, 1181–1190.

106. Kolb, H. C., Finn, M. G., and Sharpless, K. B. (2001). Click chemistry: diverse chemical function from a few good reactions, *Angew. Chem. Inter. Edt .*, **40**, 2004–2021.

107. Huisgen, R. (1984). In 1,3-*Dp olar Cycloaddt ion Chemś try*, Vol. 1 (ed. Padwa, A.), Wiley, New York, pp 1–176.

108. Rostovtsev, V. V., Green, L. G., Fokin, V. V., and Sharpless, K. B. (2002). A stepwise Huisgen cycloaddition process: copper(i)-catalyzed regioselective "ligation" of azides and terminal alkynes, *Angew. Chem. Inter. Edt .*, **41**, 2596–2599.

109. Hawker, C. J., and Wooley, K. L. (2005). The convergence of synthetic organic and polymer chemistries, *Scė nce*, **309**, 1200–1205.

110. Chikae, M., Fukuda, T., Kerman, K., Idegami, K., Miura, Y., and Tamiya, E. (2008). Amyloid-[beta] detection with saccharide immobilized gold nanoparticle on carbon electrode, *Bὸ electrochemiṡ try*, **74**, 118–123.

111. Byrne, J. D., Betancourt, T., and Brannon-Peppas, L. (2008). Active targeting schemes for nanoparticle systems in cancer therapeutics, *Adv. Drug Deli̇ er. Rev.*, **60**, 1615–1626.

112. Leamon, C. P., and Reddy, J. A. (2004). Folate-targeted chemotherapy, *Adv. Drug Deli̇ er. Rev.*, **56**, 1127–1141.

113. Hayashi, K., Moriya, M., Sakamoto, W., and Yogo, T. (2009). Chemoselective synthesis of folic acidâ-functionalized magnetite nanoparticles via click chemistry for magnetic hyperthermia, *Chem. Mat.*, **21**, 1318–1325.

114. Hayashi, K., Ono, K., Suzuki, H., Sawada, M., Moriya, M., Sakamoto, W., *et al.* (2010). High-frequency, magnetic-field-responsive drug release from magnetic nanoparticle/organic hybrid based on hyperthermic effect, *ACS Aʃ ʃ l. Mat. Interf.*, **2**, 1903–1911.

115. Santra, S., Kaittanis, C., Grimm, J., and Perez, J. M. (2009). Drug/dye-loaded, multifunctional iron oxide nanoparticles for combined targeted cancer therapy and dual optical/magnetic resonance imaging, *Small*, **5**, 1862–1868.

116. Majumdar, S., and Siahaan, T. J. (2011). Peptide-mediated targeted drug delivery, *Med. Res. Rev.*, doi: 10.1002/med.20225.

117. von Maltzahn, G., Ren, Y., Park, J.-H., Min, D.-H., Kotamraju, V. R., Jayakumar, J., *et al.* (2008). *In vi̇ o* tumor cell targeting with "click" nanoparticles, *Bὸ conjugate Chem.*, **19**, 1570–1578.

118. Martin, A., Hickey, J., Ablack, A., Lewis, J., Luyt, L., and Gillies, E. (2008). Synthesis of bombesin-functionalized iron oxide nanoparticles and their specific uptake in prostate cancer cells, *J. Nanoʃ art. Res.*, **12**, 1599–1608.

119. Devaraj, N. K., Keliher, E. J., Thurber, G. M., Nahrendorf, M., and Weissleder, R. (2009). 18F labeled nanoparticles for *ṅ vi̇ o* PET-CT imaging, *Bὸ conjugate Chem.*, **20**, 397–401.

120. Brennan, J. L., Hatzakis, N. S., Tshikhudo, T. R., Razumas, V., Patkar, S., Vind, J., *et al.* (2006). Bionanoconjugation via click chemistry: the creation of functional hybrids of lipases and gold nanoparticles, *Bὸ conjugate Chem.*, **17**, 1373–1375.

121. Zhang, M.-X., Huang, B.-H., Sun, X.-Y., and Pang, D.-W. (2010). Clickable gold nanoparticles as the building block of nanobioprobes, *Langmuṙ*, **26**, 10171–10176.

122. Cutler, J. I., Zheng, D., Xu, X., Giljohann, D. A., and Mirkin, C. A. (2010). Polyvalent oligonucleotide iron oxide nanoparticle "click" conjugates, *Nano Lett.*, **10**, 1477–1480.

123. Thorek, D., Elias, D., and Tsourkas, A. (2009). Comparative analysis of nanoparticle-antibody conjugations: carbodiimide versus click chemistry, *Mol. Imaging*, **8**, 221–229.

124. Canalle, L. A., van Berkel, S. S., de Haan, L. T., and van Hest, J. C. M. (2009). Copper-free clickable coatings, *Adv. Funct. Mater.*, **19**, 3464–3470.

125. Rana, S., Yeh, Y.-C., and Rotello, V. M. (2010). Engineering the nanoparticle-protein interface: applications and possibilities, *Curr. Opin. Chem. Biol.*, **14**, 828–834.

126. Hild, W. A., Breunig, M., and Goepferich, A. (2008). Quantum dots—Nano-sized probes for the exploration of cellular and intracellular targeting, *Eur. J. Pharm. Biopharm.*, **68**, 153–168.

127. Katz, E., and Willner, I. (2004). Integrated nanoparticle–biomolecule hybrid systems: synthesis, properties, and applications, *Angew. Chem. Inter. Edt.*, **43**, 6042–6108.

128. Charalambous, A., Andreou, M., and Skourides, P. (2009). Intein-mediated site-specific conjugation of Quantum Dots to proteins *in vivo*, *J. Nanobiotechnology*, **7**, 9.

129. Moghimi, S. M., Hunter, A. C., and Murray, J. C. (2005). Nanomedicine: current status and future prospects, *FASEB J.*, **19**, 311–330.

130. Xia, Y., Xiong, Y., Lim, B., and Skrabalak, S. E. (2009). Shape-controlled synthesis of metal nanocrystals: simple chemistry meets complex physics? *Angew. Chem. Inter. Edt.* **48**, 60–103.

131. Peer, D., Karp, J. M., Hong, S., Farokhzad, O. C., Margalit, R., and Langer, R. (2007). Nanocarriers as an emerging platform for cancer therapy, *Nat. Nano*, **2**, 751–760.

132. Janssen, X. J. A., van Ijzendoorn, L. J., and Prins, M. W. J. (2008). On-chip manipulation and detection of magnetic particles for functional biosensors, *Bio sens. Bio electron.*, **23**, 833–838.

133. Graham, D. L., Ferreira, H. A., Freitas, P. P., and Cabral, J. M. S. (2003). High sensitivity detection of molecular recognition using magnetically labelled biomolecules and magnetoresistive sensors, *Bio sens. Bio electron.*, **18**, 483–488.

134. Huang, S., Yang, H., Lakshmanan, R. S., Johnson, M. L., Wan, J., Chen, I. H., *et al.* (2009). Sequential detection of Salmonella typhimurium and Bacillus anthracis spores using magnetoelastic biosensors, *Bio sens. Bio electron.*, **24**, 1730–1736.

135. Lenglet, L., Nikitin, P., and Péquignot, C. (2008). Magnetic immunoassays: a new paradigm in POCT, *IVD Technology*, July /Aug.: 43–49.

136. Lawaczeck, R., Menzel, M., and Pietsch, H. (2004). Superparamagnetic iron oxide particles: contrast media for magnetic resonance imaging, *Appl. Organomet. Chem.*, **18**, 506–513.

137. Motte, L., Benyettou, F., de Beaucorps, C., Lecouvey, M., Milesovic, I., and Lalatonne, Y. (2011). Multimodal superparamagnetic nanoplatform for clinical applications: immunoassays, imaging & therapy, *Faraday Discuss.*, **149**, 211–225.

138. Tanimoto, A., and Kuribayashi, S. (2006). Application of superparamagnetic iron oxide to imaging of hepatocellular carcinoma, *Eur. J. Radiol.*, **58**, 200–216.

139. Enochs, W. S., Harsh, G., Hochberg, F., and Weissleder, R. (1999). Improved delineation of human brain tumors on MR images using a long-circulating, superparamagnetic iron oxide agent, *JMRI J. Magn. Reson. Im.*, **9**, 228–232.

140. Perrault, S. D., Walkey, C., Jennings, T., Fischer, H. C., and Chan, W. C. W. (2009). Mediating tumor targeting efficiency of nanoparticles through design, *Nano Lett.*, **9**, 1909–1915.

141. Chang, E., Alexander, H. R., Libutti, S. K., Hurst, R., Zhai, S., Figg, W. D., *et al.* (2001). Laparoscopic continuous hyperthermic peritoneal perfusion, *J. Am. Coll. Surgeons*, **193**, 225–229.

142. Feldman, A. L., Libutti, S. K., Pingpank, J. F., Bartlett, D. L., Beresnev, T. H., Mavroukakis, S. M., *et al.* (2003). Analysis of factors associated with outcome in patients with malignant peritoneal mesothelioma undergoing surgical debulking and intraperitoneal chemotherapy, *J. Clin. Oncol.*, **21**, 4560–4567.

143. van der Zee, J. (2002). Heating the patient: a promising approach?, *Ann. Oncol.*, **13**, 1173–1184.

144. Wust, P., Hildebrandt, B., Sreenivasa, G., Rau, B., Gellermann, J., Riess, H., *et al.* (2002). Hyperthermia in combined treatment of cancer, *Lancet Oncol.*, **3**, 487–497.

145. Jain, S., Coulter, J. A., Hounsell, A. R., Butterworth, K. T., McMahon, S. J., Hyland, *et al.* (2011). Cell-specific radiosensitization by gold nanoparticles at megavoltage radiation energies, *Inter. J. Radiat. Oncol.*, **79**, 531–539.

Chapter 5

Propagating and Standing-Wave Plasmonic Modes in Low-Dimensional Atomic-Scale Objects

Tadaaki Nagao

International Center for Materials Nanoarchitectonics (MANA),
National Institute for Materials Science (NIMS),
1-1 Namiki, Tsukuba 305-0044, Japan

nagao.tadaaki@nims.go.jp

Plasmons can be confined and propagated in ultimately thin media down to the scale of atomic sheets and atomic wires. This new class of plasmons exhibits resonance frequency in the infrared (IR) spectral region and shows extremely small phase velocity compared with conventional surface plasmons in larger systems. Recently, such atomic-scale slow plasmons are also found to support standing-wave plasmons similar to cases with thicker and longer nanoantennas prepared by electron beam lithography. In this chapter, dispersion relations of propagating plasmon modes and plasmonic standing-wave modes in some atomic-scale systems are briefly overviewed.

Nanoantenna: Plasmon-Enhanced Spectroscopies for Biotechnological Applications
Edited by Marc Lamy de la Chapelle and Annemarie Pucci

ISBN 978-981-4303-61-3 (Hardcover), 978-981-4303-62-0 (eBook)
www.panstanford.com

5.1 Introduction

The first theoretical work on surface plasmons was on energy loss spectroscopy for thin metallic foils.[1,2] The thickness of the film was in the order of tens of nanometers, and thickness below this level was mostly difficult for direct comparison with the experiments partly because of a lack of ideal films. Owing to development in the molecular beam epitaxy (MBE) technique in recent years, films down to a thickness of a few atomic layers are now readily achievable with perfect crystallinity and atomic flatness. Since the screening lengths and Fermi wavelength of typical metals are in the scale of 1 nm or less, size and shape effects are expected to manifest themselves most strongly in a scale of this length (Fig. 5.1).

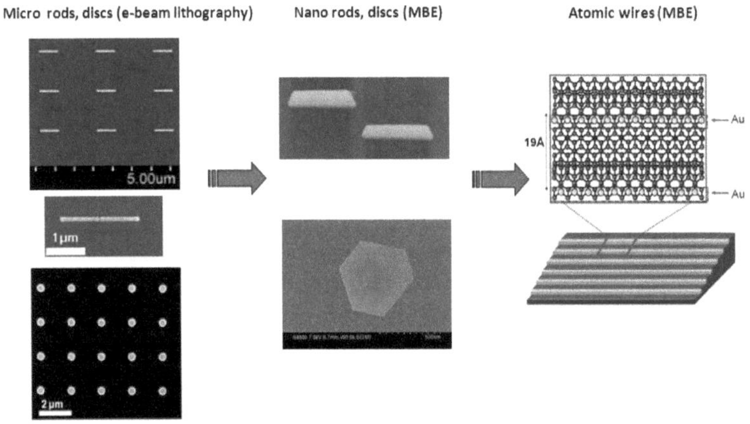

Figure 5.1 Miniaturization of plasmonic antenna structures—from lithographically fabricated Au microrods/discs, to self-assembled metallic (Pb, Ag) nanorods/disks, and finally to atomic wires made of Au.

As a consequence, a naive question arises: Do plasmons still exist in such small, atomic-scale systems? If they exist, what are the differences from the surface plasmons on semi-infinite surfaces and larger objects? Are they confined to forming standing-wave resonators in closed systems? As prototype systems, one can imagine systems such as two-dimensional (2D) electron gas made of atomic sheets or one-dimensional (1D) electron gas made of atomic wires. These ultimately thin, low-dimensional (low-D) systems are prepared via self-organization of metallic overlayers, for example, via epitaxial growth on silicon high-index substrates. The especially high crystallinity of silicon ensures perfect ordering of such low-D objects.[3–6] Another advantage of using Si is that since the plasmonic excitations in these low-D systems are in the mid-IR region, these plasmons are free from hybridizing with the near-IR interband transitions from the substrate. Such a situation makes these systems clean and ideal arenas of opportunity for the study of electronic excitation in low-D atomic-scale systems.[6–16]

In this chapter, we show some recent examples of plasmonic excitations in atomic sheets and atomics wires by both electron spectroscopic and optical spectroscopic methods.[17–20] In the first part, the dispersion relations of propagating plasmonic modes measured by electron energy loss spectroscopy (EELS) are introduced. In the second part, the first example of the measurement of plasmonic resonators in atomic wires is introduced together with nanoscale structure characterization by electron diffraction.

5.2 Plasmon Dispersion of Atomic Sheets and Atomic Wires

In this section, we discuss the dispersion relation of propagating modes of plasmons in atomic-scale systems. Figure 5.2 shows schematically the dispersion relations of surface plasmon polaritons and low-D plasmons in atomic sheets or atomic wires. One can notice easily that the slope of the low-energy mode of surface plasmon polaritons, that is, the Fano mode, is rather close to that of the light line at the small wave number **q**. This situation makes this mode couple relatively easily with visible light by using a grating coupler or a prism with an index of refraction $n > 1$.[21,22] On the other hand, plasmon dispersion for atomic-scale, low-D structures is rather different from that of surface plasmons. For example, the slope of the dispersion curve is far smaller than that of the light line, and this situation makes it less possible for these modes to be coupled with the light. Their resonance energy mostly lies in the mid-IR region, and they damp above the near-IR region.

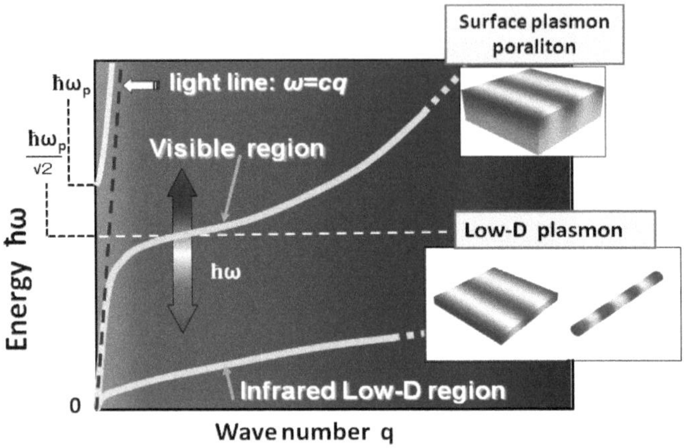

Figure 5.2 Schematic dispersion curves of surface plasmon polaritons and low-D atomic-scale plasmons. ω_p denotes the plasma frequency. The upper two branches show the two surface plasmon polaritons at the semi-infinite metallic surface (Brewster and Fano modes). The lowest one represents the IR plasmon in low-D atomic-scale systems. See also Color Insert.

In the region next to the light line, the propagation wavelength becomes smaller and charge density oscillation as well as associated electromagnetic waves become strongly localized near the metallic object. As the wave number becomes larger, and consequently the propagation wavelength approaches the atomic scale, the plasmon dispersion is expected to reflect short-range interaction of the electrons, such as exchange correlation (XC) effects. To extract such features, we need high energy and wave number resolutions together with a large scanning range in the wave number. Optical spectroscopy mostly does not have enough probing range in the wave number. On the other hand, spectroscopy by charged- particle probes, such as EELS, is a powerful tool that gives access to a large momentum to fully examine the feature of the dispersion relation of plasmons.[6,14,15] In the following section some examples of the propagating modes of atomic-scale plasmons are shown.

In the case of a nanometer-thick metallic slab, surface charge distribution associated with plasmons that are remote from one another changes character markedly as the thickness of the slab becomes smaller. For example, the Fano mode of the surface plasmon polariton splits into two modes. The higher-energy mode has polarization perpendicular to the surface (normal mode, antisymmetric mode), and the lower-energy mode has polarization parallel to the surface (tangential mode, symmetric mode). The latter mode becomes lower in energy, and its phase velocity also becomes smaller as the thickness becomes smaller. When the thickness reaches one-monoatomic layer, the latter symmetric mode finally transforms into 2D plasmons.

In Fig. 5.3 an example of such a 2D plasmon in an atomic sheet is shown.

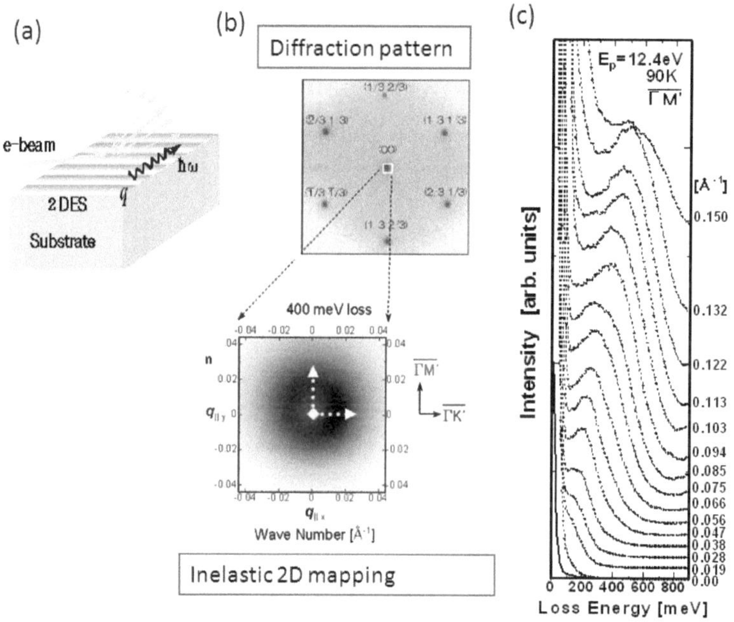

Figure 5.3 (a) Schematic illustration of the excitation of 2D charge density wave oscillation by an electron beam. (b) (Top) Electron diffraction pattern taken from the Ag monolayer on the Si(111) surface. (Bottom) "Inelastic 2D scan" of the beam scattered from the same sample with an energy loss of 400 meV. Electron primary energy was 50.3 eV and 12.3 eV, respectively, for the two patterns. (c) Wave number–resolved EELS spectra from the 2D plasmon of the Ag monolayer.

Figure 5.3a shows a schematic illustration of the reflection EELS from the 2D charge density wave supported on a dielectric substrate. When one monolayer of Ag was deposited onto the Si(111) surface in ultrahigh vacuum and annealed at 800 K, a sharp diffraction pattern, as shown in Fig. 5.3b, is observed. In the loss mode, one can see the mapping of the inelastic intensity in this 2D reciprocal space, as the lower figure in Fig. 5.3b shows (recorded at the loss energy of 400 meV). The intensity distribution exhibits a ring-like feature with respect to the azimuthal orientation, which indicates the 2D isotropic nature of the excitation.

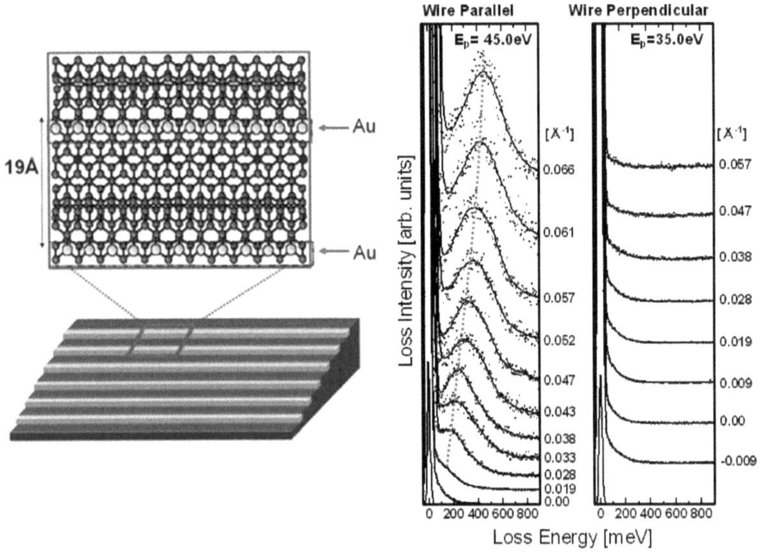

Figure 5.4 (Left) Atomic structure of a self-assembled Au atomic wire on a 1D template substrate—a high-index Si(557) surface. The wire is nearly one atom wide (0.3 nm), and the interwire distance is 1.9 nm. (Right) EELS spectra taken at different values of momentum in wire-parallel and wire-perpendicular directions. A monotonically increasing sound-wave-like dispersion is clearly shown. Electron incident energy was 45.0 eV and 35.0 eV, respectively.

Figure 5.3c shows typical momentum-resolved spectra obtained by scanning the **q** along a one-symmetry direction (the ΓM′ direction). The excitation energy goes down to 0 at small **q** values. Nearly identical slopes are also observed in other directions (such as in the ΓK′ symmetry direction), which again evidences the 2D nature of the excitation. Here, the quickly increasing line width as a function of **q** is clearly perceived; it is less possible to assign such a feature to single-particle excitations. All these features are consistent with 2D-type plasmonic excitation and evidence the confinement of the electromagnetic wave in atomically thin media.[7,8,15] The next example is from the 1D plasmon in an Au atomic wire array fabricated on a high-index silicon substrate. This is one of the most prototypical atomic-scale 1D system: self-assembled Au atom chains on the Si(557) surface (Fig. 5.4, left).

The high-index Si(557) surface serves as an ideal anisotropic template with high crystallinity to sustain perfectly ordered atom chains that can be viewed as the ultimate 1D electron system (1DES). Its electronic band near the Fermi level is parabolic along the wires, and it is flat and nearly nondispersive in the direction perpendicular to the wires. Such a prototype 1D electronic band offers a diversity of intriguing topics, such as spin-charge separation, Peierls instability, and the Rashba-type spin-orbit (SO) coupling effect of the 1D band.[23,24] The surface of the Si(557) wafer was cleaned by resistive heating using direct current fed through the wafer. A 0.2 monolayer of Au was then evaporated while the substrate Si(557) wafer was heated at 800 K in ultrahigh vacuum. Prior to the EELS measurement, the diffraction pattern was observed to check the crystallinity and crystallographic orientation of the atom chains. After the formation of the atomic wires, sharp Bragg spots with full width at half maximum less than < 0.01 $Å^{-1}$ were observed.

The right side of Fig. 5.4 shows the wave number–resolved EELS spectra taken in two different directions, wire parallel and wire perpendicular. In the wire-parallel direction, a single loss peak is clearly observed to disperse rapidly from the mid- to near-IR region as a function of the wave number accompanied by monotonic line width broadening. This feature is very similar to the case of the 2D plasmon in an Ag monolayer shown above and clearly evidences that the plasmon propagates in an "atomically narrow" 1D system. On the other hand, in the wire-perpendicular direction, no prominent feature was observed. This means that the observed excitation freely propagates along the wires, but it is strongly confined in each atomic wire and does not propagate across the wires.

In Fig. 5.5, plasmon dispersion curves for the Ag atomic sheet and Au atomic wire are shown. These curves are constructed by the EELS spectra, as shown in Figs. 5.3 and 5.4. The impact energy of the electron beam was varied to ascertain that the excitation does not have probing depth dependence of the electron beam: the excitation takes place only at the topmost surface and not in the subsurface region. Both curves shows very similar features except the strong anisotropy in the latter case. The plasmon frequency starts from practically zero from the origin of the graph and increases monotonically as a function of the wave number.

In the case of the 2D plasmon (Fig. 5.5a), the dispersion curve follows the $\sqrt{q}$ curve at a small q. The approximation is as follows:

$$\omega_{2D}(\mathbf{q}) = [4\pi N_{2D} e^2 m^{*-1}(1 + \varepsilon_{Si})^{-1}\mathbf{q}]^{1/2} \tag{5.1}$$

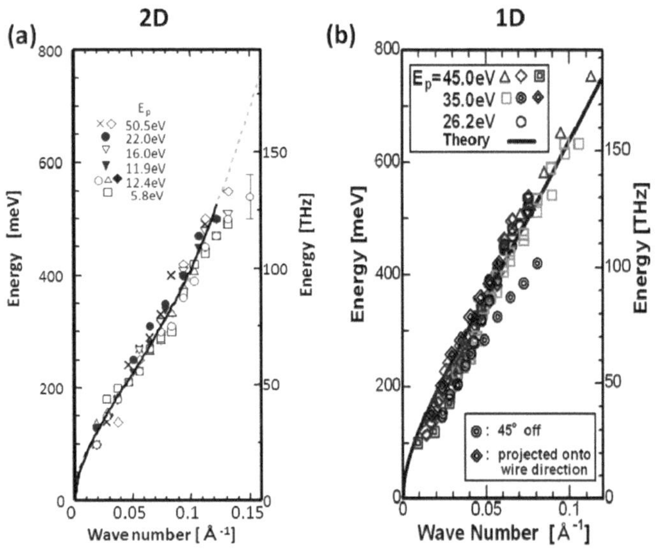

Figure 5.5 Examples of plasmon dispersion curves for 2D and 1D atomic-scale systems. (a) Result of the monolayer Ag is shown. The bold curve shows the theoretical result based on nearly-free-electron theory (RPA theory). The thin, dashed curve is the region where the plasmon is expected to decay by Landau damping. (b) Result of the Au atomic wires on the Si(557) surface is shown. The bold curve shows the theoretical result, including the LFC (electronic correlation effects). The dispersion curve measured for the wave numbers azimuthally rotated 45° from the wire-parallel direction is also shown by double circles to show the anisotropic nature of the plasmon. *Ab evū tō n*: LFC, local-field correction.

Here, $\omega_{2D}(\mathbf{q})$ is the wave number–dependent 2D plasma frequency, N_{2D} is the areal density of electrons in the 2D electron system (2DES), ε_{Si} is the dielectric constant of the Si substrate, which is nearly dispersionless within the frequency range of interest ($\varepsilon_{Si} \approx 11.5$),[25] m^* is the electron's effective mass, and e is the elementary electric charge. At a large q, the dispersion curve then deviates significantly from the $\sqrt{q}$ curve, that is, it is determined by solving the nonanalitic formula. The theoretical fit shown in this figure is based on nearly-free-electron theory with random phase approximation (RPA). This simulation does not take into account XC effects.

The black solid curve in Fig. 5.5a is the best fit to the non-analytic full RPA dispersion. The overall fit is excellent. The electron density and the electron's effective mass are determined to be $N_{2D} = 1.9 \times 10^{13}/\text{cm}^2$ and $m^* = 0.30\ m_e$, respectively (m_e is the vacuum electron mass). It should be noted that the effective density parameter, or the effective Wigner-Zeitz radius $r_s^* = (\pi N_{2D})^{-1/2} a_B^{*-1}$, of this system is very small, that is, 1.2. Here the effective Bohr radius a_B^* is defined by $a_B^* = \{(\varepsilon_{Si} + 1)/2\}(m_e/m^*)a_B$ with the Bohr radius $a_B = 0.529$ Å. It should be noted that this value is rather small compared with most of the free-electron metals like bulk aluminum, sodium, potassium, etc. (in the case of aluminum, it is still 2.1).

RPA theory predicts that the 2D plasmon decays into single-particle excitation by Landau damping above $q = 0.125$ Å^{-1}, and in the experiment also, the intensity becomes rather weak and disappears completely above $q = 0.15$ Å^{-1}. This situation can be seen in Fig. 5.5a. In addition, it should be also noted that at $q < 0.125$ Å^{-1}, we can see monotonic line width broadening that cannot be explained by RPA theory. One of the possible explanations will be the decay via two (and higher multiple) electron-hole pairs, induced by Coulomb collision or electron-electron scattering.[26,27] The presence of this type of damping was strongly suggested for volume plasmons in free-electron metals, where the Wigner-Zeitz radius r_s values are more or less of the same order as r_s^* of the present system.[28,29] Thus, the observed line width dispersion strongly suggests that for this 2D plasmon, the electronic correlation effect is more operative in the aspect of plasmon damping than in the energy dispersion.

Now we move on to the energy dispersion curve for 1D plasmons (Fig. 5.5b). As is clearly seen, the peak position approaches zero as the wave number q tends toward zero, which is the expected sound-wave-like feature first discussed by Tomonaga.[30] At a small momentum **q**, the plasmon dispersion in an isolated wire is expressed approximately as follows:

$$\omega = \omega_0 qa \left[\ln\left(\frac{2\sqrt{2}}{qa} \right) \right]^{1/2} + O(q^2) \;\; (qa \ll 1) \tag{5.2}$$

$$\omega_0 = \left(\frac{4N_{1D}e^2}{(1+\varepsilon_{Si})m^*a^2} \right)^{1/2} \tag{5.3}$$

Here $2a$ and ε_{Si} denote the wire width and the dielectric constant of the substrate Si (ε_{Si} = 11.5), respectively. N_{1D} and m^* are the linear electron density and the electron's effective mass, respectively. The initial slope of the dispersion is almost linear in contrast to the case for a 2D plasmon where the dispersion is proportional to the $\sqrt{q}$.[7,15,20,31] However, the real system here is composed of arrayed wires and the crosstalk between the neighboring wires (with 1.9 nm separation) will be operative. In the case of wires excited by a planar electromagnetic wave, the excited charge density waves in all the wires has the same phase (phase-locked) and the repulsive Coulombic interaction consequently lifts up the plasmonic frequency. This effect becomes stronger at a smaller q since the electromagnetic field from the oscillating charge becomes more extended, and thus the ensemble of wires should give rise to plasmonic dispersion similar to the $\sqrt{q}$ dispersion of the 2D plasmon. This situation can be seen in the theoretical calculation (solid curve) in Fig. 5.5b, which assumes plane wave excitation. In real EELS experiments this tendency might be smaller since the excitation field is a spherical one associated with individual electrons that arrive randomly. Thus the phase relation between the wires is random, and the Coulombic repulsion is cancelled out.

As q becomes larger, another issue emerges: The effect of higher-order terms becomes significant and electronic correlation effects, such as dynamic XC effects, become more operative. Such an effect, which is normally inaccessible using conventional optical probes, can be examined in detail by EELS with a wide range of q scanning and with a high q resolution. In contrast to a 2DES, the electrons in a 1D system are more tightly packed and elude each other with difficulty. This situation causes the electronic correlation effects to be more operative in 1D than in 2D. Such a problem in 1DESs has been one of the intriguing topics in condensed matter physics from the early theory of Tomonaga.[30] The effective density parameter (electron density) of this system is as small as $r_s^* = 1/(2N_{1D}a_B^*) = 0.7$ (using the value of $m^* = 0.60\ m_e$). This indicates a very high average velocity of electrons, and seemingly, the electron-electron coupling is negligible. To check this effect in more detail, we analyzed the data with a nonlocal dielectric-response theory, which includes a self-consistent LFC developed by Singwi *et al.*[32], which was first applied to the 1D system by Friesen *et al.*[33] In this scheme the XC effects can be described by the XC holes that make the electrons steer clear of each other within the distance of the XC hole radius $r_{xc} \approx k_F^{-1} = 2.4$ Å. The best theoretical fit to the experimental data is shown as the bold solid curve in Fig. 5.5b. We can obtain markedly good agreement over the entire range of the observed q up to 0.11 Å^{-1}. The parameters m^* and $2a$ are adjusted to $m^* = 0.60\ m_e$ and $2a = 4$ Å, which are in good agreement with reported works.[34]

In Fig. 5.5b, double circles indicate the data taken by using the excitation wave vector $\mathbf{q}$ azimuthally rotated (45° off) from the wire-parallel direction. When the plasmon energy is plotted against $q = |\mathbf{q}|$ the dispersion curve lies significantly lower than the one measured with $\mathbf{q}$ vectors in the wire-parallel excitation. However, if one replots the plasmon energy against the component of $\mathbf{q}$ projected onto the wire-parallel direction, that is, against $q\cos(45°)$, they fall again onto the same dispersion curve as for $\mathbf{q}$ in the wire-parallel direction (shown by double diamonds). This behavior again evidenced that this excitation has a strong anisotropic nature as expected for 1D plasmons.

5.3 Standing-Wave Plasmons in Atomic Wires

As we have seen earlier, propagating modes of plasmons are measured successfully by using EELS with high resolutions in both wave number and energy. On the contrary since the plasmon dispersion curves of atomic sheets and wires locate far away from the light line, and also the thickness of the systems is extremely small, it is expected that these modes are difficult to be detected by optical spectroscopy. In this section, a trial experiment for detecting plasmons in atomic wires by using IR absorption spectroscopy is introduced.[18]

The prototype 1D systems we chose are indium (In) atomic wires self-assembled on the Si(111) surface (Fig. 5.6). This 1D structure can be prepared in a ultrahigh vacuum condition by depositing In onto a heated clean Si(111) substrate at 350°C with a deposition rate of 0.2 monolayers/minute. The In wire array grows in the direction of atomic steps on the substrate and forms large domains (ca. 100 nm or larger) with 4 × 1 periodicity with respect to the 1 × 1 lattice of the Si surface.[4] The In wire width is 0.9 nm, and the distance between the wires is 0.4 nm (Fig. 5.6). At room temperature, three electronic bands (m_1, m_2, m_3) cross the Fermi level and exhibit three large electron pockets in the wire-parallel direction. In the wire-perpendicular direction, these bands are nondispersive and make this electronic structure strongly one dimensional. At low temperatures (below 120 K), the structure doubles its periodicity along the wire due to Peierls instability and its electronic band structure consequently opens a large gap near the Fermi level.

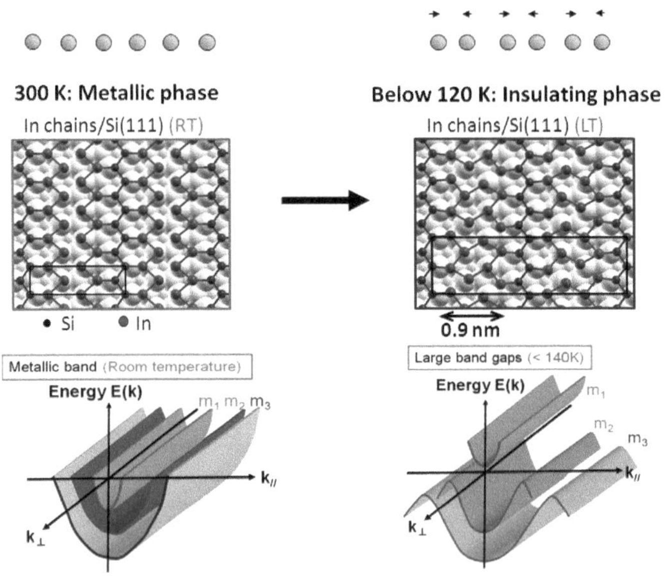

Figure 5.6 (Top) Conceptual diagram of the Peierls transition. (Middle) Self-assembled In atomic wires on the Si(111) surface. Below 120 K the structure transforms from 4 × 1 to 8 × 2 periodicity by small-lattice displacement of In atoms (models in Ref. 18). (Bottom) The 1D electronic bands, m_1, m_2, and m_3, transform from strongly metallic ones to insulating ones by opening the gaps.

First, electron energy loss experiments were performed to determine the plasmon dispersion in this 1D multiband system. The EELS spectra exhibit a strongly dispersing loss peak in the wire-parallel direction (Fig. 5.7, right). The wire-perpendicular direction shows no loss feature except the broad Drude tail from the elastic peak. The observed feature is basically the same as in the case for Au atomic wires on Si(557) (see Fig. 5.5b). However, one interesting fact is that we observed only one plasmonic band, although the system has three large electron pockets: m_1, m_2, and m_3. Each of

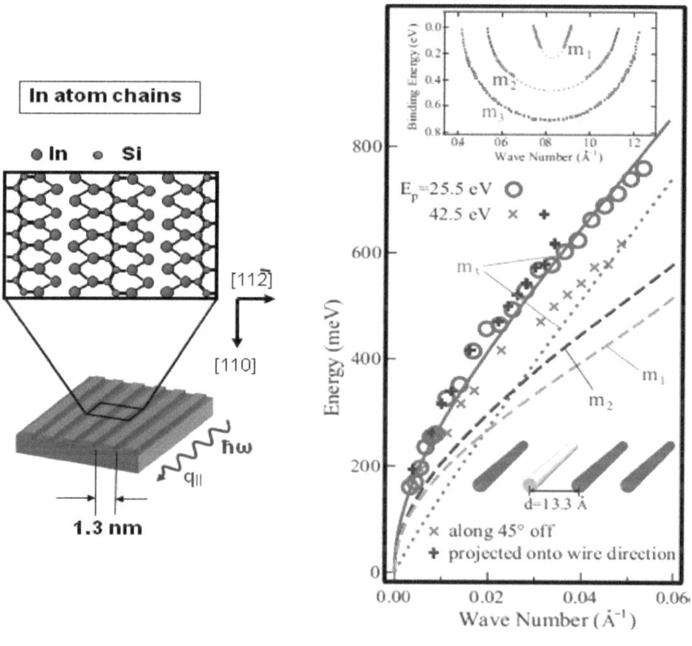

Figure 5.7 (Left) Schematic illustration of In atomic wires on the Si(111) surface. The wire width is about 0.9 nm, and the wire distance is about 0.4 nm. (Right) Plasmon dispersion curve of In wires. The open circles are obtained from EELS spectra along the In wire direction, while the crosses (×) are from those along 45° off from the wire direction, and the pluses (+) are their projection onto the wire direction. The curves are the calculated plasmons corresponding to the three metallic bands: m_1, m_2, and m_3. For the m_3 band, both calculated results based on single- and multiwire models are displayed as dotted and solid curves, respectively. The electronic band structure is shown in the upper inset, and the multiwire model is shown in the lower inset.

them is expected to sustain plasmons with different frequencies. To determine which one of these bands is responsible for the observed plasmonic band, simulations of the plasmons for these three electron pockets were carried out to compare with the experimental result.

In simulation, the plasmons for m_1, m_2, and m_3 bands were calculated by employing the same nonlocal dielectric response theory with a self-consistent LFC.[10,32,33] The electrons' effective mass m^* and the Fermi wave number k_F for the three bands used in the calculation were determined from the reported band structure with binding energy less than 0.1–0.3 eV.[11] A small binding energy was chosen because the electronic states close to the Fermi lever E_F are mostly responsible for the low-energy plasmonic excitations. The simulated results are shown in the graph in Fig. 5.7. It is clear that the plasmon dispersion calculated for the m_3 band agrees perfectly with the experiment. On the other hand, plasmons for the m_1 and m_2 bands locate far from the experimental one. From this result, we can conclude that the observed plasmon is assigned to the biggest electron pocket m_3. Nonobservation of the m_1 and m_2 plasmons is due possibly to the dynamic screening of these plasmons from the large amount of m_3-state electrons. A 3D analog of this phenomenon was known for two-plasma systems in the early theory by Pines[35,36] and in the examples of acoustic plasmons, such as in bulk Bi and at the Be surface.[36,37] This type of excitation can be regarded as the low-energy plasmonic excitation dynamically screened by the charge density oscillations in the coexisting metallic bands. Because of this screening, the intensity of the plasmon becomes very weak and is difficult to be observed. Here in the In system, plasmons in m_1 and m_2 are well screened by the plasmons in m_3.

The dotted line m_1 in Fig. 5.7 is the simulation for a single isolated wire and the solid curve m_1 is the simulation with an interacting wire array with a periodicity of 1.3 nm. It is clear that the single-wire simulation does not reproduce the experimental curve, but in the case of interacting multiwires, the calculation for the m_3 band agrees very well with the experiment. It should be noted that the dispersion curve rises up rapidly with $\sqrt{q}$-law, assuming the feature similar to 2D plasmons at a small q. Compared with the case of Au wires shown earlier (Figs. 5.4 and 5.5b), the $\sqrt{q}$-like feature at a vanishing q is much stronger for these In wires. This is because for the In wires, the wire width is approximately three times that of the Au wires; on the other hand, the separation between the In wires is several times smaller compared with the Au wires (see Figs. 5.4 and 5.7). These differences make the In system somewhat closer to the 2DES than the Au wires and will make its plasmon dispersion closer to 2D plasmons.

Now, in the next few pages, optical spectroscopic characterization of 1D plasmons in the same In atomic wire is introduced. In contrast to EELS, where plasmonic excitation is possible for the wide wave number region, optical spectroscopy for plasmons in atomic sheets and wires appears difficult as their phase velocity is far smaller than that of the light line. Although it is less possible to detect the propagating modes of plasmons, localized standing-wave modes of plasmons in finite-sized objects can give optical signals, for example, via dipole resonance. In this context, IR spectroscopy is one of the most effective tools for detecting and characterizing metallic conductivity as well as low-energy electronic excitation down to the nanometer scale. Here the In system was chosen because of its very high electron density supported by the large electron pockets m_1, m_2, and m_3, which ensure strong optical absorption. Also the metal-to-insulator transition of this system can be used to distinguish plasmonic excitation from single-particle excitation.

Figure 5.8 shows the preliminary result from the IR spectroscopy for In atomic wires. The spectra are taken in the transmission geometry with polarization in the wire-parallel direction. The top spectrum (grey circles) shows 100% of the line showing spectrometer stability. The middle one (filled triangles) shows the spectrum taken from the In wires in the insulating phase at 88 K. The lowest one (filled diamonds) corresponds to the spectrum taken in the metallic phase at room temperature. A clear absorption feature is seen for the wires measured at room temperature as well as at 88 K. The latter case corresponds to the wires in the insulating phase, and thus the IR spectrum unambiguously corresponds to the interband transition in the gap-opened 1D bands. On the other hand, the room-temperature spectrum has a prominent broad absorption feature at around 0.2 eV. Surprisingly, the overall feature of the room-temperature spectrum is very similar to the case of antenna resonance in larger objects such as nanoantenna made by lithographic fabrication.[20,38–40] The observed spectral change almost synchronizes with the metal-to-insulator transition: it is clearly assigned to the excitation in the In atomic wires reflecting their change in metallicity. The metallicity here means strong alternating current conductivity that can give rise to collective oscillation of the electrons in the wires.[41]

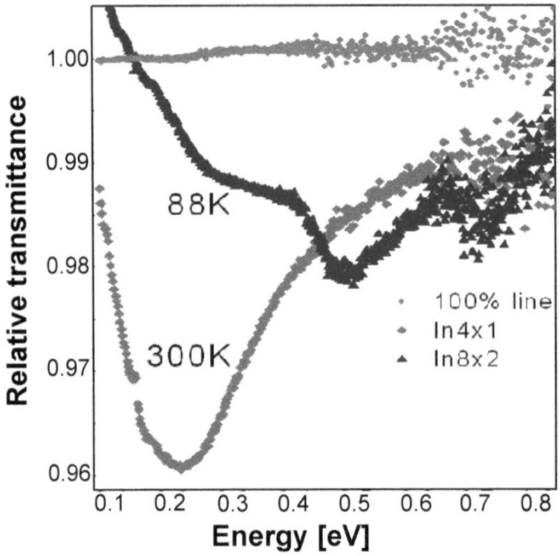

Figure 5.8 The metal-to-insulator transition of In wires supported on an Si substrate measured by IR spectroscopy. The spectra are taken with polarization parallel to the wires. The top spectrum (grey circles) shows 100% of the line showing spectrometer stability. The middle one (filled triangles) shows the spectrum in the insulating phase taken at 88 K. The lowest one (filled diamonds) corresponds to the spectra taken in the metallic phase.

As already seen in the EELS measurement shown earlier, we know that IR 1D plasmons can be excited and propagated freely in In atomic wires. Therefore the observed resonance-like feature at room temperature is most possibly assigned to the excitation of the same 1D plasmons. One way to judge whether this is the case is to see the size effect on the excitation spectra. To our knowledge, such experiments have not been carried out yet for atomic-scale IR plasmons. To examine this, we prepared atomic wires by gradually changing the preparation condition to tune the crystallinity of the In atomic wires in a systematic manner. This was done by changing the annealing current fed through the Si(111) wafer to control the annealing temperature. The current was gradually changed from 0.02 A to 0.16 A, which corresponds to the temperature change from 260°C to 440°C.

Figure 5.9 shows the low-energy electron diffraction (LEED) pattern and the IR transmittance spectra recorded by changing the temperature in this range.

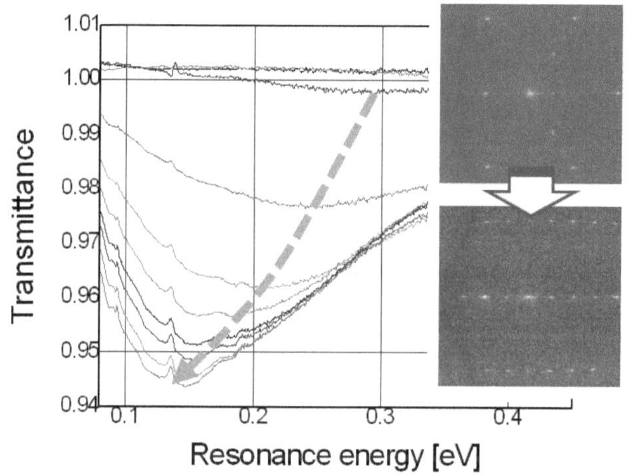

Figure 5.9 (Right) From top to the bottom: LEED patterns taken after annealing at 260°C and 440°C, respectively. (Left) IR transmittance spectra of In atomic wires taken with annealing temperature gradually changed from 260°C to 440°C. The top two straight spectra show 100% of the spectrum and the In as-deposited spectrum taken at room temperature. They are nearly the same in the IR region. After annealing at 260°C, a weak and broad feature emerges around 0.3 eV because of the nucleation of the In wire structure. As the annealing temperature increases, the resonance frequency red-shifts dramatically. Arrows indicate the direction of the temperature change from 260°C to 440°C in LEED and IR measurements. See also Color Insert.

The LEED pattern shows a change from the three-domain to the single-domain pattern and also exhibits clear sharpening in the 4 × 1 super-reflection spots from the In wires. This means that at 260°C (where the In wire nucleation starts), the In wires are short and oriented in the three equivalent directions with respect to the three-fold Si(111) substrate. As the temperature rises, they grow into longer wires and start to align into one of the three possible directions to form large domains. The direction of the wires coincides with the direction where the surface atomic steps align (the [110] direction). The IR transmission spectra first show a very shallow, broad absorption feature at around 0.3 eV at 260°C. And this feature evolves into a deeper and sharper absorption feature with its resonance position continuously red-shifted as the temperature increases. One can learn from these results that the excitation is strongly correlated with the size of the In wires, that is, the longer the wire length, the lower the excitation frequency. The change in the broadness of the IR spectra can be assigned to the size distribution of the wires; the initial randomly segmentized small wires become larger and more monodispersed as the annealing temperature increases.

The above behavior is consistent with the plasmon scenario where the strong size effect is expected to appear as the crystallinity and length of the In wires change. In fact, by taking a closer look at the data, one can find that the observed resonance frequencies are in good agreement with the propagating mode of the IR 1D plasmon observed by EELS. If we assign this plasmon to the standing-wave 1D plasmon confined in the wire direction, we can roughly estimate the wire length by using the plasmon dispersion relation shown in Fig. 5.7.

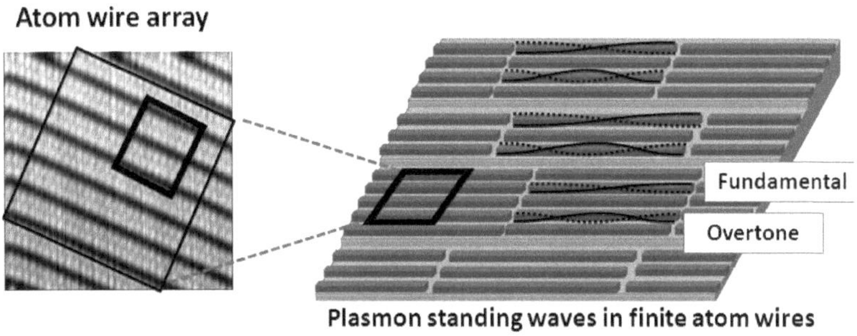

Figure 5.10 An STM image and a schematic illustration of In atomic wires self-assembled on the Si(111) surface. Finite-length In atomic wires can be prepared by controlling the annealing temperature. The wires are segmentized by domain boundaries, and 1D plasmons are confined to form standing-wave resonating modes. Dotted and solid curves schematically illustrate the standing-wave 1D plasmons. (Only the fundamental is observed, and the overtone is not identified in our study.) *Abbreviation*: STM, scanning tunneling microscopy.

Let us take the sample that shows the plasmon resonance frequency around ~0.2 eV (fifth from the top). If one looks at the dispersion curve in Fig. 5.7, the corresponding wave number of the propagating plasmon with an energy of 0.2 eV reads $q \approx 0.0057$ [1/Å]. By using the relation $q = 2\pi/\lambda = \pi/L$, where λ is the plasmon wavelength (of the propagating mode) and L is the size of the confining media (i.e., the In wire length), we can estimate the average wire length as $L \approx 55$ nm for this particular sample. That is, this length corresponds to the half wavelength of the fundamental mode of the propagating plasmon with an energy of 0.2 eV. The estimated confinement length of this In wire falls in the same order of the value estimated from the LEED diffraction spot analysis. A systematic LEED experiment with detailed spot-profile analysis is currently underway. The above assignments are based on fundamental-mode plasmons, and overtone plasmons are not observed in our experiment, possibly due to insufficient sensitivity or the broad size distribution of In wires (Fig. 5.10).

5.4 Summary and Future Perspective

A brief overview of the fundamental aspects of the propagating modes of plasmons as well as the first example of the detection of standing-wave antenna resonance in atomic-scale low-D objects have been given. In the first part, dimensionality effects and electronic correlation effects in relation to the atomic-scale confinement of plasmons were detailed. In the second part, an In metallic atomic wire was chosen to exemplify that atomic wires are prospective materials for supporting plasmonic resonating modes. The results presented here clearly demonstrate that the frequency of the plasmon in an atomic wire is fully tunable by controlling its width, separation, and length in the range of one to tens of nanometers. This fact suggests the great potential of atomic wires as IR antenna. The greatest advantage of this new type of antenna materials is that these materials are fabricated via simple bottom-up processes, giving object sizes one to two orders of magnitude smaller than the ones fabricated by top-down processes. Such advantages are certainly of great use in the field of nanotechnology, nano-photonics, and plasmonics. A new class of materials for future nanoscale optical circuitry, high-efficiency photovoltaic devices, and quick and high-sensitivity medical diagnosis is expected from the results and the concepts shown here.

References

1. Ritchie, R. H. (1957). Plasma losses by fast electrons in solid, *Phys. Rev.*, **106**, 874–881.
2. Ferrell, R. A. (1958). Predicted radiation of plasma oscillations in metal films, *Phys. Rev.*, **111**, 1214–1222.
3. Sato, N., Takeda, S., Hasegwa, S., and Nagao, T. (1999). Electron standing waves on the Si(111)-$\sqrt{3} \times \sqrt{3}$-Ag surface, *Phys. Rev. B*, **59**, 2035–2039.
4. Yeom, H. W., Takeda, S., Rotenberg, E., Matsuda, I., Horikoshi, K., Schaefer, J., Lee, C. M., Kevan, S. D., Ohta, T., Nagao T., and Hasegawa, S. (1999). Instability and charge density wave of metallic quantum chains on a silicon surface, *Phys. Rev. Lett.*, **82**, 4898–4902.
5. Nagao, T., Sadowski, J. T., Saito, M., Yaginuma, S., Fujikawa, Y., Kogure, T., Ohno, T., Hasegawa, Y., Hasegawa, S., and Sakurai, T., (2004). Nanofilm allotrope and phase transformation of ultrathin Bi film on Si(111)-7 × 7, *Phys. Rev. Lett.*, **93**, 105501-1–105501-4.
6. Nagao, T., Yaginuma, S., Inaoka, T., Sakurai, T., and Jeon, D. (2007). Sound-wave-like collective electronic excitations in Au atom chains, *J. Phys. Soc. Jpn.*, **76**, 114714-1–114714-8.
7. Nagao, T., Hildebrandt, T., Henzler, M., and Hasegawa, S. (2001). Dispersion and damping of a two-dimensional plasmon in a metallic surface-state band, *Phys. Rev. Lett.*, **86**, 5747–5750.
8. Nagao, T., Hildebrandt, T., Henzler, M., and Hasegawa, S. (2001). Two-dimensional plasmon in a surface-state band, *Surf. Sci.*, **493/13** , 680–686.
9. Nagao, T., Yaginuma, S., Inaoka, T., and Sakurai, T. (2006). One-dimensional plasmon in an atomic-scale metal wire, *Phys. Rev. Lett.*, **97**, 116802-1–116802-4.
10. Inaoka, T., and Nagao, T. (2007). Exchange-correlation effects on low-dimensional plasmons in a quantum-wire array, *Mater. Trans.*, **48**(4), 718–721.
11. Liu, C., Inaoka, T., Yaginuma, S., Nakayama, T., Aono, M., and Nagao, T. (2008). Disappearance of the quasi-one-dimensional plasmon at the

metal-insulator phase transition of indium atomic wires, *Phys. Rev. B*, **77**, 205415-1–205415-5.

12. Liu, C., Inaoka, T., Yaginuma, S., Nakayama, T., Aono, M., and Nagao, T. (2008). Excitation of one-dimensional plasmons in Si and Au-Si complex atom wires, *Nanotechnology*, **19**, 355204-1–355204-5.

13. Rugeramigabo, E. P., Nagao, T., and Pfnuer, H. (2008). Experimental investigation of two-dimensional plasmons in a $DySi_2$ monolayer on Si(111), *Phys. Rev. B*, **78**, 155402-1–155402-6.

14. Nagao, T. (2008). Characterization of atomic-level plasmonic structures by low-energy EELS, *Surf. Interface Anal.*, **40**, 1764–1767.

15. Nagao, T. (2010). "Low-dimensional plasmons in atom sheets and atom chains," in *Dynamics at Solid State Surfaces and Interfaces*, Volume I (ed. U. Bovensiepen, H. Petek, and M. Wolf), Wiley VCH, pp. 189–211. ISBN: 978-3-527-40937-2

16. Nagao, T., (2010). Low-dimensional plasmons in atom-scale metallic objects, *Proc. SPIE*, **7600**, 76001Q 1–76001Q 8.

17. Neubrech, F., Weber, D., Enders, D., Nagao, T., and Pucci, A. (2010). Antenna sensing of surface polaritons, *J. Phys. Chem.* C, **114**(16), 7299–7301.

18. Hoang, C. V., Kubber, C. J., Han, G., Rigamonti, S., Sanchez-Portal, D., Enders, D., Pucci, A., and Nagao, T. (2010). Optical detection of plasmonic and electronic excitations in 1 nm-wide indium atomic wires, *Appl. Phys. Lett.*, **96**, 243101-1–243101-3.

19. Hwang, C. G., Shin, S. Y., and Chung, J. W. (2007). Possible evidence of non-Fermi liquid behavior from quasi-one-dimensional indium nanowires, *New J. Phys.*, **9**, 249-1–249-9.

20. Nagao, T., Han, G., Hoang, C. V., Wi, J.-S., Pucci, A., Weber, D., Neubrech, F., Silkin, V. M., Enders, D., Saito, O., and Rana, M. (2010). Plasmons in nanoscale and atomic-scale systems, *Sci. Technol. Adv. Mater.*, **11**, 054506-1–054506-12.

21. Otto, A. (1968). Excitation of nonradiative surface plasma waves in silver by the method of frustrated total reflection, *Z. Phys.*, **216**, 398–410.

22. Raether, H., and Kretschmann., E. (1968). Radiative decay of non-radiative surface plasmons excited by light, *Z. Naturforsch.*, **23**A, 2135–2136.

23. Robinson, I. K., Bennett, P. A., and Himpsel, F. J. (2001). Structure of quantum wires in AuSi(557), *Phys. Rev. Lett.*, **88**, 096104-1–096104-4.

24. Sánchez-Portal, D., Riikonen, S., and Martin, R. M. (2004). Role of spin-orbit splitting and dynamical fluctuations in the Si(557)-Au surface, *Phys. Rev. Lett.*, **93**, 146803-1–146803-4.

25. Philipp, H. R., and Ehrenreich, H. (1963). Optical properties of semiconductors, *Phys. Rev.*, **129**, 1550–1560.

26. Sturm, K., and Oliveira, L. E. (1981). Wave-vector-dependent plasmon linewidth in the alkali metals, *Phys. Rev. B*, **24**, 3054–3062.

27. Totsuji, H., (1976). Theory of two-dimensional classical electron plasmas, *J. Phys. Soc. Jpn.*, **40**, 857–862.

28. Ichimaru, S. (1986). *Plasma Physics: An Introduction to Statistical Physics of Charged Particles*, Benjamin/Cummings.

29. vom Felde, A., Sprosser-Prou, J., and Fink, J. (1989). Valence-electron excitations in the alkali metals, *Phys. Rev. B*, **40**, 10181–10193.

30. Tomonaga, S. (1950). Remarks on Bloch's method of sound waves applied to many-Fermion problems, *Prog. Theor. Phys.*, **5**, 544–569.

31. Stern, F. (1967). Polarizabilty of two-dimensional electron gas, *Phys. Rev. Lett.*, **18**, 546–548.

32. Singwi, K. S., Tosi, M., Land, R. H., and Sjölander, A.(1968). Electron correlations at metallic densities, *Phys. Rev.*, **176,** 589–599.

33. Friesen, W. I., and Bergersen, B. (1980). Dielectric response of a one-dimensional electron gas, *J. Phys. C*, **13**, 6627–6640; Calmels, L., and Gold, A. (1995). Exchange and correlation in the quasi-one-dimensional electron gas: the local-field correction, *Phys. Rev. B*, **52**, 10841–10857.

34. Barke, I., Zheng, F., Rügheimer, T. K., and Himpsel, F. J. (2006). Experimental evidence for spin-split bands in a one-dimensional chain structure, *Phys. Rev. Lett.*, **97**, 226405-1–226405-4.

35. Pines, D. (1956). Electron interaction in solids, *Can. J. Phys.*, **34**, 1379–1394.

36. McWhorter, A. L., and May, W. G. (1964). Acoustic plasma waves in solids, *IBM J. Res. Dev.*, **8**, 285–290.

37. Diaconescu, B., Pohl, K., Vattuone, L., Savio, L., Hofmann, P., Silkin, V. M., Pitarke, J. M., Chulkov, E. V., Echenique, P. M., Farías, D., and Rocca, M. (2007). Low-energy acoustic plasmons at metal surfaces, *Nature*, **448**, 57–59.

38. Neubrech, F., Weber, D., Enders, D., Nagao, T., and Pucci, A, (2010). Antenna sensing of surface phonon polaritons, *J. Phys. Chem. C*, **114,** 7299–7301.

39. Han, G., Weber, D., Neubrech, F., Yamada, I., Mitome, M., Bando, Y., Nagao, T., and Pucci, A. (2011). Infrared spectroscopic and electron microscopic characterization of gold nanogap structure fabricated by focused ion beam, *Nanotechnology*, **22**, 275202.

40. Weber, D., Albella, P., Alonso-González, P., Neubrech, F., Han, G., Nagao, T., Hillenbrand, R., Aizpurua, J., and Pucci, A. (2011). Longitudinal and transverse coupling in infrared gold nanoantenna arrays: long range versus short range interaction regimes, *Opt. Express*, **19**, 15047.

41. Wendler, L., and Grigoryan, V. G. (1991). Optical and acoustic plasmons in cylindrical quantum-well wires, *Phys. Rev. B*, **49**, 14531–145341.

Chapter 6

Antenna Interaction in the Infrared

Daniel Weber and Annemarie Pucci

Kirchhoff Institute for Physics, Heidelberg University, Im Neuenheimer Feld 227, 69120 Heidelberg, Germany

dweber@kip.uni-heidelberg.de

6.1 Introduction

Because the resonant excitation of metallic nanostructures by light can yield considerable electromagnetic near-field enhancement, the optical properties of metal nanoparticles have been the subject of many studies.[1,2] Local field enhancement can be exploited for sensing applications like surface-enhanced infrared (IR) spectroscopy (SEIRS)[3–8] or surface-enhanced Raman scattering (SERS)[3,9–11] up to single molecule detection.[12–14] Today the highly developed standard in nanostructure production, with methods such as electron-beam lithography,[15] allows the preparation of all kinds of plasmonic nanostructures with a high degree of precision. Gold is the usual metal of choice because it combines chemical stability with excellent plasmonic behavior in the visible range. For many applications it is important to have very little

Nanoantenna: Plasmon-Enhanced Spectroscopies for Biotechnological Applications
Edited by Marc Lamy de la Chapelle and Annemarie Pucci

ISBN 978-981-4303-61-3 (Hardcover), 978-981-4303-62-0 (eBook)
www.panstanford.com

damping of plasmon polaritons. This equates to low electronic damping and thus a low defect density in the nanostructures and smooth surfaces on the atomic scale, which are important issues in gold nanostructure production.

Some what like radio frequency antennas, resonantly excited nanostructures, especially in the infrared, act as optical antennas that confine energy of electromagnetic radiation to a restricted volume of subwavelength scale. Hence, nanorods with micrometer-sized lengths L, which show so-called antenna resonances in the IR spectral range,[16–18] can be termed *nanoantennas*. However, the simple $\lambda/2$-dipole behavior associated with radio frequency antennas does not really hold for nanoantennas at optical (including IR) frequencies because the finite penetration depth of the light into the metal as well as the non-negligible diameter D of the antenna lead to a modified relation:[16,19]

$$2L = c_1 + c_2\,[\lambda_{\text{res}}/\lambda_{\text{p}}]. \tag{6.1}$$

In Eq. 6.1, λ_{res} denotes the photon wavelength of the antenna resonance and λ_{p} denotes the plasma frequency of the antenna's material. The coefficients c_1 and c_2 depend on D and the refractive index n_{med} of the surrounding medium. Basic assumptions in this model are a high aspect ratio of the antenna ($D << L$) and that the metal can be described by a free-electron gas Drude-type dielectric function with the electronic damping neglected. Since these conditions are sufficiently fulfilled for crystalline gold nanorods with $L/D > 10$ in the mid and near IR,[20] Eq. 6.1 can be used to describe the resonance behavior of isolated or at least noninteracting nanoantennas. For gold in the mid to near IR, the parameter c_1 can be negligibly small.

Interaction between particles in multimers and arrays of nanoantennas affects the optical properties of the system. They therefore strongly depend on the separation distances, especially between nearest neighbor particles. The interaction effects have been broadly analyzed experimentally and theoretically for many kinds of arrangements and different shapes of particles. Besides the investigation of coupling in dimers, consisting of, for example, ellipsoids,[21] spheres,[22,23] nanodisks,[24] nanorods,[25,26] or nanoantennas,[27,28] many particle systems like nanorod assemblies,[29] linear arrays of nanocylinders[30] or two-dimensional nanoparticle arrays[31–36] have also been subjects of interest. It is important to note that all of these studies focused on the visible spectral range, while there are only few that focused on the IR range,[37–39] where retardation is especially important and the line shape of resonances is mainly determined by resonant scattering, not only by electronic damping.

6.2 Infrared Spectroscopy of Individual Nanoantennas

This section reports on the microscopic IR spectroscopy of individual gold nanoantennas supported by IR transparent substrates. Cylindrical nanorods are prepared by electrochemical deposition in etched ion-track polymeric membranes, and cuboid-shaped nanorods are fabricated by electron-beam lithography.[40]

Spectroscopic investigation of individual gold nanoantenna resonances in the far-field was performed with a commercial IR microscope (Bruker IR scope II) coupled to a Fourier-transform infrared (FTIR) spectrometer (Bruker IFS 66 v/S). Since signals of individual nanoantennas are rather small, a high brilliance IR light source is recommended for acceptable accumulation periods. In the studies shown here, the ANKA light source at the Karlsruhe Institute of Technology was chosen for the measurements because it provides synchrotron light with sufficiently high intensity in the IR range.

Prior to an IR spectroscopic measurement can be done, an individual nanorod has to be located by white light microscopy. Subsequently, the IR radiation is focused on the nanoantenna, collected by a Schwarzschild objective, and finally detected by a liquid nitrogen-cooled mercury cadmium telluride (MCT) detector. All measurements shown in this section were performed in transmittance geometry under normal incidence of light. Since the collection lens has a numerical aperture of 0.52, a negligibly small part of light scattered away from the normal direction is also detected. However, most of the power scattered by the antenna goes to directions outside the cone of collected light. In addition, a circular aperture was used to select an area, here with a diameter of 8.3 μm in the focal plane, which was centred on an individual nanorod for the IR measurement.

IR spectra recorded at the position of the nanorod (sample position) were always normalized to spectra of the bare substrate (background) taken at least 10 μm away from any nanorod. When the ratio between the two spectra is calculated, the relative transmittance is obtained in which all spectral features of the substrate, light source, and so on are eliminated. Sample and background measurements were recorded by acquisition of at least 10 scans in the spectral range from 600 to 7000 cm^{-1} with a resolution of at least 16 cm^{-1}. To allow polarization-dependent measurements, an IR polarizer has to be inserted into the beam path.

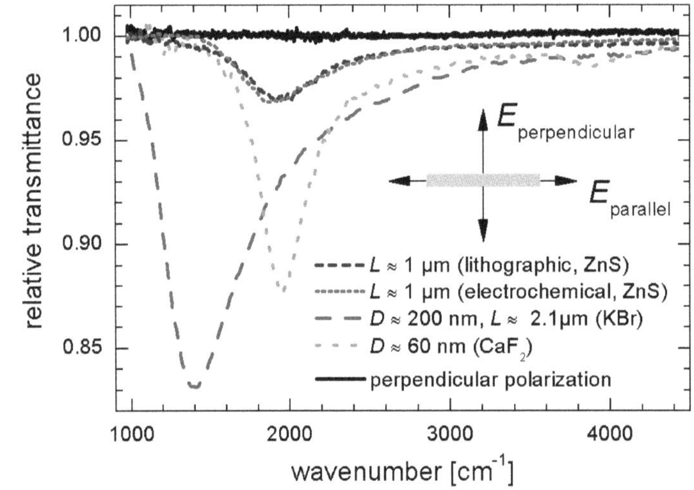

Figure 6.1 Typical relative transmittance spectra of electrochemically (red, green, and orange) and lithographically (blue) prepared individual nanoantennas with different lengths and diameters on different substrates. The nanorods on ZnS feature similar cross sections ($\pi D^2/4 \approx w\ h$, with width w and height h). Polarization of the incident light was parallel to the long rod axis (except for the black curve). For the nanorod on CaF_2, a higher order excitation at around 3900 cm^{-1} is observed (see Section 6.4 for details). Note that its length L could not be determined by white light microscopy since the diameter was too small for such measurement. See also Color Insert.

Figure 6.1 shows typical relative transmittance spectra (plotted versus photon wavenumber $1/\lambda$) of different individual nanorods. The fundamental antenna resonances can be observed if the incident light is polarized parallel to the long rod axis. In contrast, polarization perpendicular to it yields no signal since the longitudinal plasmonic resonance (oscillation of the electrons along the long rod axis) cannot be excited. Note that the transverse resonance is located somewhere in the visible range[41] and hence cannot be detected in the IR range.

The spectral resonance position is determined by different parameters (see Section 6.1) such as rod length L, diameter D, refractive index n_{sub} of the substrate, and the antenna's material properties (dielectric function). The influence of n_{sub} on the resonance intensity is indicated in Fig. 6.1: Nanorods placed on substrates with a higher refractive index ($n_{ZnS} = 2.2$ compared to $n_{CaF_2} = 1.4$ and $n_{KBr} = 1.53$) show lower signals. However, a systematic comparison here is difficult because L and D also influence signal intensity.[17]

Although a high-brilliance light source is recommended for the investigation of individual nanorods, it is possible to use a conventional thermal light source (e.g., globar), but measurements may take many hours (about 15 hours compared to some minutes if synchrotron light is used) and are therefore rather laborious. Furthermore, the investigation of individual electrochemically prepared nanorods features some additional inconveniences. Due to the preparation process, they are available as a suspension and have to be transferred on a substrate for the measurement. Hence, they are randomly distributed, which makes their localization rather difficult with white light microscopy. Additionally, it can be hard to distinguish between individual rods and randomly appearing dimers (two closely spaced rods which are aligned parallel to each other) formed during deposition of the nanorod solution. Besides that, the rod length L cannot be exactly designed by the electrochemical preparation process and its determination with the optical microscope is afflicted with a rather big error (up to 8%). In fact, other methods offer a higher degree of accuracy (e.g., scanning electron microscopy (SEM)), but here the problem arises that it is almost impossible to find the investigated nanorod again in order to measure L. Note that this problem can be overcome by evaporating a grid on the substrate or by nanomanipulation methods that are being developed at many places.

In contrast, lithographically fabricated structures have some advantages. First, they can be prepared in well-arranged patterns that feature a much more intense signal. Hence, the convenient investigation of the nanorods with a thermal light source becomes possible. Second, the geometric dimensions of the rods can be reproducibly adjusted to desired values. In addition, the rods can be easily located and measured by SEM with smaller errors (2%–3%). Finally, the lower crystalline quality of the lithographic nanorods does not seem to be a disadvantage: no significant differences in the far-field properties have been observed between polycrystalline lithographic and crystalline electrochemically prepared IR nanoantennas.[18] One example is shown in Fig. 6.1, where similar optical responses of a cuboid-like lithographic rod (blue curve) and a cylindrical crystalline rod (red curve) with nearly the same length and similar cross sections are observed.

6.3 IR Spectroscopy of Individual Nanoantennas Arranged in Arrays

As indicated at the end of Section 6.2, nanoantenna arrays are better suited for potential sensing applications because they feature a much more intense signal per substrate area. That is why measurements with a common thermal light source become possible with satisfying accumulation times. The measuring principle that was introduced in Section 6.2 stays nearly the same. Only small differences are given, since a larger number of scans (typically 1000–4000) is needed when a thermal light source and bigger apertures (16.7–33.3 μm diameter in the focal plane) are used. The overall measuring time for one array can then be compared to individual nanorod measurements with synchrotron light.

The geometry of the arrays is as follows: First of all, the rods that belong to one array should all have approximately the same length. At this point, it is important to state that inhomogeneous length distributions would certainly influence the optical properties of arrays compared to individual rods, but EBL provides a sufficiently homogenous preparation. The rods are arranged in patterns that typically have dimensions between (50 × 50) and (80 × 80) μm^2.[42] The inset in Fig. 6.2 shows a schematic drawing of a small part of such an array with the characteristic separation distances in longitudinal (d_x) and transverse (d_y) direction.

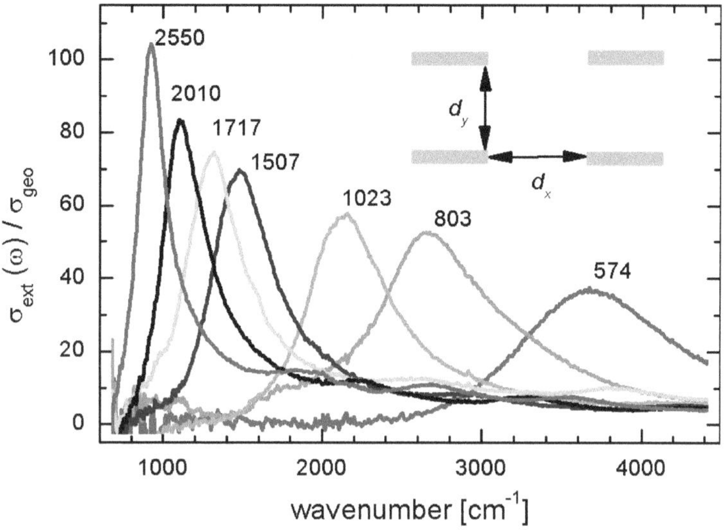

Figure 6.2 Extinction cross section σ_{ext} related to geometric cross section σ_{geo} and normalized to one rod (see Eq. 6.2) for gold nano-antenna arrays on ZnS. All rods in the arrays have quadratic cross sections ($w \approx h \approx 60$ nm) and the lengths in nm are given in the figure. The inset shows the schematic arrangement of the rods in the arrays with the corresponding longitudinal (d_x) and transverse (d_y) separation distances, which are both approximately 5 μm for the measurements shown here.

From relative transmittance spectra (shown exemplary for individual rods in Fig. 6.1), the extinction cross section σ_{ext} can be estimated from the transmittance change $1 - T_{rel}$ in the spectrum by assuming $1 - T_{rel} \approx 2\ (\sigma_{ext}/A_0)/(n_{sub} + 1)$, where A_0 denotes the illuminated spot size and n_{sub} the refractive index of the substrate.[17,40] Since many rods contribute to the measured signal in the case of array measurements, the signal has to be divided by their number, N, to obtain the extinction cross section of an individual rod. In addition, it is beneficial to relate σ_{ext} to the rod's geometric cross section σ_{geo}, which is simply the product of length L and width w or diameter D, respectively. Taking these all together, we arrive at:

$$\frac{\sigma_{ext}}{\sigma_{geo}}(\omega) \approx \frac{A_0}{N}[1 - T_{rel}(\omega)]\frac{n_{sub} + 1}{2Lw}. \qquad (6.2)$$

Any value of this ratio above 1 means an extinction of intensity above geometric shadowing. This is a clear indication of the nanorods' ability to confine light and therefore an indication for an enhanced local field in the vicinity of the nanorods.[17] However, direct conclusions about near-field behavior are not possible with the measuring technique presented here, since only the far-field response is monitored in that way.

Figure 6.2 shows values of $\sigma_{ext}/\sigma_{geo}$ for different gold nanoantenna arrays on a ZnS substrate. Since the separation distances in both directions are relatively large ($d_x \approx d_y \approx 5$ µm), interaction between the nanorods can be excluded[36] and the nanorod ensemble can be treated as noninteracting.[42] (Experimental evidence of this claim is given later in this section, and the influence of interaction between nanorods for smaller values of d_x or d_y is discussed in Section 6.5.) It is obvious from Fig. 6.2 that a longer antenna length leads to resonances at lower wavenumbers. Later, we see that this dependence is quite linear in the IR spectral range. Furthermore, longer rods feature a more intense signal, which can be explained by the fact that a bigger dipole moment can be induced.

The value of $\sigma_{ext}/\sigma_{geo}$ can also be estimated out of relative reflectance measurements at normal incidence, which are analogous to the relative transmittance measurements described in Section 6.2. The only difference is that the sample is illuminated from above and that the normally reflected light is collected and analyzed. In Fig. 6.3, the values of $\sigma_{ext}/\sigma_{geo}$ for two different nanorod arrays are determined from relative transmittance (solid curves) as well as from relative reflectance measurements (dotted curves). As in Eq. 6.2, the influence of the substrate on reflectance measurements is taken into account by a factor of $[(n_{sub})^2 - 1]/4$, which follows analogously to transmittance from thin film approximations of Fresnel's equations at normal incidence. Therefore, $\sigma_{ext}/\sigma_{geo}$ can also be derived by using

$$\frac{\sigma_{ext}}{\sigma_{geo}}(\omega) \approx \frac{A_0}{N}[R_{rel}(\omega)-1]\frac{n_{sub}^2-1}{4Lw} \tag{6.3}$$

with R_{rel} being the relative reflectance. Since significant differences are not observed in the spectra of Fig. 6.3, the approximations made in Eqs. 6.2 and 6.3 seem to be reasonable.

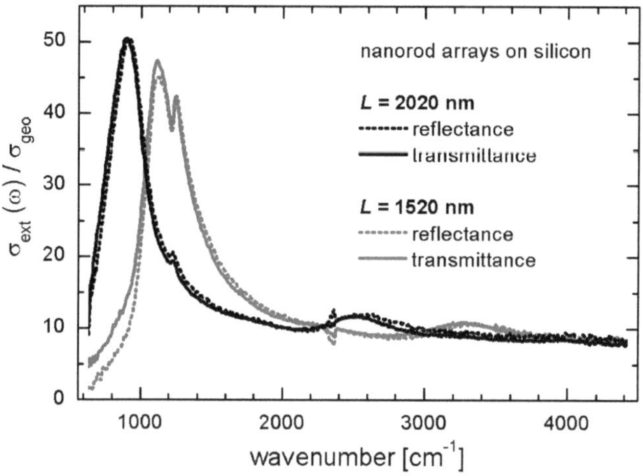

Figure 6.3 Calculated values of $\sigma_{ext}/\sigma_{geo}$ normalized to one nanorod for two different nanorod arrays ($w \approx 120$ nm, $h \approx 100$ nm) on silicon. The dotted curves result from relative reflectance measurements (see Eq. 6.3), and the solid curves from relative transmittance measurements (see Eq. 6.2), respectively. The Fano-like feature at around 1230 cm^{-1} originates from the excitation of a surface phonon-polariton.[43] Besides the fundamental resonances, higher order excitations at 2600 and 3400 cm^{-1}, respectively, are observable (see Section 6.4 for details). The dips at around 2350 cm^{-1} are due to measurement artefacts (CO_2-gas absorption in the optical beam path). See also Color Insert.

Now we want to turn to the dependency of spectral resonance position on antenna length, L. In Fig. 6.4, the measured resonant photon wavelength λ_{res} of noninteracting ($d_x \approx d_y \approx 5$ μm) gold nanoantenna arrays on different substrates (open symbols) is shown in dependence on L. Obviously, there exists a strong influence of the supporting substrate on the exact resonance position. For a fixed value of L, the resonance shifts to longer wavelengths as the refractive index increases from glass (SiO_2) to zinc sulphide (ZnS) and silicon (Si), respectively (values of the refractive indices are given in Fig. 6.4). Besides that, the linear relation between λ_{res} and L is valid in all three cases. In addition, a comparison between the open squares (array measurements) and the full squares (individual rod measurements) supports the previously made assumption that the nanorods do not interact, since no significant differences are present in the experiments. A more detailed comparison between the optical properties of individual rods and rod arrays can be found in Neubrech, *et al.*[18]

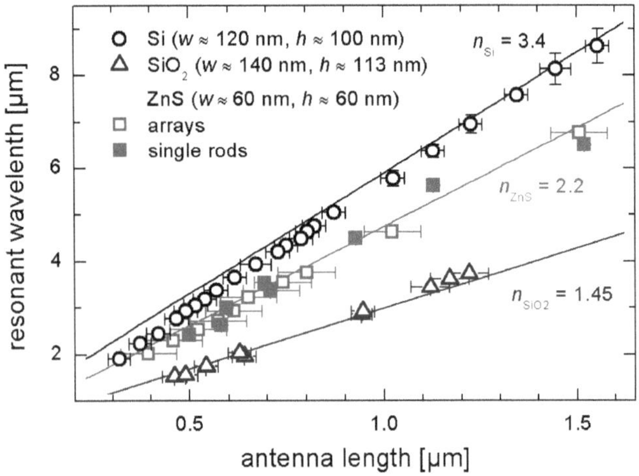

Figure 6.4 Resonance photon wavelength vs. antenna length for nanoantenna arrays (open symbols) on silicon (Si), quartz glass (SiO_2), and zinc sulphide (ZnS) with $d_x \approx d_y \approx 5$ µm in each case. The filled red squares are results from measurements of individual gold nanorods on ZnS and the solid lines are fits to Eq. 6.1. The literature values of the refractive indices of the corresponding substrates are given in the figure. See also Color Insert.

The experimental resonance positions can be reasonably fitted with the analytical approximation of Eq. 6.1, which considers an individual cylindrical nanoantenna embedded in a dielectric surrounding. The solid lines in Fig. 6.4 represent such linear fits with n_{med} being a free parameter and describe the experimental data quite well. Only small differences for shorter rod lengths (especially on silicon) are observed, which can be attributed to the fact that the model's assumption of a large rod-aspect ratio (L/D) does not hold for these shorter rods.

The fit results of n_{med} (not shown here) considerably differ from the literature values of the substrate materials (given in Fig. 6.4). The reason for that discrepancy certainly is that in the experiment the rods are not surrounded by a homogenous medium but are supported by the substrate. Therefore n_{med} has to be considered as an effective refractive index. The simplest approximation for n_{med} would it be to average over the dielectric constants of substrate and air, $\varepsilon_{med} = ½ (\varepsilon_{sub} + 1)$, as proposed in Neubrech *et al.*[17] However, recent numerical simulations show a big difference between a nanorod embedded in such an effective dielectric medium and a nanorod supported by a substrate.[44] This indicates the complexity of substrate effects.

6.4 Even and Odd Higher-Order Excitations

Until now, we mainly focused on the fundamental antenna resonance mode that is analogous to the resonance of RF antennas. Besides that strong fundamental ($l = 1$) mode, nanoantennas also exhibit multipolar ($l > 1$) resonances at λ_{res}/l, as could just about be seen in Figs. 6.2 and 6.3. Furthermore, Fig. 6.5 shows $\sigma_{ext}/\sigma_{geo}$ for a nanorod array on ZnS where these higher-order excitations (marked with $l = 2, 3, 4$) can be observed more clearly. These higher orders also feature a linear dependence on L, as the inset in Fig. 6.5 shows.

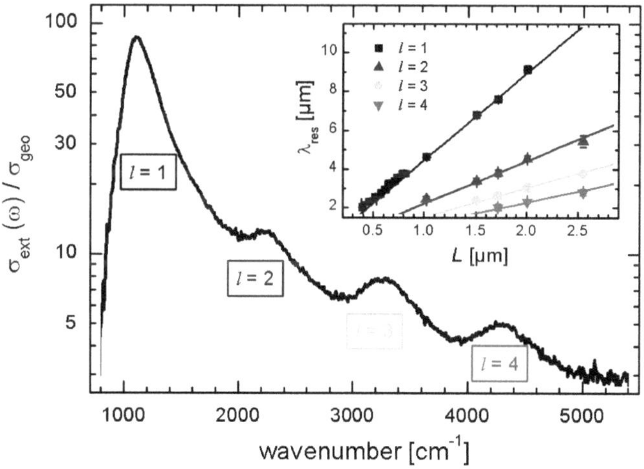

Figure 6.5 Extinction cross section σ_{ext} related the geometric one and normalized to one nanorod for a nanorod array on ZnS ($L \approx 2010$ nm, $w \approx h \approx 60$ nm, $d_x \approx d_y \approx 5$ μm). The measurement was performed at normal incidence. Besides the fundamental ($l = 1$) resonance, higher-order dipolar resonances ($l > 1$) can be observed. The inset shows the dispersion relation of the fundamental mode as well as those of higher-order modes (straight lines correspond to linear fits to the experimental data).

It is known from classical antenna theory[45] that only multipolar modes with an antisymmetric charge distribution (odd *l*) can be optically excited at a normal incidence of light because their total dipole moment does not vanish.[46,47] In contrast to these bright modes, plasmonic excitations with a centrosymmetric charge distribution (even *l*) cannot be excited under such conditions and are therefore referred to as dark modes. Thus, the appearance of the $l = 2$ and $l = 4$ mode in Fig. 6.5 is surprising since the measurement was performed at normal incidence. However, dark modes become visible when some type of symmetry-breaking is present,[23] such as when the nanorod is excited at oblique incidence.[48,49] Another possibility for the activation of dark modes is a structural defect of the nanorod.[20] This is illustrated in Fig. 6.6, where normalized extinction cross sections for two individual nanorods on ZnS are shown. The nanorods were investigated by atomic force microscopy and a structural defect[20] was found for one of the nanorods (black curve). And in fact, the even-order resonance $l = 2$ only occurs for this imperfectly prepared nanorod, whereas it is not detected for the other nanorod (red curve). Numerical simulations also support that dark modes can be activated at normal incidence

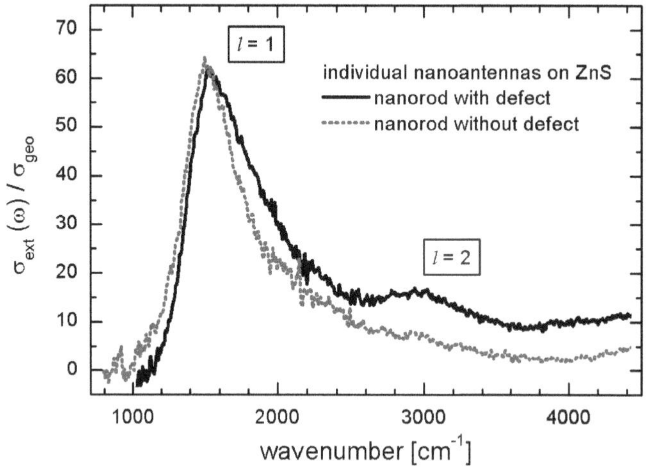

Figure 6.6 $\sigma_{ext}/\sigma_{geo}$ for two individual nanorods on ZnS with similar geometric dimensions ($L \approx 1.5$ μm, $w \approx 60$ nm, $h \approx 100$ nm), measured at normal incidence. The even order ($l = 2$) excitation is clearly visible for the nanorod that has a structural defect (black curve). See also Color Insert.

when the excited nanostructure features some kind of asymmetry such as a structural defect.[20] Such lithographic defects may be induced by inhomogeneities in the substrate's surface (scratches, for example) or imperfections arising from the preparation process of the nanorods.

6.5 Interaction Effects in Nanoantenna Arrays

In Section 6.3 it was mentioned that no significant differences in the optical properties between individual rod and array measurements were observed. This is true if the rods in the arrays have large separation distances (5 μm seems to be enough) but it changes if the distance between adjacent rods is decreased. This consequently leads to interaction effects,[31,35–37] which are discussed in this section.

In the following, only regularly arranged arrays, which were introduced in Section 6.2, are considered. A scheme of the rods' geometric arrangement was shown in the inset in Fig. 6.2, where the longitudinal and transverse separation distances d_x and d_y, respectively, were defined. In the subsequent analysis, only one of them is varied while the other is constantly kept to a value of 5 μm. Hence, the possible interaction mechanisms in longitudinal and transverse direction can be investigated separately.

At first, we will have a look on the effect of reducing the distance between the rod's tip ends (d_x). Figure 6.7 shows $\sigma_{\text{ext}}/\sigma_{\text{geo}}$ for three nanoantenna arrays on ZnS with similar geometric dimensions. Here, d_y is kept to approximately 5 μm to exclude any interaction in this direction. Reducing the longitudinal distance from 5 μm (black solid curve) to 105 nm (red dashed curve) and 65 nm (blue dotted curve) clearly leads to a shift of the fundamental resonance to smaller photon wavenumbers. This spectral shift for decreasing d_x can be understood by looking at a coupled dipole model including retardation.[39] Strong attraction between opposite charges across a small gap leads to a reduced intra-rod restoring force, which consequently decreases the resonance frequency. Numerical simulations also predict a decline of the far-field intensity and a broadening of the resonance if nanorods are interacting in this way,[39] which is in agreement with the findings from Fig. 6.7. The origin of this interaction is a coupling of the near fields of the nanorods, which have their highest intensities at the tip ends (lightning rod effect[50]). Since this direct dipolar interaction features a $1/d^3$ dependence on the particle separation d,[30,34] it only becomes important for small d_x. For instance, no significant changes of the optical properties were observed between nanorod arrays with d_x = 1 μm and d_x = 5 μm.[42] However, the fact that the far-field intensity can be strongly reduced (see Fig. 6.7) does not imply that the near field behaves in the same manner. In fact, calculations[39] and experiments[8] indicate that a strong enhancement of the near-field intensity can be obtained with such coupled structures. This makes them particularly interesting for sensing applications (such as, for example, surface-enhanced infrared spectroscopy, see respective chapters in this book).

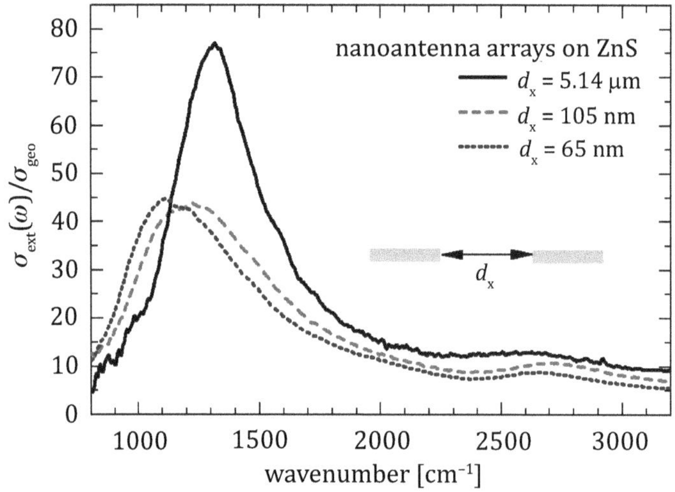

Figure 6.7 $\sigma_{ext}/\sigma_{geo}$ for three nanorod arrays on ZnS with similar geometric dimensions ($L \approx 1.7$ μm, $w \approx h \approx 60$ nm, $d_y \approx 5$ μm) but different separation distances, d_x. For decreasing d_x, a shift of the fundamental resonance to lower wavenumbers is observed. Additionally, the extinction decreases and the resonances become broader, both of which are indications of interaction between neighboring nanorods. See also Color Insert.

At last, interaction effects in transverse direction (d_y) are shortly discussed. Figure 6.8 shows $\sigma_{ext}/\sigma_{geo}$ for nanorod arrays prepared on a silicon substrate. All the rods feature similar geometric dimensions and are separated in longitudinal direction for 5 μm

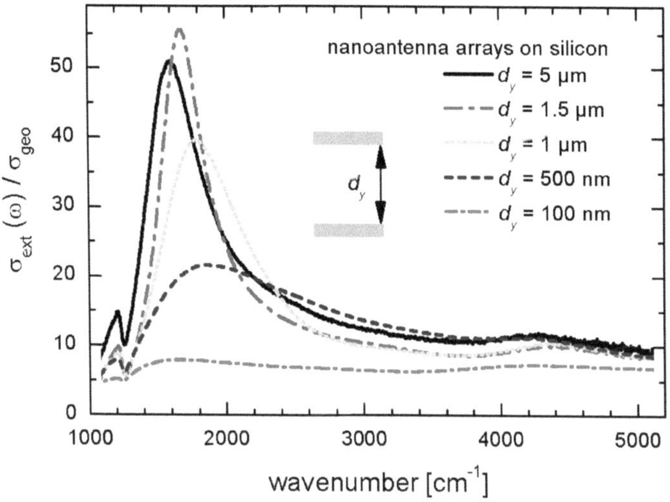

Figure 6.8 $\sigma_{ext}/\sigma_{geo}$ for five nanorod arrays on silicon with similar geometric dimensions ($L \approx 1$ µm, $w \approx 90$ nm, $h \approx 60$ nm, $d_x \approx$ 5 µm) but different separation distances d_y. The Fano-like feature at around 1230 cm^{-1} originates from the excitation of a surface phonon-polariton.[43] See also Color Insert.

to exclude interaction in this direction. Beginning from d_y = 5 µm (black curve), d_y is gradually decreased down to 100 nm. In contrast to the variation of d_x, changes in the optical properties already occur for relatively large distances (in the µm range). For example, a shift of the resonance to higher wavenumbers and an increased extinction is observed for d_y = 1.5 µm. Furthermore, the far-field intensity drops down dramatically if d_y is continuously decreased: For d_y = 100 nm, only a broad background can be detected. These effects are attributed to the long-range radiative coupling, which varies with distance as as e^{ikd}/d,[30,34] with k being the wavenumber of the emitted radiation. Here, retardation and therefore the phase difference between the emitted fields of neighboring particles become very important. In fact, it was found that an onset of interaction effects on the resonance condition occurs if the resonant wavelength in the substrate (λ_{res}/n_{sub}) roughly equals the separation distance d_y.[42] When this condition is fulfilled, constructive interference between the nanoparticles leads to shaper resonances. Summarized, it can be stated that the demonstrated anisotropy of the far-field interaction in nanorod arrays is mainly influenced by the anisotropic emission pattern of an oscillating dipole: Most of the electromagnetic power is radiated perpendicular to the oscillation axis and thus the change in d_y gives stronger interference effects, whereas interaction in a tip-to-tip direction only becomes relevant for small gaps (below 100 nm) and originates from near-field interaction. A detailed analysis of such interaction effects can be found in Weber, *et al.*[42]

References

1. Maier, S. A., and Atwater, H. A. (2005). Plasmonics: localization and guiding of electromagnetic energy in metal/dielectric structures, *J. Appl. Phys.*, **98**(1), 011101.
2. Pelton, M., Aizpurua, J., and Bryant, G. (2008). Metal-nanoparticle plasmonics, *Laser & Photon. Rev.*, **2**(3), 136–159.
3. Aroca, R. (2006). *Surface-Enhanced Vb atio nal Spectroscopy*, Wiley, Chichester, England.
4. Enders, D., and Pucci, A. (2006). Surface-enhanced infrared absorption of octadecanethiol on wet-chemically prepared Au nanoparticle films, *Appl. Phys. Lett.*, **88**(18), 184104.
5. Le, F., Brandl, D. W., Urzhumov, Y. A., Wang, H., Kundu, J., Halas, N. J., Aizpurua, J., and Nordlander, P. (2008). Metallic nanoparticle arrays: a common substrate for both surface-enhanced Raman scattering and surface-enhanced infrared absorption, *ACS Nano*, **2**(4), 707–718.
6. Levin, C. S., Kundu, J., Barhoumi, A., and Halas, N. J. (2009). Nanoshell-based substrates for surface-enhanced spectroscopic detection of biomolecules, *Analyst*, **134**(9), 1745–1750.
7. Neubrech, F., Pucci, A., Cornelius, T. W., Karim, S., García-Etxarri, A., and Aizpurua, J. (2008). Resonant plasmonic and vibrational coupling in a tailored nanoantenna for infrared detection, *Phys. Rev. Lett.*, **101**(15), 157403.
8. Pucci, A., Neubrech, F., Weber, D., Hong, S., Toury, T., and Lamy de la Chapelle, M. (2010). Surface-enhanced infrared spectroscopy using gold nanoantennas, *Phys. Stat. Sol. B*, **247**(8), 2071–2074.
9. Billot, L., Lamy de la Chapelle, M., Grimault, A.-S., Vial, A., Barchiesi, D., Bijeon, J.-L., Adam, P.-M., and Royer, P. (2006). Surface-enhanced Raman scattering on gold nanowire arrays: evidence of strong multipolar surface plasmon resonance enhancement, *Chem. Phys. Lett.*, **422**(4–6), 303–307.

10. Grand, J., Lamy de la Chapelle, M., Bijeon, J.-L., Adam, P.-M., Vial, A., and Royer, P. (2005). Role of localized surface plasmons in surface-enhanced Raman scattering of shape-controlled metallic particles in regular arrays, *Phys. Rev. B*, **72**(3), 033407.

11. Lal, S., Grady, N. K., Kundu, J., Levin, C. S., Lassiter, J. B., and Halas, N. J. (2008). Tailoring plasmonic substrates for surface-enhanced spectroscopies, *Chem. Soc. Rev.*, **37**(5), 898–911.

12. Kneipp, K., Wang, Y., Kneipp, H., Perelman, L. T., Itzkan, I., Dasari, R. R., and Feld, M. S. (1997). Single molecule detection using surface-enhanced Raman scattering (SERS), *Phys. Rev. Lett.*, **78**(9), 1667–1670.

13. Kneipp, J., Kneipp, H., and Kneipp, K. (2008). SERS—a single-molecule and nanoscale tool for bioanalytics, *Chem. Soc. Rev.*, **37**(5), 1052–1060.

14. Nie, S., and Emory, S. R. (1997). Probing single molecules and single nanoparticles by surface-enhanced Raman scattering, *Science*, **275**(5303), 1102–1106.

15. Tseng, A. A., Chen, K., Chen, C. D., and Ma, K. J. (2003). Electron beam lithography in nanoscale fabrication: recent development, *IEEE Trans. Electron. Pack. Manuf.*, **26**(2), 141–149.

16. Bryant, G., García de Abajo, F. J., and Aizpurua, J. (2008). Mapping the plasmon resonances of metallic nanoantennas, *Nano Lett.*, **8**(2), 631–636.

17. Neubrech, F., Kolb, T., Lovrincic, R., Fahsold, G., Pucci, A., Aizpurua, J., Cornelius, T. W., Toimil-Molares, M. E., Neumann, R., and Karim, S. (2006). Resonances of individual metal nanowires in the infrared, *Appl. Phys. Lett.*, **89**(25), 253104.

18. Neubrech, F., Weber, D., Lovrincic, R., Pucci, A., Lopes, M., Toury, T., and Lamy de La Chapelle, M. (2008). Resonances of individual lithographic gold nanowires in the infrared, *Appl. Phys. Lett.*, **96**(16), 163105.

19. Novotny, L. (2007). Effective wavelength scaling for optical antennas, *Phys. Rev. Lett.*, **98**(26), 266802.

20. Neubrech, F., Garcia-Etxarri, A., Weber, D., Bochterle, J., Shen, H., Lamy de la Chapelle, M., Bryant, G. W., Aizpurua, J., and Pucci, A. (2010). Defect-induced activation of symmetry forbidden infrared resonances in individual metallic nanorods, *Appl. Phys. Lett.*, **96**(21), 213111.

21. Su, K. H., Wei, Q.-H., Zhang, X., Mock, J. J., Smith, D. R., and Schultz, S. (2003). Interparticle coupling effects on plasmon resonances of nanogold particles, *Nano Lett.*, **3**(8), 1087–1090.

22. Atay, T., Song, J.-H., and Nurmikko, A. V. (2004). Strongly interacting plasmon nanoparticle pairs: from dipole-dipole interaction to conductively coupled regime, *Nano Lett.*, **4**(9), 1627–1631.

23. Nordlander, P., Oubre, C., Prodan, E., Li, K., and Stockman, M. I. (2004). Plasmon hybridization in nanoparticle dimers, *Nano Lett.*, **4**(5), 899–903.

24. Jain, P. K., Huang, W., and El-Sayed, M. A. (2007). On the universal scaling behavior of the distance decay of plasmon coupling in metal nanoparticle pairs: a plasmon ruler equation, *Nano Lett.*, **7**(7), 2080–2088.

25. Funston, A. M., Novo, C., Davis, T. J., and Mulvaney, P. (2009). Plasmon coupling of gold nanorods at short distances and in different geometries, *Nano Lett.*, **9**(4), 1651–1658.

26. Gluodenis, M., and Foss, C. A. (2002). The effect of mutual orientation on the spectra of metal nanoparticle rod-rod and rod-sphere pairs, *J. Phys. Chem. B*, **106**(37), 9484–9489.

27. Fromm, D. P., Sundaramurthy, A., Schuck, P. J., Kino, G., and Moerner, W. E. (2004). Gap-dependent optical coupling of single "bowtie" nanoantennas resonant in the visible, *Nano Lett.*, **4**(5), 957–961.

28. Muskens, O. L., Giannini, V., Sánchez-Gil, J. A., and Gómez Rivas, J. (2007). Optical scattering resonances of single and coupled dimer plasmonic nanoantennas, *Opt. Express*, **15**(26), 17736–17746.

29. Jain, P. K., Eustis, S., and El-Sayed, M. A. (2006). Plasmon coupling in nanorod assemblies: optical absorption, discrete dipole approximation simulation, and exciton-coupling model, *J. Phys. Chem. B*, **110**(37), 18243–18253.

30. Hicks, E. M., Zou, S., Schatz, G. C., Spears, K. G., Van Duyne, R. P., Gunnarsson, L., Rindzevicius, T., Kasemo, B., and Käll, M. (2005). Controlling plasmon line shapes through diffractive coupling in linear arrays of cylindrical nanoparticles fabricated by electron beam lithography, *Nano Lett.*, **5**(6), 1065–1070.

31. Auguié, B., and Barnes, W. L. (2008). Collective resonances in gold nanoparticle arrays, *Phys. Rev. Lett.*, **101**(14), 143902.

32. Félidj, N., Aubard, J., Lévi, G., Krenn, J. R., Schider, G., Leitner, A., and Aussenegg, F. R. (2002). Enhanced substrate-induced coupling in two-dimensional gold nanoparticle arrays, *Phys. Rev. B*, **66**(24), 245407.

33. Félidj, N., Laurent, G., Aubard, J., Lévi, G., Hohenau, A., Krenn, J. R., and Aussenegg, F. R. (2005). Grating-induced plasmon mode in gold nanoparticle arrays, *J. Chem. Phys.*, **123**(22), 2140699.

34. Haynes, C. L., McFarland, A. D., Zhao, L. L., Van Duyne, R. P., Schatz, G. C., Gunnarsson, L., Prikulis, J., Kasemo, B., and Käll, M. (2003). Nanoparticle optics: the importance of radiative dipole coupling in two-dimensional nanoparticle arrays, *J. Phys. Chem. B*, **107**, 7337–7342.

35. Lamprecht, B., Schider, G., Lechner, R. T., Ditlbacher, H., Krenn, J. R., Leitner, A., and Aussenegg, F. R. (2000). Metal nanoparticle gratings: influence of dipolar particle interaction on the plasmon resonance, *Phys. Rev. Lett.*, **84**(20), 4721–4724.

36. Zhao, L., Kelly, K. L., and Schatz, G. C. (2003). The extinction spectra of silver nanoparticle arrays: influence of array structure on plasmon resonance wavelength and width, *J. Phys. Chem. B*, **107**, 7343–7350.

37. Adato, R., Yanik, A. A., Amsden, J. J., Kaplan, D. L., Omenetto, F. G., Hong, M. K., Erramilli, S., and Altug, H. (2009). Ultra-sensitive vibrational spectroscopy of protein monolayers with plasmonic nanoantenna arrays, *Proc. Natl. Acad. Sci.*, **106**(46), 19227–19232.

38. Adato, R., Yanik, A. A., Wu, C.-H., Shvets, G., and Altug, H. (2010). Radiative engineering of plasmon lifetimes in embedded nanoantenna arrays, *Opt. Express*, **18**(5), 4526–4537.

39. Aizpurua, J., Bryant, G. W., Richter, L. J., and García de Abajo, F. J. (2005). Optical properties of coupled metallic nanorods for field-enhanced spectroscopy, *Phys. Rev. B*, **71**(23), 235420.

40. Pucci, A., Neubrech, F., Aizpurua, J., Cornelius, T., and Lamy de la Chapelle, M. (2008). Electromagnetic nanowire resonances for field-enhanced spectroscopy, in *One-Dimensional Nanostructures* (ed. Wang, Z. M., Salamo, G., Waag, A., and Kishimoto, N.), Springer, New York, pp. 175–213.

41. Link, S., Mohamed, M. B., and El-Sayed, M. A. (1999). Simulation of the optical absorption spectra of gold nanorods as a function of their aspect ratio and the effect of the medium dielectric constant, *J. Phys. Chem. B*, **103**(16), 3073–3077.

42. Weber, D., Albella, P., Alonso-Gonzalez, P., Neubrech, F., Gui, H., Nagao, T., Hillenbrand, R., Aizpurua, J., and Pucci, A. (2011). Longitudinal and transverse coupling in infrared gold nanoantenna arrays: Long range versus short range interaction regimes, *Opt. Express*, **19**(16), 15047-15061.

43. Neubrech, F., Weber, D., Enders, D., Nagao, T., and Pucci, A. (2010). Antenna sensing of surface phonon polaritons, *J. Phys. Chem. C*, **114**(16), 7299–7301.

44. Albella, P., and Aizpurua, J., private communication.

45. Jackson, J. D. (1999). *Classic al Electrodynamic s*, Wiley, New York.

46. Krenn, J. R., Schider, G., Rechberger, W., Lamprecht, B., Leitner, A., Aussenegg, F. R., and Weeber, J. C. (2000). Design of multipolar excitations in silver nanoparticles, *Appl. Phys. Lett.*, **77**(21), 3379–3381.

47. Schider, G., Krenn, J. R., Hohenau, A., Ditlbacher, H., Leitner, A., Aussenegg, F. R., Schaich, W. L., Puscasu, I., Monacelli, B., and Boreman, G. (2003). Plasmon dispersion relation of Au and Ag nanowires, *Phys. Rev. B*, **68**(15), 155427.

48. Dorfmüller, J., Vogelgesang, R., Weitz, R. T., Rockstuhl, C., Etrich, C., Pertsch, T., Lederer, F., and Kern, K. (2009). Fabry-Pérot resonances in one-dimensional plasmonic nanostructures, *Nano Lett.*, **9**(6), 2372–2377.

49. Laurent, G., Félidj, N., Aubard, J., Lévi, G., Krenn, J. R., Hohenau, A., Schider, G., Leitner, A., and Aussenegg, F. R. (2005). Evidence of multipolar excitations in surface enhanced Raman scattering, *Phys. Rev. B*, **71**(4), 045430.

50. Liao, P. F., and Wokaun, A. (1982). Lightning rod effect in surface-enhanced Raman scattering, *J. Chem. Phys.*, **76**(1), 751–752.

Chapter 7

Enhanced Spectroscopies and Surface Plasmon Thin-Film Sensors

Ibrahim Abdulhalim

Department of Electro Optic Engineering and the Ilse Katz Institute for Nanoscale Science and Technology, Ben Gurion University of the Negev, Beer Sheva 84105, Israel

abdulhlm@bgu.ac.il

Surface plasmon resonance (SPR)-based sensors have seen tremendous developments during the past decade, and some of these sensors are becoming a mature technology now, although improvements in their sensitivity and detection limit are ongoing. Together with the recent investigations in localized SPR (LSPR) phenomena, extraordinary optical transmission through nanoapertures in metals, and surface-enhanced spectroscopies, drastic developments are expected to revolutionize the field of optical sensing. Sensitivity enhancement techniques based on SPR phenomena are reviewed, focusing both on physical transduction mechanisms and system performance. It is shown that in the majority of cases, sensitivity enhancement is associated with the enhancement of the electromagnetic (EM) field overlap integral describing the EM interaction energy within the analyte.

Nanoantenna: Plasmon-Enhanced Spectroscopies for Biotechnological Applications
Edited by Marc Lamy de la Chapelle and Annemarie Pucci

ISBN 978-981-4303-61-3 (Hardcover), 978-981-4303-62-0 (eBook)
www.panstanford.com

Examples are given starting from the well-known Kretschmann configuration to the addition of high-index dielectric thin films to metals, the addition of gratings, the use of nanosculptured thin films (nano-STFs), and enhanced transmission of nanoslits. Special attention will be given to nano-STFs, which are assemblies of shaped, parallel, and tilted nanorods, prepared using many variants of the basic oblique angle deposition (OAD) technique. Because of the special shapes and nanoscale dimensions of STFs, they exhibit a great potential in the biosensing field. The fact that the nanofeatures composing these films exhibit LSPR, which falls in the visible–near infrared (Vis-NIR) range, gave these structures the property of enhancing significantly the EM field in their vicinity. Enhancing the field by some orders of magnitude leads to amplification of many interesting optical phenomena such as surface-enhanced fluorescence (SEF), surface-enhanced Raman scattering (SERS), and surface-enhanced infrared absorption (SEIRA). A comparison is presented between different materials, such as silver, gold, copper, and silicon, and between various shapes, such as columns, screws, spheres, and helices, which are deposited on different substrates, with different porosities and orientations.

7.1 Introduction

Thin-film sensors based on surface plasmon (SP)-related technology have been developed in various configurations and formats for sensing a variety of target samples, including pesticides, pollutants, biological pathogens, toxins, and diseased tissue. The past decade has seen tremendous advancements in the optics of nanostructured metals, and a new research area called nanoplasmonics has thereby emerged. The sensitivity and detection limit of SPR-based sensors continues to improve so that samples of ever smaller volumes can be detected with enhanced reliability and specificity, in particular when combined with Raman, fluorescence and infrared (IR) spectroscopies. Nanoplasmonics is the field of research that deals with the interaction of EM waves with metallic or, more generally, conducting nanostructures. It is a subfield of the wider research field that emerged recently called *plasmoni̇ s*,[1-3] which deals with SPR-related science and technology.[4]

The coupling between the EM field and the collective charge density oscillations on the interface between a conductor and a dielectric gives rise to a wealth of phenomena that are of interest from both fundamental and applied research points of view. SPR is a quantum EM phenomenon arising from the interaction of light with free electrons at a conductor–dielectric interface emerging as a longitudinal EM wave in a two-dimensional (2D) gas of charged particles such as free electrons in metals. Under certain conditions the energy carried by the photons is transferred to collective excitations of free electrons, called SPs, at that interface. This transfer of energy occurs only at a specific resonance wavelength of light when the momentum of the photon along the surface matches that of the plasmon. The SPs excited are strongly localized across the interface and may be considered, classically, EM surface waves that propagate along the interface and decay exponentially with distance normal to the interface. SPR is responsible for a dip in reflectance at the specific wavelength, the dip resulting from the absorption of optical energy in the metal. Since SP waves are tightly bound to metal–dielectric interfaces penetrating around 10–50 nm into the metal (the so-called skin depth) and typically more than 100–200 nm into the dielectric (depending on the wavelength), they concentrate EM waves in a region that is considerably smaller than their wavelength, a feature that enhances the sensitivity of sensors, particularly with nonlinear optical processes such as Raman scattering and second-harmonic generation (SHG).

There are two main types of SPs with respect to their propagation characteristics along the interface—extended or propagating and localized. Propagating SPs occur at the interface of a dielectric and an infinite sheet of conductor and are considered more classical since they are known for a longer time. To excite them the wave vector along the interface, k_x, should be high enough (Fig. 7.1a) to match the plasmon wave vector, k_{SP}. To achieve this, one should use a special coupling medium, such as a prism, fiber, waveguide, or grating, above the first metal interface. The latest advancements in nanotechnology have made the fabrication of structures with nanometer-scale features feasible; thus LSPR has become a subject of immense interest during the past two decades.[5] Localized SPs (LSPs) are excited at metallic structures with lateral dimensions less than half the wavelength of the exciting EM wave. The clearest manifestation of LSPs are the resonances of metallic nanoparticles (NPs) that are famous for their beautiful coloring of some stained glass windows. The SP wave frequency is different from the bulk one, depending on the dielectric environment of the metal and its geometry. In general it might be expressed as $\omega_{SP} = g\omega_p$ when the dielectric is air, where g is a geometrical factor equal to $1/\sqrt{2}$ for propagating the SP at a flat surface, $1/\sqrt{3}$ for a sphere, and determined by the ratio of the ellipsoid axis for an ellipsoidal particle. A third type of plasmon may be mentioned, called long-range SPR (LRSPR), which exists in thin metal films or stripes characterized by low attenuation and travelling along the surface for distances up to a few millimeters in the Vis or even a few centimeters in the IR. This latter type might have applications in active photonic components and highly sensitive sensors; however, the research on LRSPR is still in its infancy.

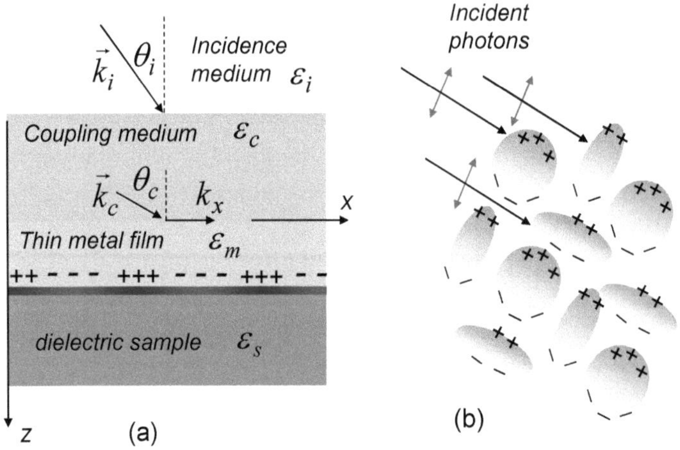

Figure 7.1 Schematic showing the geometries for SPR excitation (a) for extended or propagating SP waves at the interface of an infinite thin metal sheet and a dielectric and (b) for LSP waves at the interface of NPs of different shapes and orientations.

In the case of propagating SPR (PSPR), plasmons propagate along the interface between a metal and a dielectric for distances on the order of microns to tens and even hundreds of microns and decay evanescently in the z direction normal to the interface with $1/\varepsilon$ decay length on the order of half the wavelength (≈200 nm for wavelengths in the Vis range). The interaction between the metal surface-confined EM waves and the molecular layer of interest leads to shifts in the plasmon resonance, which can be observed in five main modes: (a) angle resolved, (b) wavelength shift, (c) intensity variation, (d) phase shift or polarization variation, and (e) imaging. In the first two modes, one measures the reflectivity of light from the metal surface as a function of either wavelength (at a constant incidence angle) or the incidence angle (at a constant wavelength). The modes c–e use light of both constant wavelength and incidence angle to measure intensity or phase shift or polarization change upon reflection or interrogate a 2D region of the sample, mapping the reflectivity of the sample as a function of position. To increase the system reliability one can also measure the phase, intensity, or polarization change as a function of wavelength or angle.

In the case of LSPR, light interacts with particles or structures on a scale much smaller than the incident Vis wavelength (nanoscale usually defined within the 100 nm range). This leads to a plasmon that oscillates locally around the NP with a frequency known as the LSPR. Similar to PSPR, LSPR is sensitive to changes in the local dielectric environment. Typically one measures changes in the local dielectric environment through LSPR wavelength-shift measurements, although a variant of angle-resolved sensing for LSPR is also possible. Both PSPR and LSPR can provide thermodynamic and real-time kinetic data for binding processes. Moreover, although PSPR spectroscopy provides much higher sensitivity to changes in the bulk refractive index than LSPR spectroscopy, the responses of the two techniques become comparable when measuring short-range (in the nanometer vicinity of the metal surface) changes in the refractive index, owing to a molecular adsorption layer. This is a result of the much smaller sensing volume offered by the LSPR sensors, as the EM decay length is 40–50 times shorter than that of the PSPR sensors. Moreover, advances in synthetic and lithographic fabrication techniques allow researchers to tune the localized resonance wavelength through the Vis, to the NIR, and into the IR regions of the EM spectrum by varying the shape, size, and material of the NPs that support LSPR.[6] This offers additional flexibility when designing LSPR-sensing experiments.

PSPR biosensors have been widely applied in a diverse range of fields,[7] including molecular recognition and disease immunoassays. Even though conventional SPR biosensors are more sensitive than other label-free devices, they are still unable to achieve the direct detection of small molecular (few hundreds of Daltons) interactions or low molecular concentrations (physiological concentration) on the surface of the biosensor. Consequently, various proposals have been developed to enhance the sensitivity or resolution of biosensors by using different SPR modes or detection methods.[8] Also, various LSPR biosensors have been proposed, which employ the strong ultraviolet (UV)-Vis absorption band of the metal NPs to yield an area mass detection limit of 100–1,000 pg/mm^2. However, this detection capability is poorer than that of conventional PSPR biosensors by an order of at least 100 times. On the other hand the local EM field enhancement near nanostructures causes huge enhancement of other spectroscopic signals, such as SERS and SEF. These spectroscopic techniques provide another important sensing parameter, which is specificity. The enhancement factor depends strongly on the shape of the nanostructure and the type of the metal.

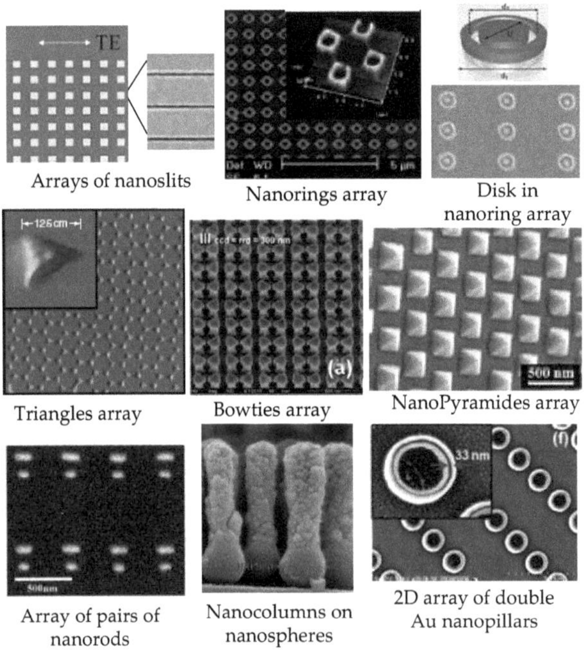

Figure 7.2 Variety of nanostructured thin-film images reproduced with permission from different references, starting from the top left: a 2D array of squares of 1D nanoslits,[9] a nanorings array,[10] a disk in a nanorings array,[11] a triangles array,[12] a bowties array,[13] a nanopyramides array,[14] an array of pairs of nanorods,[15] nanocolumns,[16] and double nanopillars.[17] Other geometries exist, such as a 2D array of nanoholes and a film on nanospheres.[18] *Ab evü tö n*: 1D, one dimensional.

Single particles are important for single-molecule detection; however, for more practical sensor schemes, platforms of nano-features in the form of a thin film on a substrate are preferable. This is important as well for biochip applications in which a 2D array of small pixels (0.1 mm × 0.1 mm) on a substrate are used to sense a large number of analytes in parallel. Nanostructured thin films come in a variety of forms, as shown in Fig. 7.2. These include perforated metal films for enhanced optical transmission (EOT) and other periodic arrays of NPs. An important class of nanostructured thin films is called sculptured thin films (STFs), which are prepared in large areas easily using glancing angle deposition. The last section of this chapter is dedicated to sensing with STFs.

7.2 Plasmons Excitation

7.2.1 Bulk Plasmons

Within a bulk conductor, plasmons are collective oscillations of charged particles with density *N*, charge *q*, and mass *m*, having a characteristic frequency $\omega_p = (Nq^2/\varepsilon_0 m)^{1/2}$ (in SI units). For an ideal free electron gas without damping, the Drude model gives:

$$\varepsilon(\omega) = 1 - \frac{\omega_p^2}{\omega^2} \tag{7.1}$$

which becomes negative when $\omega < \omega_p$, meaning the metal becomes absorptive but with a large imaginary part; hence it is also reflective. For $\omega > \omega_p$ the metal becomes more transmissive, which explains why metallic mirrors are not good enough for the UV part of the spectrum. When $\omega = \omega_p$, bulk plasmons are excited and the dielectric function is null as expected in ideal plasma, which implies from the constitutive relation $\vec{D} = \varepsilon\vec{E}$ that there is only a longitudinal or compressional EM wave. The dispersion relation for light inside the metal can be easily shown from Maxwell's equations and the ideal Drude model to be $\omega = \sqrt{\omega_p^2 + c^2k^2}$, showing that it is always above the light line $\omega = ck$ and therefore the EM waves propagation is only allowed for $\omega > \omega_p$ when the metal is not absorptive.

7.2.2 Surface Plasmons at the Semi-Infinite Metal–Dielectric Interface

As mentioned earlier, to excite SP waves at a semi-infinite conductor–dielectric interface, there is a need for a coupling medium at the other metal interface that can provide a high enough wave vector along the interface to match the plasmon wave vector. There are several ways to achieve this, for example, using a prism, a grating, or in general evanescent wave coupling, which can be launched using fibers, waveguides, or near-field microscopy probes (Fig. 7.3).

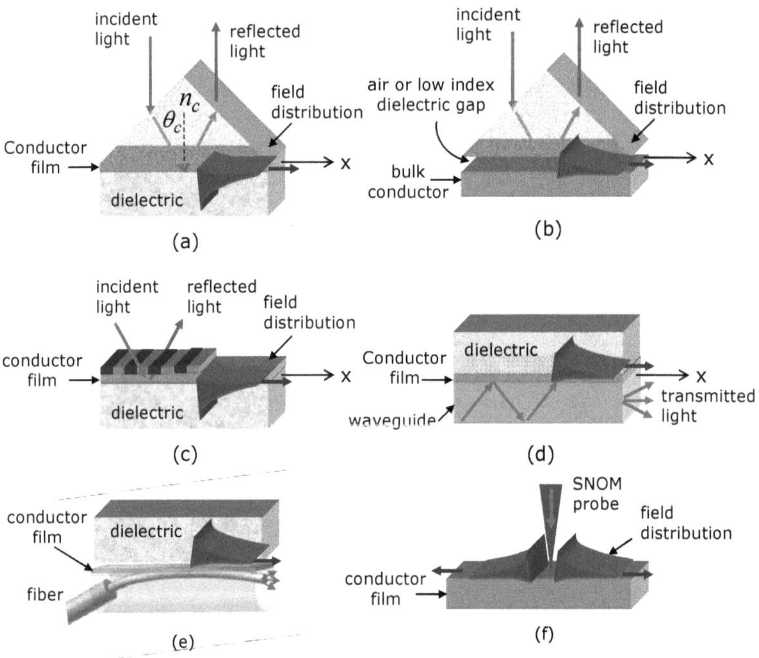

Figure 7.3 Extended SPR excitation schemes using (a) prism coupling in Kretschmann configuration, (b) prism coupling via Otto configuration, (c) grating coupling, (d) waveguide coupling, (e) fiber coupling, and (f) coupling using a SNOM nanoprobe. *Abbreviation*: SNOM, scanning near-field optical microscopy.

Propagating SP waves are excited with transverse magnetic (TM)-polarized EM waves when the component of the **k**-vector along the metal–dielectric interface matches the SP **k**-vector. The condition of TM polarization is needed to generate the charge distribution on the metal–dielectric interface. In the mentioned case the electric field has a component perpendicular to the interface. Note that in Fig. 7.3 there are two cases of prism coupling—Kretschmann and Otto configurations, where in the first the coupling is direct, while in the Otto configuration there is a thin layer of air or a low-index material between the metal and the prism, thinner than the decay length of the totally reflected wave. The SP phenomenon can easily be understood, and its main characteristics can be determined by solving Maxwell's equations to the boundary-value problem.

We start by considering the boundary problem between the metal and the dielectric sample shown in Fig. 7.1a having dielectric constants ε_m and ε_s and ignoring the interface Fresnel reflections and the finite thickness of the conducting layer. By assuming TM plane and harmonic EM fields in the metal and the sample with the appropriate waves:

$$\begin{aligned} E_j(r,t) &= (E_{xj}, 0, E_{zj}) e^{-k_{zj}|z|} e^{i(k_{xj}x - \omega t)} \\ H_j(r,t) &= (0, H_{yj}, 0) e^{-k_{zj}|z|} e^{i(k_{xj}x - \omega t)} \end{aligned} \tag{7.2}$$

where (j = metal m, sample s), k_{zj} the z component of the wave vector (**k**-vector) in a medium j, k_{xj} the x component of the wave vector, and ω the radial frequency of the radiation. Substituting the fields from Eq. 7.2 into Maxwell's equations and applying the continuity conditions $H_{ym} = H_{ys}$; $E_{xm} = E_{xs}$ and using the relation $k_{zj}^2 + k_{xj}^2 = (\varepsilon_j k_0)^2$ lead to:

$$\varepsilon_s k_{zm} + \varepsilon_m k_{zs} = 0 \tag{7.3}$$

where ε_j is the relative permittivity of a medium j and c is the speed of light in vacuum. Using this relation and the phase-matching condition $k_{xm} = k_{xs} = k_x$, the following expression is found for k_x, which is equal to the SP wave vector k_{SP} for the SP wave to be excited:

$$k_x = k_{SP} = \frac{2\pi}{\lambda}\sqrt{\frac{\varepsilon_m \varepsilon_s}{\varepsilon_m + \varepsilon_s}} \tag{7.4}$$

This is also known as the dispersion relation of the propagating SP. The value on the left-hand side of Eq. 7.4 represents the SP wave vector. Since in the range of optical frequencies, $\text{Re}(\varepsilon_m) < 0$

and assuming that the imaginary parts of both ε_m and ε_s are small, a purely real SP wave vector can be obtained when $\mathrm{Re}(\varepsilon_m) < -\varepsilon_s$ allowing LRSP propagation. Hence to conclude, the conditions for SPR excitation are as follows:

(a) Incident light is TM polarized.
(b) The real part of the dielectric constant of the metal and the dielectric are of opposite signs and satisfying $\mathrm{Re}(\varepsilon_m) < -\varepsilon_s$.
(c) The wave vector of the incident light is large enough to satisfy the momentum matching $k_s = k_{SP}$.

To derive the dispersion relation explicitly and present it graphically, the simplified Drude model for metal dispersion may be substituted in Eq. 7.4 to give the dispersion equation in normalized form:

$$\omega_n^2 = \frac{1}{2}\left[1 + \left(1 + \frac{1}{\varepsilon_s}\right)k_n^2 - \sqrt{1 + 2k_n^2\left(1 - \frac{1}{\varepsilon_s}\right) + k_n^4\left(1 + \frac{1}{\varepsilon_s}\right)^2}\,\right] \qquad (7.5)$$

where we used the normalized parameters $\omega_n = \omega/\omega_p$ and $k_n = k_x/(\omega_p/c)$, representing a normalized frequency and wave vector, with the normalization being to the plasma frequency and plasmon wave vector $k_p = \omega_p/c$.

On the other hand, the wave vector of light falling on the metallic film satisfies the dispersion line of a photon in free space, $\omega = ck_0$ (where k_0 is the **k**-vector in vacuum) or in terms of the normalized quantities $\omega_n = k_n/(\sin\theta_c)$, where n_c and θ_c are the refractive index of the coupling medium and the propagation angle in it, respectively. The slope of this line is $1/(n_c \sin\theta_c)$, which for light incident from vacuum always lays on the left of the dispersion line of the plasmon for any frequency (see Fig. 7.4).

Hence a light beam that hits a metal surface from air or through a low-index medium with a flat surface will never excite the plasmon at the metal interface unless the photon momentum is enhanced to fulfill the matching with the SP momentum and an intersection between the two lines is obtained. The intersection represents the resonance phenomenon and defines an operating point at which the frequency and the wave vector of both the exciting light and the plasmon are determined. Note that the SP dispersion curve asymptotically approaches $1/\sqrt{1 + \varepsilon_s}$. Hence the frequency $\omega_p/\sqrt{1 + \varepsilon_s}$ represents the SP frequency, which is smaller by $\sqrt{1 + \varepsilon_s}$ than the bulk plasma frequency.

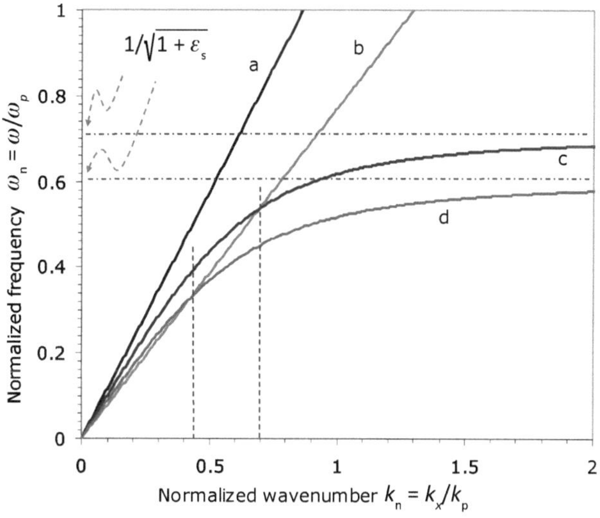

Figure 7.4 (a) Dispersion curve for a free photon incident onto the metal from vacuum at an incidence angle of 60°, (b) dispersion curve for a photon emerging from a coupling medium with a refractive index of 1.73 and with an incidence angle of 60°, (c) dispersion relation of an SP at the interface between the metal and air (as a sample), and (d) dispersion relation of an SP generated at the interface between the metal and water as a sample with a refractive index of (water) 1.33. For a fixed incidence angle, changes in the sample refractive index demand a corresponding change in the wavelength to maintain resonance or alternatively a modification in the incidence angle for a fixed wavelength. The frequency is normalized to the plasma frequency ω_p, and the x component k_x of the light wave vector is normalized to the plasma wave number k_p.

Perturbation in the substrate refractive index causes a change in the intersection point and consequently a shift in the resonance wavelength for a fixed incidence angle or a shift in the incidence angle for a fixed wavelength because $k_x = 2\pi n_c \sin\theta_c/\lambda$. Figure 7.4 demonstrates clearly the sensing mechanism in the Kretschmann configuration. A small increase in the analyte (sample) refractive index modifies the plasmon's dispersion curve, and the resonance is obtained at a larger wavelength when the incidence angle is fixed by the light dispersion line. On the other hand if we want to keep the wavelength fixed, the slope of the incident light should be changed by increasing the incidence angle.

The two important parameters affecting performance are the penetration depth of the EM field inside the dielectric or the metal and the propagation length along the surface. The field penetration depth is determined by $1/\mathrm{Im}(k_{zj})$, which from the previous treatment is given by:

$$\delta_{s,m} = \frac{\lambda}{2\pi} \cdot \sqrt{\frac{|\varepsilon_{mr}| - \varepsilon_s}{\varepsilon^2_{s,mr}}} \tag{7.6}$$

As an example Fig. 7.5 shows the penetration depth inside water as a sample material on silver metal.

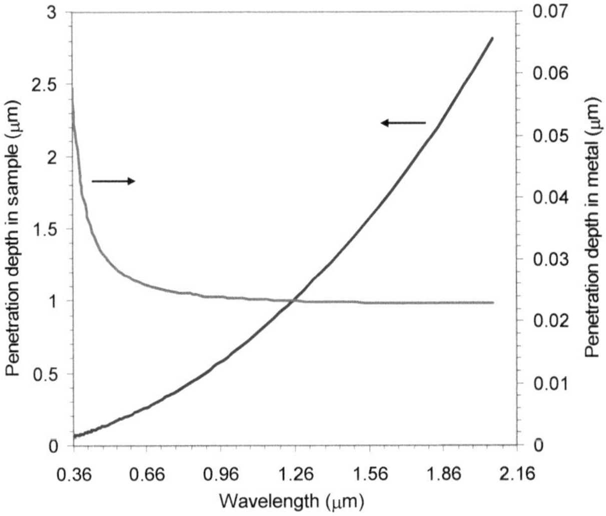

Figure 7.5 Penetration depth of an SP wave inside water and silver, showing that in the IR the penetration inside the sample increases with the wavelength more than linearly. The dispersion of water was taken into account in the calculation.

Hence the penetration depth in the NIR range is larger by a factor of 8 than that in the Vis range, although the wavelength ratio is only 2.5. The reason for that is the difference in the real part of the metal-dielectric function. Note that the penetration depth inside the metal is smaller than that in the sample by the factor $|\varepsilon_{mr}|/\varepsilon_s$, which is very large; hence the penetration depth in the metal is only few to tens of nanometers (see Fig. 7.5). This asymmetry in the field distribution is responsible for high sensitivity to variations in the dielectric sample refractive index.

The propagation length of the extended SP along the surface is defined as the distance at which the intensity decays to $1/\varepsilon$ of its maximum value and hence equals $1/2\,\mathrm{Im}(k_x)$ and is found to be:

$$L_x = \frac{\lambda}{2\pi} \cdot \frac{\varepsilon_{\mathrm{mr}}^2}{\varepsilon_{\mathrm{mi}}} \cdot \left[\frac{\varepsilon_{\mathrm{s}} + \varepsilon_{\mathrm{mr}}}{\varepsilon_{\mathrm{s}} \cdot \varepsilon_{\mathrm{mr}}} \right]^{3/2} \tag{7.7}$$

The propagation length versus the wavelength at the silver-water interface is shown in Fig. 7.6, showing that it scales more than linearly with the wavelength.

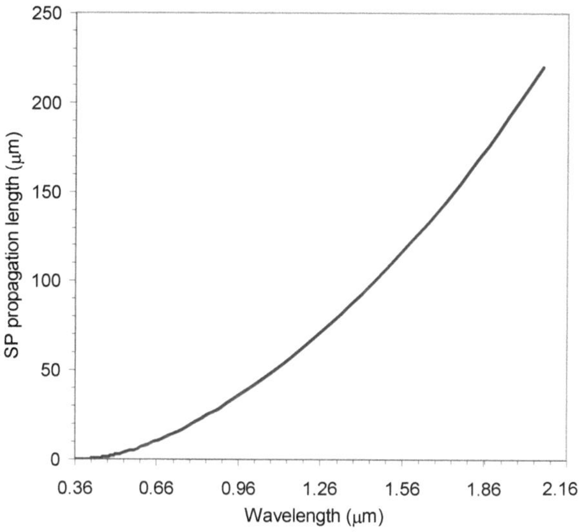

Figure 7.6 Propagation length of an SP wave along the silver–water interface, showing that in the IR it increases with the wavelength more than linearly. The dispersion of water was taken into account in the calculation.

In the UV and the Vis it is in the range of 10–15 μm, but in the NIR it is in the range 50–250 μm, although the large propagation length means less lateral resolution in sensing variations at the interface.

7.2.3 Metal Thickness Effect and Long-Range Surface Plasmons

The earlier treatment is valid when the thickness of the metal layer is ignored; however, the metal films used have a finite thickness of the order of 40–50 nm. To see the effect of the thickness one can simply write down the reflectivity function from a single layer using the Airy formulae. Considering for example the Kretschmann configuration in Fig. 7.3a, with the prism being the ambient and the dielectric sample being the substrate, we get the following formulae for the Airy reflectivity function for a single layer between two semi-infinite media:

$$R_{\mathrm{TM}} = \left| \frac{r_{\mathrm{pm}} + r_{\mathrm{ms}} \exp(2i\beta)}{1 + r_{\mathrm{pm}} r_{\mathrm{ms}} \exp(2i\beta)} \right|^2 \tag{7.8}$$

where $\beta = k_{zm} d_{\mathrm{m}}$ is the single-pass-phase factor through the metal film and r_{lj} are the Fresnel reflection coefficients from a medium j = m, s when the light is incident from medium l = c, m, with the subscript c standing for the coupling medium. For TM waves the Fresnel coefficients may be expressed as:

$$r_{\mathrm{lj}} = \frac{k_{z\mathrm{l}} / \varepsilon_{\mathrm{l}} - k_{z\mathrm{j}} / \varepsilon_{\mathrm{j}}}{k_{z\mathrm{l}} / \varepsilon_{\mathrm{l}} + k_{z\mathrm{j}} / \varepsilon_{\mathrm{j}}} \tag{7.9}$$

where $k_{z\mathrm{j}} = \sqrt{\varepsilon_{\mathrm{j}} k_0^2 - k_x^2}$. As it is well known from the optics of thin films, multiple interference effects are important and Fabry–Perot-type resonances in the transmission appear. To obtain the optimum thickness for SPR excitation, the condition $R_{\mathrm{TMmin}} = 0$ should be satisfied. However, since total internal reflection (TIR) takes place, there is no transmission and a surface wave is excited. The dispersion relation of such a surface wave is determined by the poles of the reflectivity function; hence it is given by:

$$1 + r_{\mathrm{pm}} r_{\mathrm{ms}} \exp(2i\beta) = 0 \quad \text{or} \tag{7.10}$$

$$\begin{aligned}(k_{zp} / \varepsilon_{\mathrm{c}} + k_{zm} / \varepsilon_{\mathrm{m}})(k_{zm} / \varepsilon_{\mathrm{m}} + k_{zs} / \varepsilon_{\mathrm{s}}) + \\ (k_{zc} / \varepsilon_{\mathrm{c}} - k_{zm} / \varepsilon_{\mathrm{m}})(k_{zm} / \varepsilon_{\mathrm{m}} - k_{zs} / \varepsilon_{\mathrm{s}}) \exp(2i\beta) = 0\end{aligned} \tag{7.11}$$

Hence at SPR both the numerator and the denominator of the reflectivity vanish; however, the former vanishes more rapidly than the latter. By equating the imaginary and real parts of Eq. 7.11 to zero, two dispersion relations are obtained describing two SPs excited at both sides of the metal film. Since k_{zm} is mainly imaginary, β is mainly imaginary, which for a thick enough metal yields the two equations, each similar to Eq. 7.3, hence giving a characteristic SP wave frequency at the two metal interfaces:

$$k_{zc}/\varepsilon_c + k_{zm}/\varepsilon_m = 0 \Rightarrow \omega_{SPc} = \frac{\omega_p}{\sqrt{1+\varepsilon_c}}$$
$$k_{zm}/\varepsilon_m + k_{zs}/\varepsilon_s = 0 \Rightarrow \omega_{SPs} = \frac{\omega_p}{\sqrt{1+\varepsilon_s}} \quad (7.12)$$

For the case of a thin metal film surrounded by the same medium of dielectric constant ε_s, the two equations in Eq. 7.12 become:

$$\omega^+: \quad \varepsilon_{mr}\,\mathrm{Im}(k_{zs}) + \varepsilon_s\,\mathrm{Im}(k_{zm})\tanh(\mathrm{Im}(k_{zm})d_m/2) = 0$$
$$\omega^-: \quad \varepsilon_{mr}\,\mathrm{Im}(k_{zs}) + \varepsilon_s\,\mathrm{Im}(k_{zm})\coth(\mathrm{Im}(k_{zm})d_m/2) = 0 \quad (7.13)$$

Note that even though the structure is symmetric, the fact that the metal is thin causes the two plasmons not to have the same SPR frequency. This is because the wave excited at the first surface does not gain the same phase as the wave excited at the second interface. The upper branch (ω^+) is an asymmetric mode (with respect to the center of the metal film, ($z = d_m/2$) while the lower branch is symmetric. The asymmetric mode then' has zero energy at the center of the metal film, while the symmetric mode has its maximum at the center of the metal film at $z = d_m/2$. As a result the latter gets absorbed quickly, while the former propagates for a longer distance along the interface, forming what is called LRSP. The expressions for the two frequencies in the limit of a large wav evector and $\varepsilon_{mi} \approx 0$ are:[5]

$$\omega_+ = \omega_{SPs}\sqrt{1 + \frac{2\varepsilon_s \exp(-\mathrm{Re}(k_x)d_m)}{1+\varepsilon_s}},$$
$$\omega_- = \omega_{SPs}\sqrt{1 - \frac{2\varepsilon_s \exp(-\mathrm{Re}(k_x)d_m)}{1+\varepsilon_s}} \quad (7.14)$$

These two equations show that the asymmetric mode has a higher frequency, and they degenerate when the metal film is thick enough.

In the prism coupling an additional layer of the same index as the prism (or higher) between the prism and the metal film is simply a continuation of the prism medium and therefore an additional SP cannot be excited at the first surface. This can also be verified by drawing the dispersion relation for the two plasmons on the basis of the transcendental equations in Eq. 7.13. The symmetric plasmon has a curve $\omega_-(k_x)$, which is asymptotic to the light line. To excite it an additional dielectric layer has to be inserted between the prism and the metal film, having an index less than that of the prism, as shown in the inset in Fig. 7.7. This layer breaks the symmetry of the structure, and therefore the consequences are slightly different from the symmetric case. This configuration is known as the insulator-metal-insulator (IMI) heterostructure. Note that the single-SPR dip location is not sensitive to the insertion of the buried dielectric layer. This is because for this plasmon the energy is concentrated mainly on the metal–analyte interface. The field distribution for the double-SPR case is shown in Fig. 7.8, showing that for the first resonance (smaller resonance angle) the field is enhanced at the metal–analyte interface, while for the second resonance it is enhanced at the buried dielectric–metal interface.

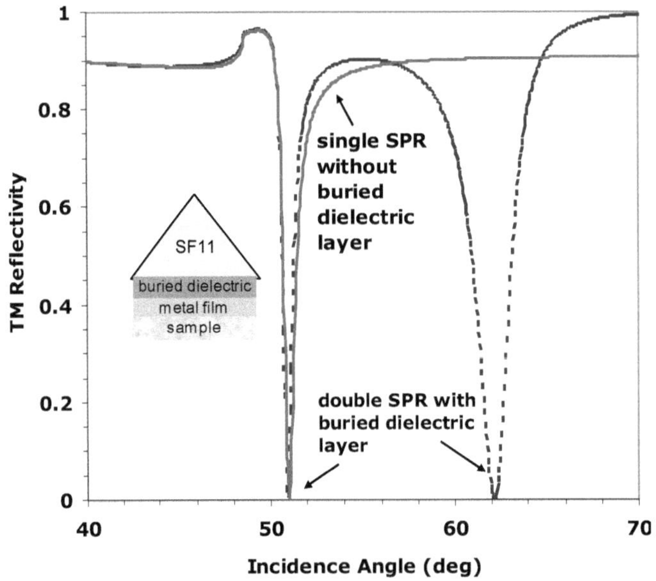

Figure 7.7 TM reflectivity vs. incidence angle for the Kretschmann configuration, as shown in the inset, without and with the addition of a buried dielectric layer, showing the excitation of single and double SPR, respectively. Parameters are as follows: the wavelength is 788 nm, the prism index is 1.772, the water analyte (substrate) index is 1.33, the buried dielectric is PMMA with an index of 1.49 and thickness of 480 nm, the silver thickness is 45 nm, and the dielectric constant is –27.28 +1.794i. *Ab evů tò n*: PMMA, poly(methyl methacrylate).

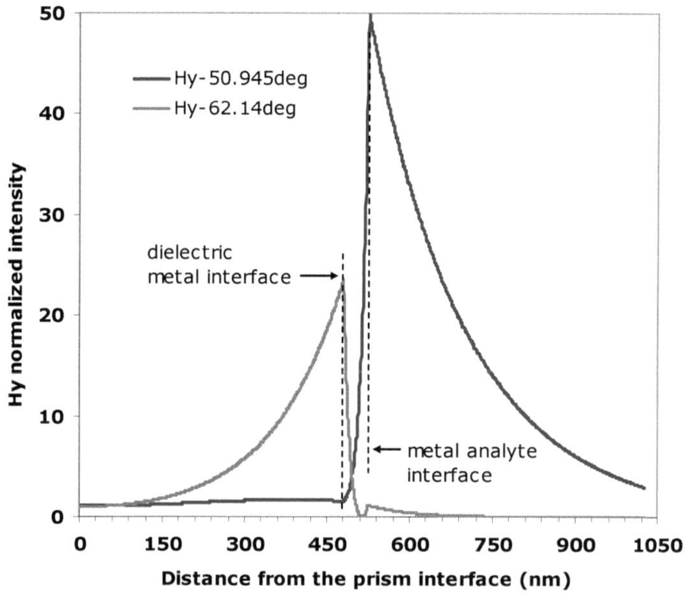

Figure 7.8 The normalized intensity of the y component of the magnetic field, corresponding to the Kretschmann configuration with double-SPR excitation by adding a buried dielectric layer, as in Fig. 7.7. The resonance angles for each SP wave are indicated in the figure. Parameters are the same as in Fig. 7.7. See also Color Insert.

To show further at which interface each of the SPs is excited, the effect of the variation of the refractive index of the analyte n_a and the buried dielectric layer index n_b on the double-SPR spectrum was calculated and shown in Fig. 7.9. In each case only one SPR dip is affected but not the other, thus proving that the first dip is excited at the metal–analyte interface (Fig. 7.9a), while the second dip is excited at the dielectric–metal interface (Fig. 7.9b).

The fact that the asymmetric SP mode can have a longer propagation range may also be seen from the fact that the corresponding dip width is narrower. The narrower the SPR dip, the smaller the corresponding value of $\text{Im}(k_x)$, and hence the larger its propagation length $1/2\,\text{Im}(k_x)$. One more interesting configuration is the *metal-insulator-metal* (MIM) heterostructure, which can also be treated using equations similar to Eqs. 7.10–7.16 by interchanging the subscripts s and m. One of the most important properties of the MIM heterostructure is its ability to support SP waves at frequencies much less than the SP frequency of a single metal–dielectric surface. MIM structures have been investigated extensively, including the fascinating properties of nanoshells.[5]

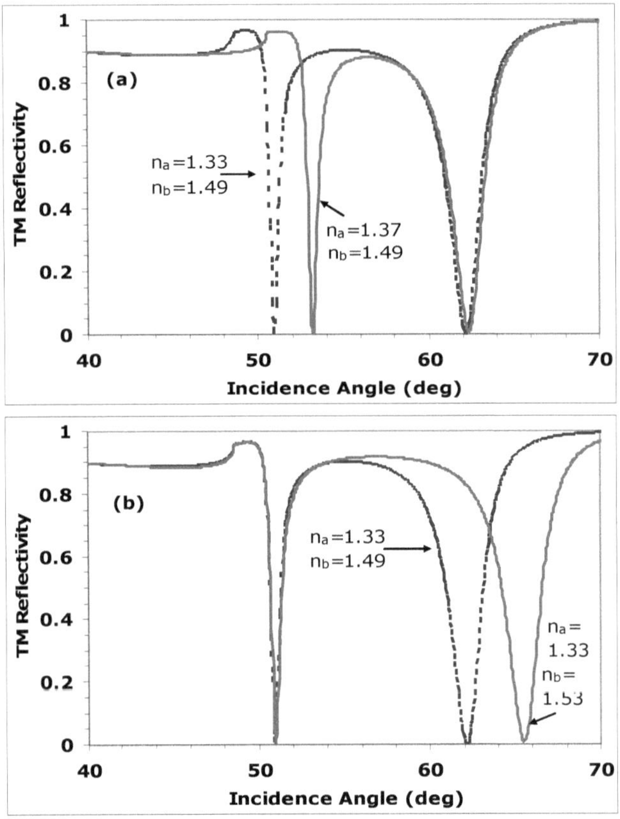

Figure 7.9 Effect of the variation of the index of the analyte n_a (a) and the buried dielectric layer n_b on the double-SPR spectrum. In each case only one SPR dip is affected but not the other, thus proving that the first dip is excited at the metal–analyte interface, while the second dip is excited on the dielectric–metal interface. Other parameters are the same as in Fig. 7.7.

7.2.4 Enhanced Transmission Through Periodic Nanoapertures in Metals

Resonant anomalies in metallic gratings are known since Wood[19] showed that EM effects can play a key role in the diffraction behavior of gratings. Fano[20] was the first to propose that some anomalies could be explained by surface waves nowadays, called plasmons, along the metallic–dielectric interface. Similar effects are observed when guided waves propagate along dielectric waveguides when the coupling is done using gratings.[21] The field of subwavelength grating waveguide couplers arose in the 1990s following the technological ability to produce them easily, characterized by the fact that only specular orders can propagate in the cladding and the substrate.[22] These gratings possess the remarkable property to have reflection and transmission variables from 0% to 100% within a fraction of the incident wavelength or angle-of-incidence variation.

In 1998 Ebbesen *et al.*,[23] presented experimental results showing that periodic arrays of subwavelength holes in a metal film can exhibit EOT spectra with peak maxima corresponding to a surprising amount of transmission in comparison with expectations based on Bethe–Bouwkamp (BB) theory.[24,25] BB theory is an analytical treatment of light transmission by a hole in a perfect metal that cannot support SP excitations. According to this theory, for apertures with a radius much less than the wavelength ($\alpha \ll \lambda$), the total transmission is given by:

$$T = \frac{1024\pi^2}{27}\left(\frac{a}{\lambda}\right)^4 \tag{7.15}$$

This formula predicts very little transmission as it exhibits the behavior of Rayleigh-type scattering from a small particle with the $(\alpha/\lambda)^4$ behavior. BB theory relies on two main approximations, (i) negligible thickness of the metal sheet and (ii) an opaque screen. Of course when the sheet is finite the solution requires rigorous simulations, which then can reveal wave-guiding properties, depending on the thickness and the diameter-to-wavelength ratio. This was proved by several investigators on different shapes.[26–28] The second approximation is problematic because some transparency to the metal has to be included (finite conductivity) when considering transmission properties. This adds the effects of SPs on the transmission.

Recently it was shown by Gordon[29] that in a periodic array of nanoapertures in a metal film, BB theory predicts almost 100% transmission at certain wavelengths. Hence SP excitation of SPs is not necessarily required for the existence of EOT[30,31] as cavity modes or transmission electron microscopy (TEM)-type modes[32–35] are possible to give EOT or even mixed modes.[36]

This EOT phenomenon is also very sensitive to the nature of any substrate placed over the holes. A simple theoretical model was presented[37] recently to study the effect of a substrate on the resonance of an aperture in a thin metal film. The transmitted energy through an aperture is shown to be governed by the coupling of the aperture waveguide mode to the incoming and outgoing EM waves into the substrate region. Transmission of polarized light through subwavelength slit apertures was studied[38] in the Vis- and NIR-range wavelengths. The authors examined the roles played by the slit apertures, such as length, depth, period, and number of slits. Birefringence aspects from plasmonic nanoslit arrays were also studied.[39] The phase of the transverse electric (TE) wave shifts ahead because of its low propagation constant. On the other hand, the phase of the TM wave is retarded due to the propagation of SPs. The opposite phase shift forms a giant birefringence.

In the case of a metal film with a periodic array of nanoholes, light transmission much larger than that predicted by classical diffraction theory has been achieved. The enhanced transmission can be larger than normally expected through a large number of nanoholes, suggesting that even the photons impinging between the nanoholes can be transmitted. The incident light is diffracted by the nanoholes to produce evanescent waves.[40] The evanescent waves diffract while tunneling through the nanoholes, resulting in their interference with the incident waves. Although the exact origin of the EOT is not fully understood and is still under investigation, the accepted view is that SPs enhance the field associated with the evanescent waves, thereby resulting in the enhancement of the transmitted light intensity.[41] So far two main types of EOT peaks have been identified—plasmon-type EOT and cavity mode– or TEM mode–type resonant EOT. SP wave–type EOT peaks appear only with TM waves; however, the cavity mode–type EOT peaks can appear also with TE polarization. As an example we consider the case of a 1D periodic array of nanoslits as, shown in Fig. 7.10a. Figure 7.10b demonstrates the EOT in TM polarization but not in TE polarization. Experimentally measured transmission spectra confirm the existence of the two peaks, as shown in Fig. 7.10c, together with the theoretical fit. For details on the experimental measurements, the reader is referred to our previous work.[42-44]

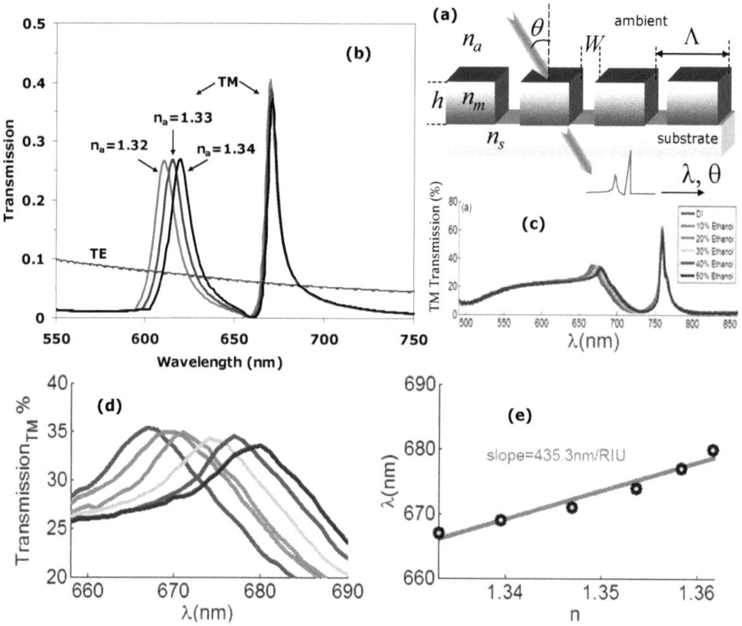

Figure 7.10 (a) Cross section of a periodic 1D array of metal nanoslits between an ambient and a substrate. (b) Theoretically calculated transmittance at normal incidence through nanoslits in silver on a glass substrate. Light is incident from an ambient with a variable index, as shown. The slits are 45 nm wide (W), the thickness (h) is 45 nm, and the period (Λ) is 450 nm. (c) Experimental transmission spectra when the fabricated silver-based grating is on an SiO_2 substrate with Λ = 500 nm, h = 45 nm, and W = 45 nm when the superstrate is deionized water and the grating is covered with a PMMA layer of thickness 15 nm. (d) Zoom on the lower EOT peak at different analyte refractive indices obtained by dissolving ethanol in water. (e) Plot of the wavelength of the EOT peak vs. the refractive index showing sensitivity of 435 nm/RIU. *Abbreviation*: RIU, refractive index unit. See also Color Insert.

The EM simulations of periodic structures are performed rigorously these days using a variety of techniques. The problem is solved numerically using several numerical approaches such as rigorous coupled wave analysis (RCWA) theory,[45–50] the C-method,[51] the Fourier factorization approach,[52] the fast Fourier factorization approach,[53] the finite difference time domain (FDTD) method,[54] the eigen waves or the exact modal method,[55,56] the S- and R-matrix methods,[57] and the scattering matrix approach.[58] Figure 7.10b shows transmission spectra through an array of nanoslits in a silver film on a glass substrate, showing two kinds of EOT peaks for TM polarization but no signs of EOT for TE polarization. The experimental verification is shown in Fig. 7.10c for different analytes. The first EOT peak around 615 nm is sensitive to variations in the ambient index (as seen in Fig. 7.10d), while the second peak around 670 nm is not sensitive to the ambient index. It was shown lately by Karabchevsky *et al.*[59] that this latter peak is sensitive to the substrate index variations, hence proposing that two SPs are involved, one on the ambient–metal interface and one on the metal–substrate interface. The presence of the grating provides the additional momentum $G = 2\pi m/\Lambda$ (m is the diffraction order) necessary to excite the SPs and the wave vector–matching condition $k_x + G = k_{SP}$ leads to:

$$\frac{2\pi}{\lambda} n_a \sin\theta \pm \frac{2\lambda}{\Lambda} m = \frac{2\pi}{\lambda}\left(\frac{n_m^2 n_{a,s}^2}{n_m^2 + n_{a,s}^2}\right)^{1/2} \tag{7.16}$$

where θ is the incidence angle, m is the diffraction order, and n_m is the real part of the refractive index of the metal. On the right-hand side n_a and n_s are used depending on whether the SP is excited on the ambient–metal interface or on the metal–substrate interface. At normal incidence the EOT peaks are then expected to follow this relation:

$$\lambda_m = \frac{\Lambda}{m}\left(\frac{n_m^2 n_{a,s}^2}{n_m^2 + n_{a,s}^2}\right)^{1/2} \tag{7.17}$$

Karabchevsky *et al.*[43] have shown that, indeed, the two EOT peaks follow this relation, where for the first one n_a has to be used, while for the second peak n_s needs to be used, as shown in Fig. 7.11. In Fig. 7.11 the resonance locations for four plasmon peaks, P_{11}, P_{12}, P_{21}, and P_{22}, are seen. It was shown also that the Eq. 7.17 holds for higher order modes $m > 1$ as well. Hence each EOT peak may be designated by the symbol P_{mj}, where m stands for the diffraction order and $j = a, s$ or $j = 1, 2$ stands for the ambient or substrate. To demonstrate this in Fig. 7.12 the shift of the second peak P_{1s} with the substrate index is shown.

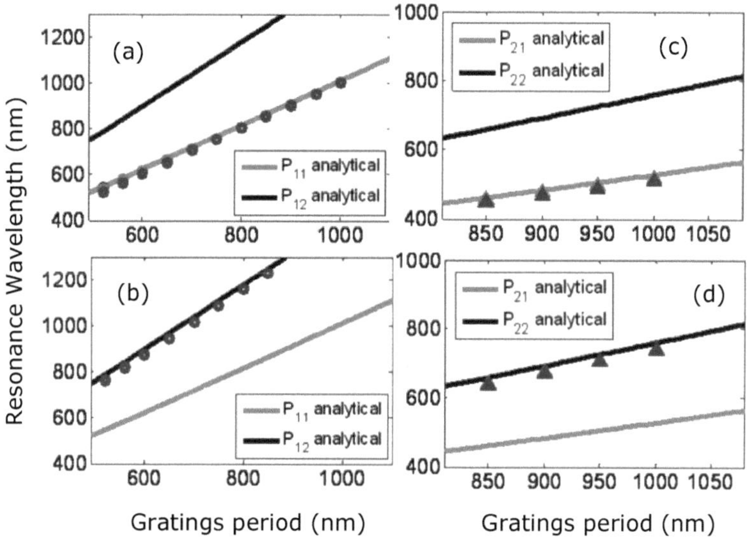

Figure 7.11 Resonance wavelength locations of first (a) and (b) and second (c) and (d) diffraction orders calculated from the rigorous simulations (circles and triangles, respectively) and the analytic expression (lines) in Eq. 7.17 vs. the grating period Λ. P_{mj} stands for the plasmon peak, with indices corresponding to the diffraction order m and the interface of excitation j (j = 1 for supertsrate, and j = 2 for substrate). Four sets of values for width, W, and depth, h, were used in the rigorous simulations to show that there was no significant dependence on the width and depth: $\{W, h\} = \{(50, 60), (60, 50), (50, 75), (70, 100)\}$ (all in nanometers). See also Color Insert.

Since Eq. 7.17 does not include metal height and slit width, resonance locations for the same periods but of different metal heights (h) and slit widths (W) were checked numerically to show that there was no significant dependence on W and h (for the thin metal structure). In Fig. 7.11a, the agreement between the rigorous results and the analytic results based on Eq. 7.17 using n_a is almost perfect for P_{11}. This is the proof that P_{11} corresponds to extended SPR excitation at the *metal-superstrate* interface. In doing a similar procedure using n_s (Fig. 7.11b), the opposite behavior is observed for P_{12}; that is, the agreement is nearly perfect for P_{12}, and there is no agreement for P_{11}. This is clear evidence that two extended SPR excitations are possible on the two sides of the metal grating for the same order of diffraction. This is the proof that P_{12} corresponds to extended SPR excitation at the metal–substrate interface. Moreover, this double-SPR excitation is occurring for each order of diffraction, as can be seen in Fig. 7.11c,d for the second order of diffraction. The small disagreement between the exact numerical result and the analytic calculation based on Eq. 7.17 is a result of the fact that the propagation length for extended SP is larger than the width of the metal lines. This means the plasmon is not exactly extended but somehow is localized or limited by the width of the metal lines. Therefore we called this type of plasmons periodically limited SPs (PLSPs). As the slit width decreases, the agreement becomes better, as it was confirmed lately by Karabchevsky *et al.*,[59] indicating that there is an interaction between the SPs excited at each neighboring metal line that becomes stronger as they come closer to each other. For each period there is an optimum slit width for the maximum EOT. On the other hand as the space width increases, the SPs become more localized since the width of the metal line is much smaller than the propagation length of an extended SP wave in silver. The role of LSPs in EOT, although small, was considered by Abajo *et al.*[60] and Chang *et al.*[61]

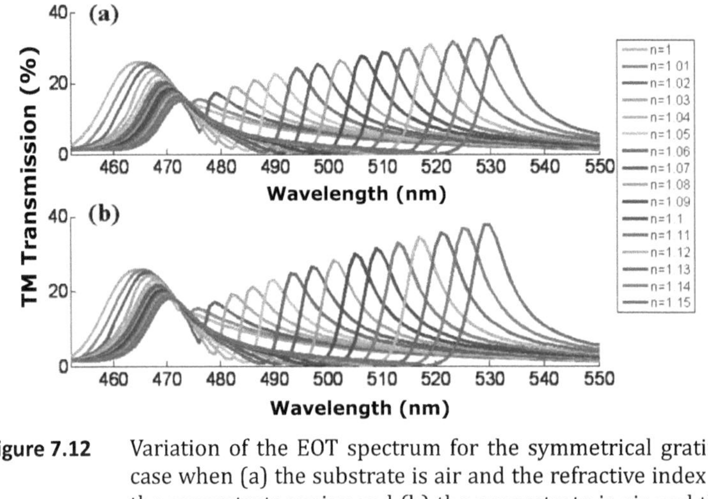

Figure 7.12 Variation of the EOT spectrum for the symmetrical grating case when (a) the substrate is air and the refractive index of the superstrate varies and (b) the superstrate is air and the refractive index of the substrate varies. Grating parameters are $\Lambda = 450$ nm, $W = 45$ nm, and metal thickness = 45 nm. See also Color Insert.

The field distribution is important to show as field enhancement has another importance for surface-enhanced spectroscopies and other properties of nanophotonic devices. Figure 7.13 shows the *y* component field distribution over a single period of a structure with a 650 nm pitch, 60 nm space width, and 50 nm silver metal height.

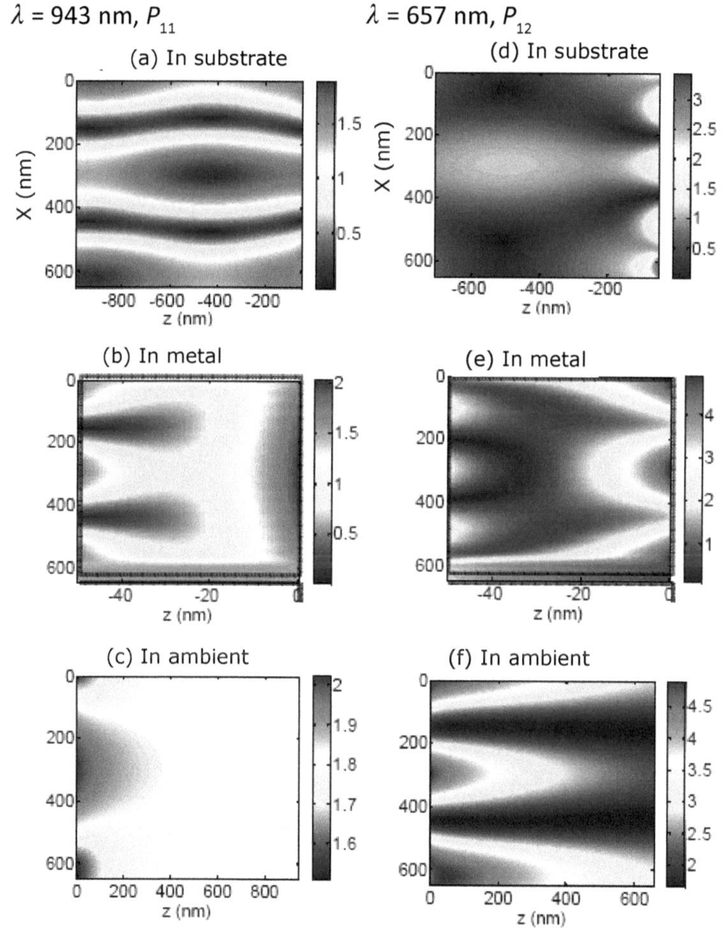

Figure 7.13 Field distribution (Hy) in the ambient, metal, and substrate calculated for the structure with the following parameters: ambient—air; grating—silver lines of 50 nm height, 60 nm space, and 650 nm period on an SiO_2 substrate. (a–c) For P_{11}, λ = 943 nm and (d–f) P_{12}, λ = 657 nm. For (a–c) most of the energy is in the ambient, as expected, while for (d–f) a large part of the energy is in the space between the lines.

The fact that the metal film has only a 45 nm thickness plays an important role in the generation of double-SPR waves, just the same as in the PSPR case when the metal film is finite in the IMI configuration. For a large metal thickness in the PSPR case only, one SPR exists, while the other decays strongly; however, for the EOT case, the penetration of the field through the nanoapertures can excite the second SP wave on the metal–substrate interface. On the other hand the thick metal starts to act as a cavity, and additional EOT peaks appear, known as cavity-mode resonances. In Fig. 7.14, simulations are presented on the same structure used to generate Figs. 7.11 and 7.12, except that the metal height is now 200 nm. As we can see from this comparison between the two figures, the two same plasmon modes appearing in Fig. 7.12 are not affected significantly in their location; however, their height is smaller, which is understood as a result of the stronger absorption in thick metal.

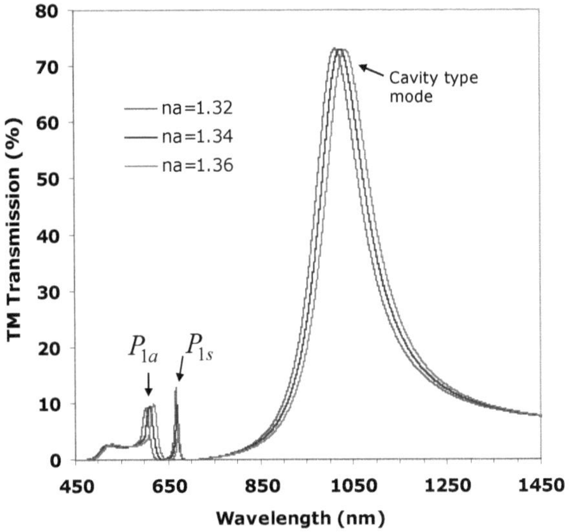

Figure 7.14 Calculated EOT spectrum from the same structure used to generate Figs. 7.11 and 7.12, except that the silver height is now 200 nm. See also Color Insert.

However, in addition to the plasmonic peaks, there is an additional EOT peak in Fig. 7.14, appearing at around 1,030 nm, which is interpreted as a cavity mode. Since the spaces are filled with a material having the same index as the ambient, varying the index of the ambient varies the peak location, following the relation $\lambda_{res} = 2hn_{eff}/m$. This relation originates from the Fabry–Perot transmission resonance condition, with n_{eff} being some effective index and m being the interference order. When the Fresnel phase is taken into account, we get:

$$\lambda_{res} = \frac{4\pi h n_{eff}}{2\pi m - \phi_{Fresnel}} \tag{7.18}$$

where $\phi_{Fresnel} = \tan^{-1}\left\{\frac{\text{Im}(r_s)}{\text{Re}(r_{ms})}\right\}$ is the Fresnel phase at normal incidence and $r_{ms} = \frac{n_s - n_m}{n_s - n_m}$ is the reflectivity amplitude from the metal–substrate interface. To confirm this further in Fig. 7.15 the EOT spectrum is calculated for different grating heights, showing that (i) the peak height is oscillatory and (ii) the peak location shifts linearly with the grating height. These two facts confirm the proposition that the peak is due to the cavity-type mode. Several approaches have been proposed to calculate the effective index. One of them uses the waveguide dispersion relation, considering the space as a waveguide medium between two semi-infinite metallic plates:

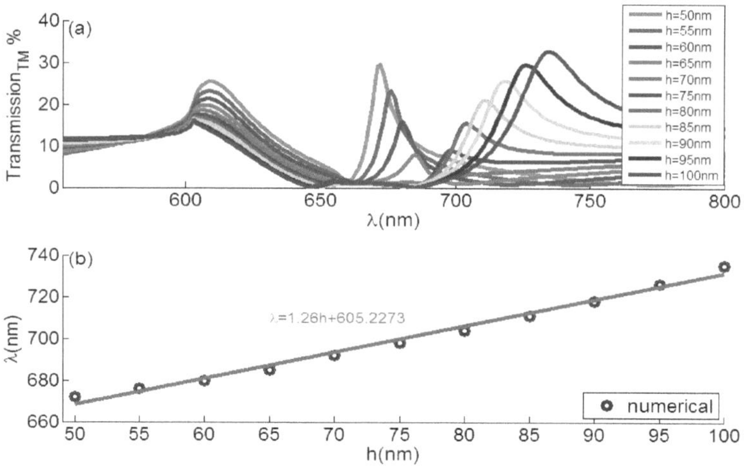

Figure 7.15 (a) Calculated EOT spectrum for a silver grating structure on an SiO_2 substrate having a period of 450 nm, a 70 nm space width, and an ambient index of 1.34 for different grating heights. The longest wavelength peak is attributed as a cavity mode exhibits oscillatory behavior with the height, as expected from a cavity-type peak. (b) Cavity-mode peak wavelength vs. the grating height. See also Color Insert.

$$\tan h\left(k_0 W\sqrt{n_{\text{eff}}^2 - \varepsilon_{\text{a}}}\right) = \frac{-2\varepsilon_{\text{a}}\sqrt{n_{\text{eff}}^2 - \varepsilon_{\text{m}}}}{\varepsilon_{\text{m}}\sqrt{n_{\text{eff}}^2 - \varepsilon_{\text{a}}}} \tag{7.19}$$

In case the space width is much smaller than the wavelength, Eq. 7.20 can be approximated to the following form:

$$n_{\text{eff}}^2 = n_{\text{a}}^2 + \frac{1}{2}\eta^2\left(1 + \sqrt{1 + \frac{4}{\eta^2}\left(n_{\text{a}}^2 - n_{\text{m}}^2\right)}\right) \tag{7.20}$$

where $\eta = 2n_{\text{a}}^2/k_0 w n_{\text{m}}^2$. For the parameters of the structure we used, the values of the effective index are $n_{\text{eff}} \approx 2 - 1.5$ in the spectral range 500–1,000 nm. The Fresnel phase at 1,000 nm is $\varepsilon_{\text{Fresnel}} \approx 0.5\text{rad}$; hence based on Eq. 7.18 and the slope of Fig. 7.15b, we get $n_{\text{eff}} \approx 2.3$. The discrepancy between the two values of the effective index is a result of two facts. One is that the mode is not exactly a planar waveguide mode, so Eq. 7.20 is not completely valid, and second, the Fresnel phase calculated for a plane wave is an approximation as the shape of the mode needs to be considered.

A note should be added on the sensitivity enhancement of EOT-based sensors. The sensitivity of the sensor is measured in nanometers per RUI (nm/RIU), defined as the slope of the variation λ_{SP} with the analyte index $n_{\text{a}} = \sqrt{\varepsilon_{\text{a}}}$, and can be derived from Eq. 7.18 in a manner similar to the grating coupled PSPR case:

$$S_\lambda = \frac{\lambda_{\text{SP}}}{\sqrt{\varepsilon_{\text{a}}}}\text{Re}\left\{\frac{\varepsilon_{\text{m}}}{(\varepsilon_{\text{m}} + \varepsilon_{\text{a}}) - 0.5\lambda_{\text{SP}}\frac{\varepsilon_{\text{a}}}{\varepsilon_{\text{m}}}\frac{\partial\varepsilon_{\text{m}}}{\partial\lambda} - 0.5\lambda_{\text{SP}}\frac{\varepsilon_{\text{m}}}{\varepsilon_{\text{a}}}\frac{\partial\varepsilon_{\text{a}}}{\partial\lambda}}\right\} \tag{7.21}$$

Hence only when the metal and the analyte dispersions are ignored, the sensitivity scales linearly with the structure period Λ and the wavelength. Since usually in the IR region $|\varepsilon_{\text{mr}}| \gg \varepsilon_{\text{a}}$, Eq. 7.21 leads to $S_\lambda \approx \lambda_{\text{SP}}/n_{\text{a}}$. Hence larger sensitivities can be obtained by designing EOT sensors with resonances in the IR range, as one can see from Fig. 7.16.

An experimental demonstration for water pollution sensing in the IR was recently presented by Krasnykov *et al.*[44]

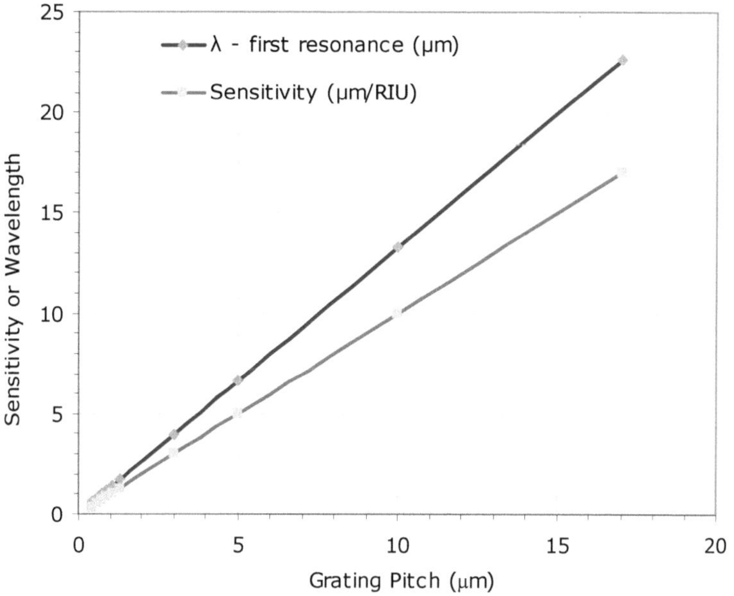

Figure 7.16 Simulation results: The blue dots represent the sensitivity vs. grating period; the pink squares represent the location of the first resonance. The structure is similar to Fig. 7.10b, but the values for the space width W and the metal thickness h were varied to obtain the maximum peak height as follows: (1) for pitches in the range Λ = 450–3,000 nm; W = 45 nm, h = 45 nm; (2) for Λ = 5,000 nm, 10,000 nm, and 17,000 nm, the width is W = 100 nm, while the thickness has to be reduced to h = 5 nm, 1 nm, and 0.5 nm, respectively. The PMMA layer thickness is 20 nm for all pitches up to 5,000 nm, while it was reduced to 5 nm for pitches of 10,000 nm and 17,000 nm. See also Color Insert.

7.2.5 Localized Surface Plasmons

The propagation length for SP waves, as it was shown in Fig. 7.6, depends on the wavelength and ranges from a few microns in the Vis range to tens of microns for the NIR. Therefore when the metal is composed of subwavelength structural units, such as NPs, or contains nanoholes, arrays of lines, etc., the SP wave cannot propagate more than the size of this structural unit; hence it becomes localized. The subject was discussed first in the context of SP waves at rough surfaces, and in the past two decades interest in this types of plasmons emerged because of the arising technological abilities of producing a variety of nanostructures as particles in solutions, on substrates or as single particles having different shapes, or as small apertures or patterns in metal films.

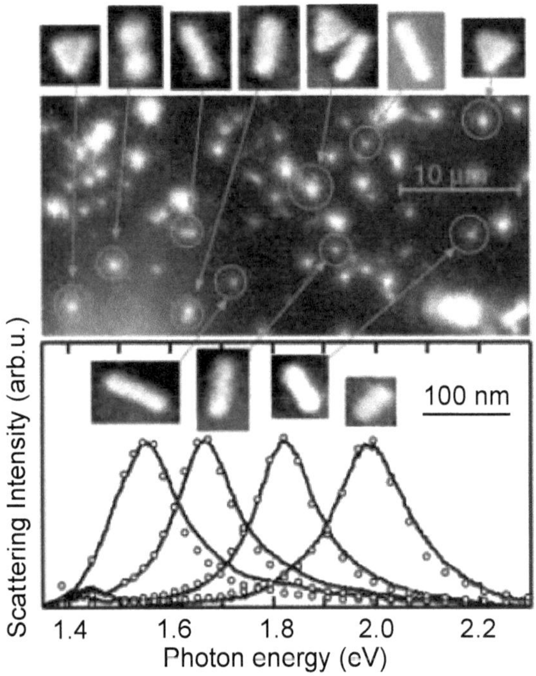

Figure 7.17 An optical dark-field image of scattered light from Au NPs, together with SEM images of the corresponding particles. Bottom: Scattering spectra from Au nanorods when the incident light is polarized parallel to the long axis; lines are experimental data, and circles are cross sections calculated by an empirical formula. Each spectrum is appropriately scaled. A higher aspect ratio gives lower resonant energy. (Reproduced from Fig. 7.2 of Kuwata *et al.*[65]) *Ab evù tò n*: SEM, scanning electron microscopy.

Optical studies of single-metal NPs started in 1903, when Zsigmondy developed what is called the "ultramicroscope" to study colloids.[62] The instrument illuminated particles in solution with an intense beam of light, and light scattered off the particles in the perpendicular direction was collected using an objective lens. Individual particles could be seen as bright flashes against a dark background as they diffused through the illuminating beam.

This technique eventually led to modern dark-field microscopy. Dark-field illumination and wavelength-resolved detection of scattered light allow for the routine measurement of the scattering spectra of single-metal NPs[63,64] (Fig. 7.17).

7.2.5.1 LSPR in isotropic, isolated metal nanoparticles: spheres

The simplest case is a metallic sphere of radius R embedded in a medium with dielectric constant ε_d. Since the magnitude of the electric field seems static around the NP in the limit of $R < \lambda$, Maxwell's equations can be solved in the quasistatic approximation. The solution of Laplace equation for the static potential inside the sphere is:[66,67]

$$\varphi(r,\theta,\phi)=\sum_{l=0}^{\infty}\sum_{m=-l}^{l}a_{lm}r^{l}Y_{lm}(\theta,\phi), \qquad 0\leq r\leq R \qquad (7.22)$$

where r, θ and ϕ are the radial, polar, and azimuthal coordinates, respectively, and $Y_{lm}(\theta,\phi)$ is the spherical harmonic function. Outside the sphere the solution is:

$$\varphi(r,\theta,\phi)=\sum_{l=0}^{\infty}\sum_{m=-l}^{l}b_{lm}\frac{1}{r^{l+1}}Y_{lm}(\theta,\phi), \qquad r\geq R \qquad (7.23)$$

The solutions in Eqs. 7.22 and 7.23 vanish at the center of the sphere and at infinity, respectively. The continuity condition on the surface for the potential and its derivative $\varepsilon\partial\phi/\partial r$ leads to the following dispersion relation:

$$\frac{\varepsilon_{\mathrm{m}}}{\varepsilon_{\mathrm{d}}}+\frac{l+1}{l}=0 \qquad (7.24)$$

Using the Drude form for metal dispersion, we get:

$$\omega_l=\omega_{\mathrm{p}}\left[\frac{l}{\varepsilon_{\mathrm{d}}(l+1)+l}\right]^{1/2} \qquad (7.25)$$

where l is a positive integer. The case with the lowest-order mode $l = 1$ describes the dipole active mode, which is the most significant for small spheres. On the other hand, for the case of a large sphere, higher-order modes are significant, that is, $l \rightarrow \infty$, and the resonance frequency becomes $\omega_\infty = \omega_p / \sqrt{1+\varepsilon_d}$, which is just the SP frequency for the case of a semi-infinite metal–dielectric interface. For the lowest-order case $l = 1$, the form of the EM field outside the sphere will be:

$$E_{\text{out}}(x,y,z) = E_0\hat{z} - \left[\frac{\varepsilon_m - \varepsilon_d}{\varepsilon_m + 2\varepsilon_d}\right] \cdot a^3 E_0 \cdot \left[\frac{\hat{z}}{r^3} - \frac{3z}{r^5}(x\hat{x} + y\hat{y} + z\hat{z})\right] \quad (7.26)$$

where E_0 is the applied field magnitude, which is polarized in the z direction in this case. The field outside the sphere is a superposition of the applied field and the dipole field produced in the center of the sphere. The EM field displaces the free-electron cloud and produces uncompensated charges near the particle surface and corresponding opposing forces (as in Fig. 7.18). The optical resonance related to these oscillations is called LSPR. The origin of the term "surface" is from the knowledge that the oscillations are caused by the polarization of the particle surface and because the generated electric field is larger near the particle surface and decays far from the surface, which makes it similar to the evanescent field at flat interface in the propagating SPR case. In the general case, the frequency of the collective oscillations does not coincide with the applied wave frequency and is determined by many factors: the electron concentration, electron effective mass, particle shape and size, interaction between particles, and influence of the environment. However, for the elementary description of the NPs' plasmon resonance, it is sufficient to use the usual dipole approximation and the Drude model.

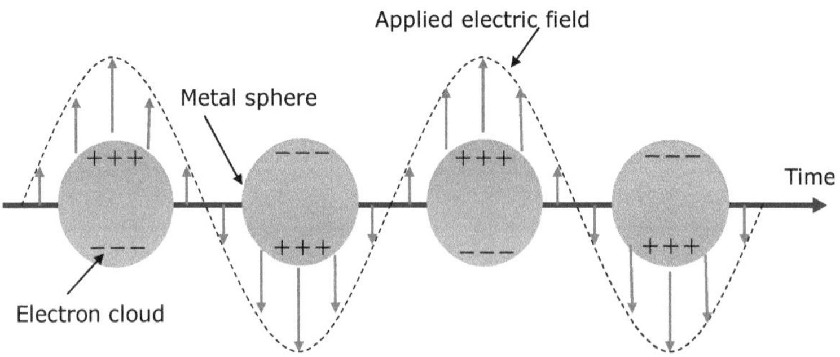

Figure 7.18 Schematic description of free-electron oscillations at the surface of a nanosphere because of an applied electric field with arbitrary polarization.

Within the dipole approximation the absorption and scattering of light by a small particle are determined by the electrostatic polarizability of the particle α_0, which can be calculated using the metal optical permittivity $\varepsilon_m(\omega)$, the medium dielectric constant ε_d, and the particle geometrical dimensions.[68] From Eq. 7.26 the polarizability is given by:

$$\alpha_0 = \frac{3V}{4\pi}\left(\frac{\varepsilon_m - \varepsilon_d}{\varepsilon_m + 2\varepsilon_d}\right) = a^3\left(\frac{\varepsilon_m - \varepsilon_d}{\varepsilon_m + 2\varepsilon_d}\right) \tag{7.27}$$

Considering the renormalized polarizability to be consistent with the energy conservation law, the integrated absorption and scattering cross sections are calculated on the basis of Mie scattering theory to be:

$$C_{ext} = C_{abs} + C_{sca} = \frac{12\pi k\, \varepsilon_d \operatorname{Im}(\varepsilon_m)}{a^3 \left|\varepsilon_m - \varepsilon_d\right|^2} \cdot |\alpha|^2 + \frac{8\pi}{3}k^4\alpha^2 \cong 4\pi k \operatorname{Im}(\alpha) \tag{7.28}$$

where $k = 2\pi\sqrt{\varepsilon_d}/\lambda$ is the wave number in the medium and the static and renormalized polarizabilities (α_0) and (α) are related by:[69]

$$\alpha = \frac{\alpha_0}{1 + \varphi(ka)\cdot a^{-3}\cdot\alpha_0} \tag{7.29}$$

where the function $\varphi(ka) \cong -(ka)^2 - 2i(ka)^3/3$ takes into account the radiative decay. The constants C_{abs}, C_{sca}, and C_{ext} are the absorption, scattering, and total cross sections, respectively. The results above are calculated with Mie scattering formalism for spherical particles. Next we shall distinguish between static polarizability and renormalized polarizability. In the static case the extinction cross section will be[70] $C_{ext} \cong 4\pi k \operatorname{Im}(\alpha = \alpha_0)$. One can see from the latter expression that the extinction spectrum has a strong resonance when:

$$\varepsilon_m(\omega = \omega_{res}) = -2\varepsilon_d \tag{7.30}$$

Although the well-known Drude model describes the permittivity of bulky metal, we will use it for estimating the resonance frequency:

$$\varepsilon_m(\omega)\, \varepsilon_\infty - \frac{\omega_p^2}{\omega(\omega + i\Gamma_b)} \tag{7.31}$$

where ε_∞ is the contribution of the interband electronic transitions and γ_b is the volume decay constant. Unlike the case of extended SPR, here we consider the exact Drude model because the metal permittivity has a main role in determining the LSPR frequency. Substituting Eq. 7.31 into Eq. 7.30 with the assumption that the significant contribution to the extinction spectrum is caused by absorption, we can express the effective absorption cross section by the resonance frequency and the metal characteristics:

$$C_{abs} = \pi a^2 \frac{12ka\varepsilon_d}{(2\varepsilon_d + \varepsilon_\infty^2)} \cdot \frac{\omega_p^2(\Gamma_b}{\omega) (\omega/\Gamma_b)^2/(\omega^2 + \Gamma_b^2 - \omega_{res}^2) + \omega_{res}^4 \Gamma_b^2/\omega^2} \tag{7.32}$$

where the resonance frequency ω_{res} and the corresponding resonance wavelength λ_{res} are given by:

$$\omega_{res} = \frac{\omega_p}{\sqrt{\varepsilon_\infty + 2\varepsilon_d}}; \quad \lambda_{res} = \lambda_p \cdot \sqrt{\varepsilon_\infty + 2\varepsilon_d} \tag{7.33}$$

where λ_p is the plasma wavelength $\lambda_p = 2\pi c/\omega_p$. It was demonstrated[71,72] that near the resonance, the absorption spectrum can be approximately reduced to a Lorenzian profile for the purpose of comparison between small metallic NPs' permittivity and the same metallic bulk permittivity.

Since the Drude model mainly uses Maxwell's equations and the boundary conditions, it is considered a classical approach to describe the permittivity function and its dependence on the angular frequency. This model is usually used for bulk samples, and it becomes problematic for very small particles. In this context, many studies were performed on adjusting the classical model to account for the small size of the NPs. A correction function was added to the permittivity function, which considers the dependence on the particle size.[73–75] A comprehensive study was done recently in several review articles.[76,77] In addition for slightly larger spheres when the Rayleigh approximation is not valid, some corrections are needed for the polarizability. Kuwata *et al.*[65] obtained the following expression:

$$\alpha_{Mie} = \frac{1 - (1/10)(\varepsilon_m + \varepsilon_d)x^2 + O(x^4)}{(1/3 + \varepsilon_d/(\varepsilon_m - \varepsilon_d)) - (\varepsilon_m - 10\varepsilon_d)x^2/30 - i4\pi^2\varepsilon_d^{3/2}\, V/3\lambda^3 + O(x^2)} \tag{7.34}$$

where $x = \pi a/\lambda$ is the size parameter.

7.2.5.2 LSPR in anisotropic metal nanoparticles

So far we have considered the case of small metallic spheres. In the general case LSPR can be excited in many other structures and shapes.[78] In most cases metal particles are not isolated and not purely spherical. The fact that metallic particles aggregate and form large and more complicated structures does not change their strong absorption and scattering properties that were mentioned earlier; the difference may be in the number of absorption peaks within the frequency range. One of the most general descriptions of a smooth and regular shape is an ellipsoid with three axes, a, b, and c, ($a \geq b \geq c$). If ($b = c$), the ellipsoid becomes a prolate spheroid (Rugby ball shape), and if ($a = b$) the ellipsoid becomes an oblate spheroid (Saturn shape). In a similar manner we can define polarizability for the structure in each direction of the major axes of the ellipsoid, namely, the polarizability that an applied electric field produces depends on the polarization direction of the external applied field, and consequently the resonance wavelength is determined for each axis i when the electric field is polarized along that particular axis:

$$\alpha_i = \frac{ab}{3}\left(\frac{\varepsilon_{\mathrm{m}} - \varepsilon_{\mathrm{d}}}{L_i(\varepsilon_{\mathrm{m}} - \varepsilon_{\mathrm{d}}) + \varepsilon_{\mathrm{d}}}\right) \tag{7.35}$$

where L_i is a form factor of axis i when the electric field is polarized along this axis and α_i is the polarizability corresponding to the same axis. The geometrical factor L_i obeys the rules $0 \leq L_i \leq 1$ and $L_a + L_b + L_c = 1$ when considering the axes a, b, and c. For a sphere $L_a + L_b + L_c = 1/3$ due to spherical symmetry, and then Eq. 7.27 is satisfied. The absorption cross section when the applied field is polarized parallel to the i axis will then be $C_{\mathrm{abs}} = k\,\mathrm{Im}(\alpha_i)$, which has peaks when[79] $\mathrm{Re}(\varepsilon_{\mathrm{m}}) = \varepsilon_{\mathrm{d}}(1 - 1/L_i)$. Since $0 \leq L_i \leq 1$, the term $1/L_i$ yields a wide range of frequencies (see Figs. 7.18–7.20). Similar behavior for nanodiscs and nanorods can be obtained because they are the limits of oblate and prolate spheroids, respectively. Furthermore, one can describe the response of structures in terms of ellipsoids with various shapes. In Figs. 7.19–7.22, several examples of the absorption cross sections are shown for spheres, prolate spheroids, and oblate spheroids, respectively.

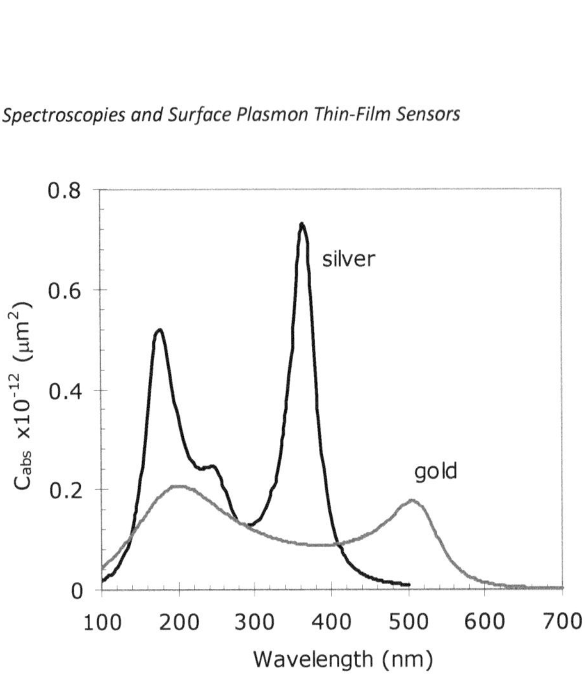

Figure 7.19 Absorption cross section for silver and gold spheres. Both silver and gold spheres have a 100 nm diameter each and are embedded in air.

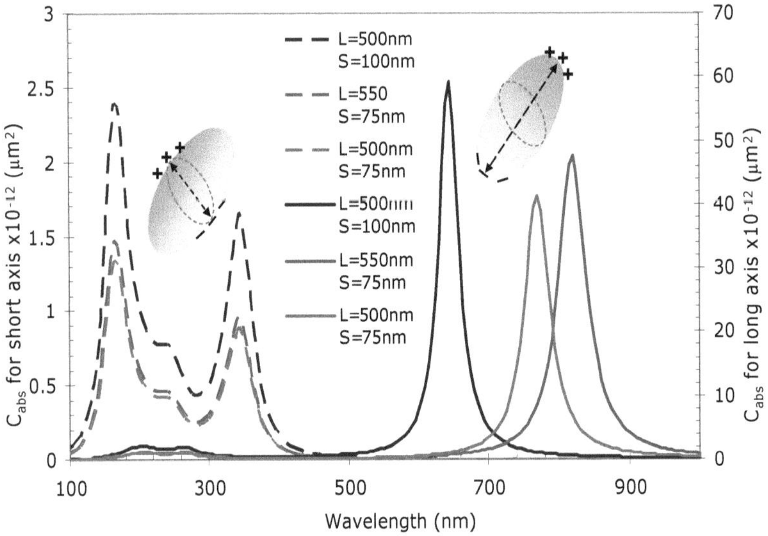

Figure 7.20 Absorption cross sections for prolate silver spheroids in air with different axes lengths. *L* indicates the long axis of the spheroid, and *S* indicates the short axis of the spheroid. For each combination of long and short axes, the absorption was calculated for the polarization both parallel to the long axis (solid curves) and parallel to the short axis (dashed curves). See also Color Insert.

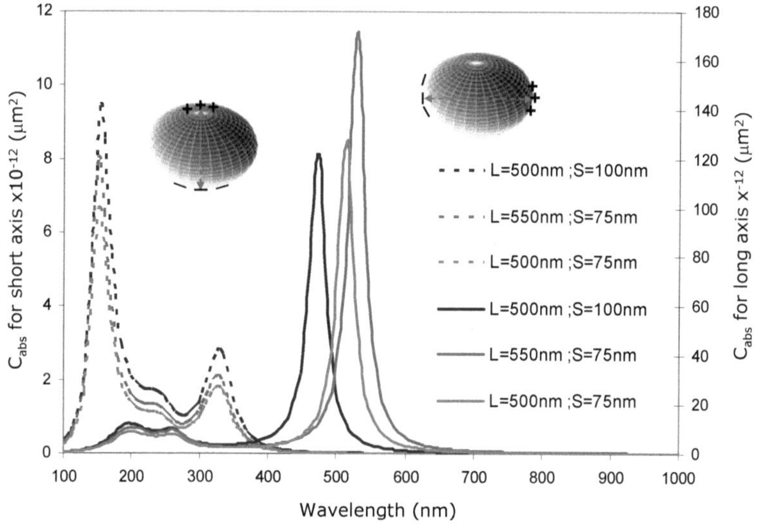

Figure 7.21 Absorption cross section for oblate silver spheroids in air. For each spheroid, the absorption was calculated for the polarization both parallel to the long axis (solid curves) and parallel to the short axis (dashed curves). The right vertical scale corresponds to the long-axis polarization, and the left vertical scale corresponds to the short-axis polarization. Other parameters are as in Fig. 7.20. See also Color Insert.

The calculation takes into account the dipolar approximation in the quasistatic limit. Although it describes qualitatively the behavior of the extinction spectra with the aspect ratio, higher-order effects, such as quadropoles, are required particularly for particles with a large size. Figure 7.22 shows the extinction spectra for different aspect ratios of Ag oblate spheroids performed using the exact Mie calculation by Kelly *et al.*[69]

By tailoring the shape and size of the NPs, one can design the desired resonance spectrum. The peak can be red-shifted or blue-shifted compared with the peak of a sphere by controlling the size and shape of the spheroid. The geometrical factor L_i can be analytically calculated for standard smooth particles like spheres, spheroids, ellipsoids, and cylinders. In the work by Stoner,[80] the geometrical factors for ellipsoids were calculated, and the elliptic integrals can be explicitly solved for spheroids. For an ellipsoid of uniform density with semiaxes a, b, and c, the principal values of the form factor are given by the general integral:

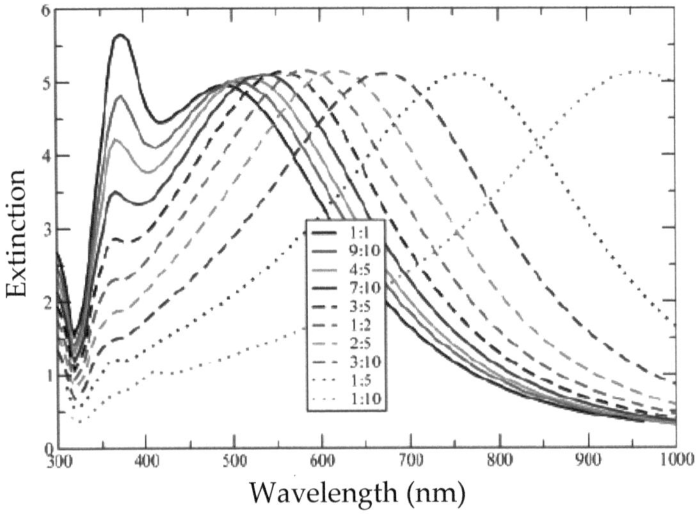

Figure 7.22 Exact electrodynamic calculation of the extinction spectra of Ag oblate spheroids, all with the same equivalent volume, corresponding to a sphere radius of 80 nm. The major-to-minor-axis ratio, *r*, is from left to right: 10, 5, 3.33, 2.5, 2, 1.67, 1.43, 1.25, 1.11, and 1. In this figure the extinction is normalized to the area of a circle with radius equal to the semimajor axis. (Reproduced from Fig. 7.5 of Kelly *et al.*[69]) See also Color Insert.

$$L_q = \frac{ab}{2}\int_0^{\infty} \frac{ds}{(x_q^2 + s)\sqrt{(a^2+s)(b^2+s)(c^2+s)}} \tag{7.36}$$

where $q = 1, 2, 3$ and $x_1 = a, x_2 = a, x_3 = a$. Normalization requires that $\sum_q L_q$. This general form of L_q for the biaxial case can be expressed in terms of elliptic integrals. In the case of a prolate spheroid with a being the major long axis and $(b = c)$ the equatorial short axes, we define $e \equiv \sqrt{1-(b/c)^2}$, and then the appropriate geometrical factors will be:

$$L_a = \frac{1-e^2}{e^2}\left[\frac{1}{2e}\ln\left(\frac{1+e}{1-e}\right) - 1\right] \text{ and } L_a = L_c = \frac{1}{2}(1 - L_a) \tag{7.37}$$

where L_a is the geometrical factor when the excitation polarization is along the major long axis, and $L_b = L_q$ are the geometrical factors when the excitation polarization is along the equatorial short axes b

and c. In the case of an oblate spheroid, when $a = b$ are the equatorial long axes and c is the short polar axis, if we define $e \equiv \sqrt{1 - (c/a)^2}$, the appropriate factors will be:

$$L_c = \frac{1}{e^2}\left[1 - \sqrt{1/e^2 - 1}\,\arcsin(e)\right] \text{ and } L_a = L_b = \frac{1}{2}(1 - L_c) \quad (7.38)$$

In addition to modeling the extinction spectrum of NPs, several equations describe how LSPR is used for sensing. For example, the LSPR extinction resonance frequency is sensitive to the medium dielectric constant ε_{d} (or refractive index, n_{d}) according to Eqs. 7.36–7.39. Thus, changes in the environment (by adding an adsorbate layer, for example) cause an appropriate shift in the resonance wavelength. The relationship between perturbations in the dielectric medium refractive index Δn_{d} and the sensitivity of the shift in the resonance wavelength Δn_{res} with respect to a change in the refractive index Δn_{d} of the adsorbate is given by:

$$S_\lambda = \frac{\Delta\lambda_{\mathrm{res}}}{\Delta n_{\mathrm{d}}} = S_{\lambda\,-\,\mathrm{bulk}}\left[1 - \exp\left(\frac{-2d}{l_{\mathrm{d}}}\right)\right] \quad (7.39)$$

where $S_{\lambda\,-\,\mathrm{bulk}}$ is the bulk refractive index sensitivity of the NPs and l_{d} is the EM field decay length. The latter equation matches most of the experiment results and forms the bases of the LSPR-sensing mechanism.

For a slightly larger anisotropic particle the Rayleigh approximation is not adequate and some correction is needed that depends on the ratio between the particle size and the wavelength. Kuwata *et al.*[65] obtained the following formula for polarizability along the principal axis j:

$$a_j \approx \frac{V}{\left(L_j + \dfrac{\varepsilon_{\mathrm{d}}}{\varepsilon_{\mathrm{dm}} - \varepsilon_{\mathrm{d}}}\right) + A_j\varepsilon_{\mathrm{d}}x^2 + B_j\varepsilon_{\mathrm{d}}^3x^2 - i\dfrac{4\pi^2 V\varepsilon_{\mathrm{d}}^{3/2}}{3\lambda^3}} \quad (7.40)$$

where $x = \pi r_j/\lambda$ and $r_j = a, b, c$, depending on along which axis the incident polarization is. From the experimental data Kuwata *et al.*[65] found the following expressions for the material-independent parameters:

$$\begin{aligned} A_j &= -0.4865L_j - 1.046L_j^2 + 0.8481L_j^3 \\ B_j &= -0.01909L_j - 0.1999L_j^2 + 0.6077L_j^3 \end{aligned} \quad (7.41)$$

Another useful expression obtained[cc] for the extinction efficiency ($Q_{ext} = C_{ext}/\pi r_{eff}^2$) up to the second order in the parameter $y = 2\pi \sqrt{\varepsilon_d} r_{eff}/\lambda$ with r_{eff} being the radius of the sphere that has the same particle volume is:

$$Q_{ext} = \frac{4x}{3} \operatorname{Im}\left\{ \frac{\varepsilon_m - \varepsilon_d}{L_i(\varepsilon_m - \varepsilon_d) + \varepsilon_d} \right\} + \frac{8y^4}{27} \left| \frac{\varepsilon_m - \varepsilon_d}{L_i(\varepsilon_m - \varepsilon_d) + \varepsilon_d} \right|^2 \tag{7.42}$$

In this expression the quadropole effect was not considered, an approximation that is justified for particle sizes less than 30 nm; however, for larger particles it has to be taken into account.[69]

7.3 Field Enhancement and Surface-Enhanced Phenomena

SPR excitation by its nature is an oscillating charge density wave localized to the surface. Hence, even with extended SPR, as we saw in Fig. 7.8, the field near the surface is enhanced. The enhancement in the case of extended SPs is by a factor of several tens up to × 100; however, because of the tight localization in the case of LSPR, field enhancement can be as high as[82,83] a factor of × 1000. In the LSPR case, according to Eq. 7.35 and as shown in Figs. 7.19–7.21, there are peaks in the absorption extinction cross section for an NP when:

$$\varepsilon_{mr} = \varepsilon_d \left(1 - \frac{1}{L_i}\right) \tag{7.43}$$

These peaks originate from the form of the polarizability that is proportional to the local EM field. Hence at the LSPR resonance there is local field enhancement. Rigorous calculations, indeed, show field enhancement by three orders of magnitude in the nanometer vicinity of the NP and then decay exponentially within $l_d \approx 10 - 100$ nm. One can visualize this phenomenon by considering the NP as localizing the electric field of a dipole field centered in the sphere, which then decays with the dipole decay law away from the surface in all directions. In this sense, the NP acts as an antenna that amplifies the intensity of the scattered light. In Fig. 7.23 field intensity distributions are shown for different cases showing that for dimmers of triangles the field enhancement is the largest. For a sphere the enhancement is less than that for anisotropic particles, prisms, and dimmers. Note the hot spots near the sharp corners.

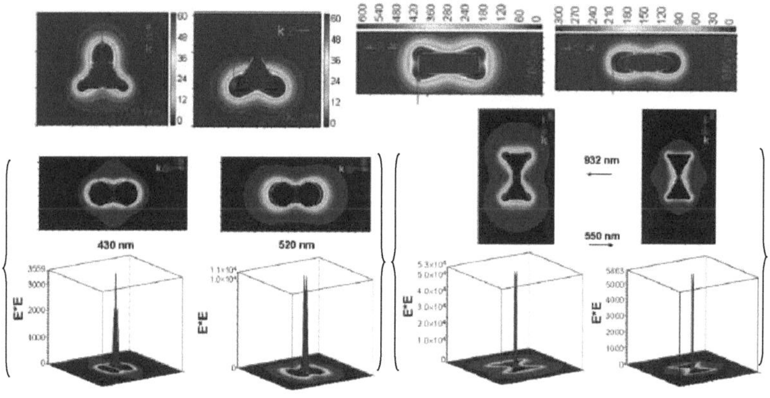

Figure 7.23 Field distribution intensities around NPs (top) and dimers of NPs. The 3D plots show the intensity distribution for the dimmers above them for different wavelengths. For the dimmers the strongest field enhancements are obtained in the center of the gap with a minimum gap of 2 nm. The wavelengths correspond to dipole and quadropole excitations, with the strongest being for the dipole excitation. (Reproduced from Figs. 7.3, 7.5, and 7.8 of Hao *et al.*[84]) *Abbreviation*: 3D, three dimensional.

These local field enhancements cause enhancements in different optical phenomena such as SHG, SEF, and SERS.[85] Since in these optical phenomena the output intensity depends on the excitation field to some power, their signal is enhanced considerably. For example, in SERS the intensity is proportional to the fourth power of the field $|E|^4$; one then can obtain enhancement on the order of 10^{12}. Fluorescence intensity is, on the other hand, proportional to $|E|^2$, so the maximum enhancement factor can be 10^6, although experiments reported much less than this factor (up to 1,000 at the most) because fluorescent dyes are chemically affected by their metallic neighborhood, in addition to bleaching, which takes place within a short time when the excitation field is enhanced.

7.3.1 SERS

SERS was originally discovered[86,87] in the 1970s when small nonresonant organic molecules, such as pyridine, were found to yield Raman signals enhanced by a factor of a million when adsorbed to a metal NP. The whole subject was revolutionized during the past two decades following emerging nanofabrication techniques and the development of surface modification protocols. The signal enhancement is so dramatic that very weak Raman peaks that are unnoticeable in spontaneous Raman spectra can appear prominently enough in SERS spectra. Some trace contaminants can also contribute additional peaks. Moreover, because of chemical interactions with metal surfaces, certain peaks that are strong in conventional Raman spectra might not be present in SERS spectra at all. The nonlinear character of signal intensity as a function of the concentration complicates things even further. Very careful consideration of all physical and chemical factors must be made while interpreting SERS spectra, which makes it extremely impractical.

Although a complete understanding of SERS has not been achieved yet, two main mechanisms are widely accepted. The first, called chemical enhancement, involves enhancement of polarizability of the analyte molecule that may occur because of a charge transfer effect or chemical bond formation between the metal surface and the analyte molecules. The second is due to the enhanced EM field produced at the surface of the metal when the wavelength of the incident light matches the SPR wavelength of the metal. Molecules adsorbed or in close proximity to the metal surface experience an exceptionally large electric field. Because the Raman effect is proportional to the fourth power of the field amplitude, the efficiency is enhanced by factors as large as 10^{14}. Molecular vibrational modes normal to the metal surface are most strongly enhanced in comparison to other vibrational modes. EM simulations confirm that the electric field can be enhanced by a factor of 10^3, and so the Raman signal is enhanced by a factor of 10^{12}.

The SPR intensity is dependent on many factors, including the wavelength of the incident light and the morphology of the metal surface. The Raman excitation wavelength should match the plasma wavelength of the metal, which is about 382 nm for a 5 μm silver particle but can be as high as 600 nm for larger, ellipsoidal silver particles. The plasma wavelength shifts to 650 nm for copper and gold, the other two metals that are used for SERS at wavelengths in the range from 350 nm to 1,000 nm. The best modality for SPR excitation is the use of either an NP (<100 nm diameter) or an atomically rough surface.

The Raman scattering from a compound (or ion) adsorbed on even within a few Ångstroms of a structured metal surface can be enhanced by a factor of 10^3 to 10^{14} compared with the case when it is in a solution. SERS is strongest on a silver surface[88,89] but is observable on gold and copper surfaces as well, and it is now known that the shape of the NP plays a crucial role in determining the enhancement factor. The triangular-pyramid shape has been found to give the strongest enhancement. SERS was discovered with pyridine. Other aromatic nitrogen- or oxygen-containing compounds, such as aromatic amines or phenols, also display strong enhancement due to SERS. The enhancement can also be seen with other electron-rich analytes, such as carboxylic acids. Although SERS allows easy observation of Raman spectra from solutions with concentration in the micromolar (10^{-6}) range, slow adsorption kinetics and competitive adsorption limit its application in analytical chemistry.

A simplified model for Raman enhancement is considering field enhancement near the surface of a single isolated particle. If we assume that the adsorbed molecules are randomly distributed on the surface of a prolate spheroid that is much smaller than the excitation wavelength and whose major semiaxis is a and equatorial semiaxis is b, the total averaged Raman enhancement EF following Schatz[90] is:

$$EF = \mathfrak{R}(\lambda_{ex}) \cdot \mathfrak{R}(\lambda_{sc}) \tag{7.44}$$

The components $\mathfrak{R}(\lambda_{ex})$ and $\mathfrak{R}(\lambda_{sc})$ refer to enhancements in the incident and the scattered fields, respectively. In doing this, we also make the simplifying approximation that the field enhancement for the Raman-emitted photon with wavelength λ_{sc} can be calculated using the same expression for the field as the incidence photon with the excitation wavelength λ_{ex}. The SERS enhancement factor is then proportional to $|E(\lambda_{ex})|^2 \cdot |E(\lambda_{ex})|^2$, which indeed was indicated by $\mathfrak{R}(\lambda_{ex})$ and $\mathfrak{R}(\lambda_{sc})$. If the spheroid is small compared with the wavelength of light, the local field $E(\lambda)$ can be obtained by solving Laplace's equation when the surrounding medium is replaced by an effective dielectric constant following Eqs. 7.37–7.39. Under the former assumptions, the field intensity enhancement at the surface is given by:

$$\mathfrak{K}(\lambda_{ex}) = |1 - g(\lambda_{ex})|^2 + \left[\frac{2\cdot \mathrm{Re}\{[1 - g(\lambda_{ex})]g^*(\lambda_{ex})\}}{Q_1(\xi_0)} + \frac{|g(\lambda_{ex})|^2}{Q_1(\xi_0)\,(\xi_0^2 - 1)}\right]$$
$$\times \left[\frac{-(\xi_0^2 - 1)^{1/2} + \xi_0^2 \sin^{-1}(1/\xi_0)}{(\xi_0^2 - 1)^{1/2} + \xi_0^2 \sin^{-1}(1/\xi_0)}\right] \qquad (7.45)$$

where the parameter $\xi_0^2 = (1 - b^2/a^2)^{-1}$ represents the aspect ratio of the spheroid (b/a) and $Q_1(\xi_0)$ is the Legendre function of the second kind, which is given by:

$$Q_1(\xi_0) = \frac{1}{2}\xi_0 \ln\left(\frac{\xi_0 + 1}{\xi_0 - 1}\right) - 1, \qquad (7.46)$$

and $g(\lambda_{ex})$ is a gain parameter, which depends on the dielectric constants of the spheroid and the medium evaluated at the excitation wavelength similar to Eq. 7.35:

$$g(\lambda_{ex}) = \frac{\varepsilon_{mi} - \varepsilon_d}{\varepsilon_{mi} + \chi_i \cdot \varepsilon_d} \qquad (7.47)$$

where $\varepsilon_{mi} = \varepsilon_{mri} + i\cdot\varepsilon_{mii}$ and ε_d are the dielectric constants of the spheroid along axis i and the surrounding medium, respectively, at the excitation wavelength and the shape-dependent parameter $\chi_i = (1/L_i) - 1$. The depolarization factor L_i is determined according to the incident field polarization state, as described in Eqs. 7.36–7.38. The component $\mathfrak{K}(\lambda_{ex})$ is calculated in the same manner as $\mathfrak{K}(\lambda_{ex})$ when the excitation wavelength is replaced by the Stokes wavelength λ_{sc}. The enhancement factor for a sphere with a radius (a) that is much smaller than the wavelength $(a \ll \lambda_{ex})$ was earlier developed by Kerker *et al.*[91] as follows:

$$EF = |(1 + 2\cdot g(\lambda_{ex}))\cdot(1 + 2\cdot g(\lambda_{sc}))|^2 \qquad (7.48)$$

The gain parameter of the sphere $g(\lambda_{ex})$ is determined as in Eq. 7.47 when the shape-dependent parameter $\chi_i = 2$ ($L_i = 1/3$ for a sphere). The enhancement appearing in Eq. 7.48 is size independent because of the first-order approximation of $g(\lambda_{ex})$ and $g(\lambda_{sc})$ used in the analysis. To extend this analysis to include small spheres with $2\pi a/\lambda_{ex}$ as large as 0.1, a higher-order approximation for the gain factors should be incorporated, and accordingly the factors $g(\lambda_{ex})$ and $g(\lambda_{sc})$ are redefined as follows:

$$g(\lambda_{ex}) = \frac{[(\varepsilon_i - \varepsilon_0) - (k_0 a)^2(\varepsilon_i^2 - 1)/10]}{[(\varepsilon_i + 2\cdot\varepsilon_0) - (k_0 a)^2(10 - 9\varepsilon_i - \varepsilon_i^2)/10]} \tag{7.49}$$

The last expression is also valid when the spheroid and the medium dielectric constants are evaluated at the scattered wavelength λ_{sc}, and the wave number k_0 is replaced by the wave number at the scattered wavelength. The expression that was given in Eq. 7.43 for spheroidal particles is also size independent and determined only by the aspect ratio of the particles. This is a result of the electrostatic limit approximation, which holds for small particles. When the spheroid dimensions become larger, the radiation of the induced dipole reduces the polarization induced by the applied field, an effect known as the radiation damping effect.[92] Moreover, the interference between the radiations emitted at different points on the surface of the particle should be taken into account for larger particles. The second effect is known as the dynamic depolarization effect. These two effects lead to dynamic correction to the Maxwell's equations solution in the static limit and can be expressed by means of a "modified induced dipole" as follows:

$$\mu_{ind} = \frac{\alpha_0 E}{1 - \left\{\dfrac{\alpha_0 k^2}{b} + \dfrac{2k^3\alpha_0}{3}\right\}} \tag{7.50}$$

Here E is the applied field, α_0 is the spheroid polarizability in the static limit, and the expression in the parentheses is the electrodynamics correction factor. To incorporate this correction into the Raman enhancement factor $\mathfrak{R}(\lambda)$, we multiply $\mathfrak{R}(\lambda)$ in Eq. 7.45 by the absolute square of the term in parentheses. In addition to the electrodynamics corrections for large particles, there is a size-dependent correction that is required when the particle is very small. It is known that the dielectric constant of a metal particle becomes size dependent whenever the particle is smaller than the conduction electron mean-free path.[93] For noble and alkali metals, this applies to particles that are roughly smaller than 20 nm in their smaller dimensions.[94] However, the main effect of the increase in plasmon width associated with decreased particle size is suppression of the field enhancement without changing its resonance wavelength or its shape dependence.

Figure 7.24 illustrates the effect of increasing the aspect ratio on the Raman enhancement factor shape. The calculations were carried for a silver prolate spheroid with a fixed minor semiaxis $b = 25$ nm, and the aspect ratio of the spheroid is varied through changing the major semiaxis a. One can see that increasing the aspect ratio of the spheroid causes a red-shift in the resonance wavelength and broadening in the enhancement shape. The red-shift is caused by increasing the rotational axis length, especially as the polarization state of the incidence field is parallel to this axis and the resonance wavelength is proportional to this length. The increase in the maximum enhancement, on the other hand, is due to the lightening rod effect in which the enhancement is maximized at the tips of the spheroid.

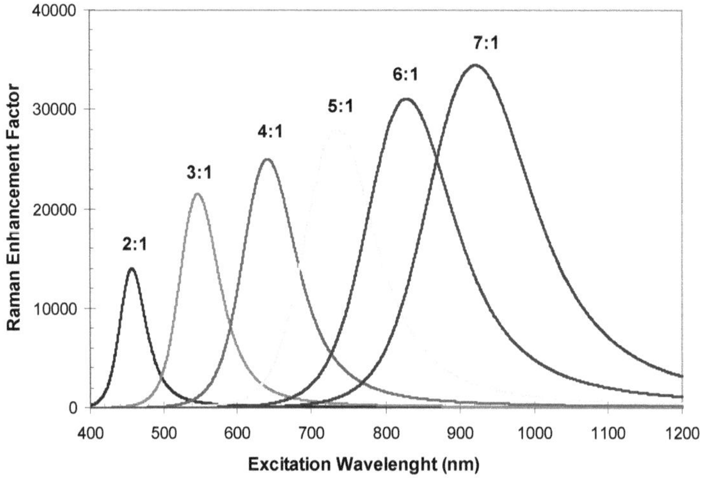

Figure 7.24 Raman enhancement factor calculated using Eqs. 7.44–7.49 when the electrodynamic correction is considered and the Raman frequency shift is 1,010 cm^{-1}. The curves correspond to a silver prolate spheroid with a fixed equatorial semiaxis equal to 25 nm and various aspect ratios. The incident field is polarized parallel to the rotational axis, and the surrounding medium is air with a dielectric constant equal to unity.

7.3.2 SEF

SEF has been known for more than a few decades now,[95] but its potential was rediscovered during the past decade because of the emerging developments in the field of optics of metallic nanostructures.[96,97] SEF is fundamentally interesting because several effects play a role in it and it can be achieved even with exciting photon energies far from the SPR. SEF is a very useful phenomenon, with significant applications in biotechnology and life sciences, alongside SERS.

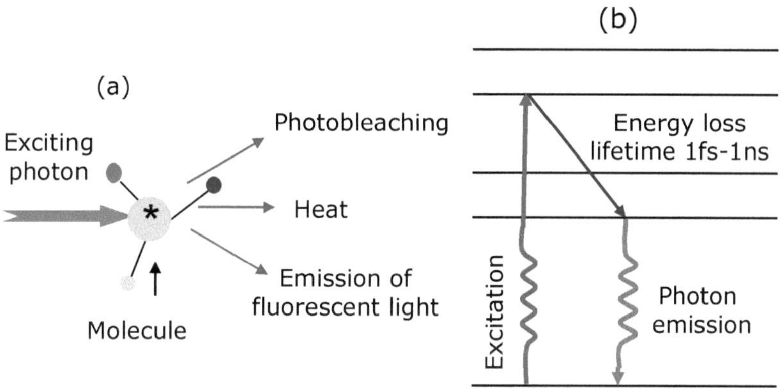

Figure 7.25 Fluorescence effect. (a) A molecule is excited by light; (b) schematic explanation of excitation and emission processes: purple: excitation from the ground state; blue: decay to the nearest energy level; and red: photon emission via returning to the ground state. See also Color Insert.

When a molecule is excited, there are a variety of processes that will return it to the ground state (Fig. 7.25). If we consider three processes for returning to the ground state (radiationless energy loss, intersystem crossing through the triplet state, and emission of photons), then the efficiency of emission will be a function of the competing rates of these processes:

$$\eta = \frac{R_f}{R_f + R_i + R_x} \tag{7.51}$$

where η is the efficiency or quantum yield, which is a dimensionless quantity given by the ratio between the emitted photon and the number of absorbed photons. The rate constants $R_{f,i,x}$ designate those for florescence emission, without any radiation energy loss, and intersystem crossing, respectively. The average lifetime of the excited state is inversely related to the fluorescence emission rate: $\tau_f = R_f^{-1}$.

SEF takes place in the proximity of metal structures. In the proximity of metals, the fluorophore radiative properties are modified and an increase in the spontaneous emission rate is observed, which is associated with a concomitant increase in the radiative quantum efficiency. This results from the shortening of the fluorescence lifetime, which enables faster cycling of the fluorophore. These are given by the following relationships:

$$\eta_{ns} = \frac{R_f + R_{ns}}{R_f + R_{ns} + R_i + R_x} \tag{7.52}$$

$$\tau_{ns} = \frac{1}{R_f + R_{ns}} \tag{7.53}$$

Here, η_{ns} is the quantum yield, modified by the nanostructure, $R_{f,i}$ are rate constants, R_{ns} is the additional rate constant induced by the nanostructure, and τ_{ns} is the fluorophore lifetime modified by the nanostructures.

SEF is a very useful phenomenon with significant applications in biotechnology and life sciences, alongside SERS. Several physicochemical effects[98–105] related to the behavior of fluorophores near NPs have been brought into focus by the emerging field of plasmonics. The effect of surfaces near an oscillating dipole on its emission properties was first investigated by Chance *et al.*,[96] which led to the development of Chance–Prock–Silbey (CPS) theory. The most important consequences of CPS theory are as follows:

(1) The reflected field from the nearby interface interferes with the emitted field and, therefore, can enhance or weaken the fluorescence, depending on the distance between the dipole and the interface.
(2) When the nearby surface is more reflective, the enhancement factor is larger.
(3) Very close to a reflective surface, the excited molecule may decay nonradiatively by coupling to lossy modes, SPs, or guided

waves. This coupling may even quench the fluorescence when the molecule is touching the interface.

(4) At small distances near a metallic surface, the lifetime decreases because of several competing relaxation processes.

(5) In SP cross emission (SPCE), one way to excite SEF, the excited SPs couple back to the radiative EM mode on the back side of a metal film[106] (Kretschmann configuration or rough metal surfaces).

On top of these effects, localization of the EM field near nano tips, corners, holes, needles, etc., has been shown to produce a large SEF by factors up to few hundreds in what is known as the lightening nanoantenna effect.[107]

The quantum efficiency of the photoluminescence (PL) from a bulk noble metal is very low (typically, 10^{-10}). However, in a similar manner to LSPR and SERS, when fluorophores (fluorescent analyte molecules) are in close proximity to metal NPs,[108] the fluorescence intensity is enhanced and the fluorescence lifetime is lowered.[109-112] These SEF effects occur because the excited fluorophores interact with freely mobile electrons in the metal, thereby resulting in increased rates of radiative decay; similar interactions with LSPR have been the subject of theoretical analysis and are closely related to the SERS mechanisms.[113-115] The effect is now known as the lightening rod effect. SEF has been demonstrated on different fluorophores and suggested as a method to enhance *in vivo* imaging.[116] Enhancement of fluorescence has also been reported on nanoscale ZnO platforms.[117] A red-shift for the PL maximum was observed with an increasing aspect ratio of the gold nanorods, and PL efficiency was found to increase with the square of the aspect ratio.[118] Rough metal surfaces can be regarded as an ensemble of randomly oriented hemispheroids of nanometer size, and SPR arising from them amplifies the local fields, thereby resulting in the enhancement of PL. It has been reported that quenching of luminescence occurs when metal NPs are in close proximity to the fluorophores (dyes or semiconductor NPs), whereas enhancement in luminescence is observed when metal NPs are located farther away from the fluorophores.[119] The fluorophore emission intensity depends on the distance between the metal and the fluorophore in a complex fashion. At very close distances (less than several nanometers), the fluorophore experiences significant nonradiative decay components, and consequently the emission is strongly quenched. At intermediate distances (e.g., for Ag around 5 nm), the nonradiative decay process subsides and the enhancement effect begins to dominate, producing the overall fluorescence amplification peak, which, for spherical metal NPs, reaches a typical value in the order of 10. For longer distances (10 nm and above), the amplification tapers off, eventually reaching unity.

7.3.3 SEIRA

SEIRA is related to SERS with a few variations in the selection rules. It was observed first by Harstein *et al.*[120] when they measured an enhancement factor of 10^4 of C–H band absorption in 4-nitrobenzoic acid on a silver thin film nanostructure on a silicon substrate. In 1997, Osawa[121] showed that it is possible to detect trace amounts using this technique, as shown in Fig. 7.26. In this experiment different enhancement factors were observed for TE- and TM-incident polarizations, thus indicating some anisotropic nature of the metal islands. *p*-nitrobenzoic acid (PNBA) was applied by dissolving it in acetone and depositing a drop onto the silver layer and allowing the solvent to evaporate. The thickness of PNBA is 1.7 nm for all samples in Fig. 7.26. With increasing mass thickness, there is an increase of the 1,390 cm^{-1} and the 1,350 cm^{-1} peaks, with a maximum enhancement at a thickness of 10 nm. The fact that not all of the peaks are shown in the SEIRA measurement will be discussed later. The 1,350 cm^{-1} band, assigned as the symmetric NO_2 stretch, shows an enhancement of about 500 for the 10 nm silver film. Yet at the 14 nm silver film, a decrease in the enhancement is observed. So there is a definite relationship between the amount of enhancement and the thickness of the metal film. It is also obvious that thicker is not always going to be better. The reason the enhancement is smaller for thicker films is because the islands are not very well separated and the field enhancement then becomes small. When the metal film becomes thick enough, then a continuous surface is formed, with all valleys and canyons between the islands having been filled.

The enhancement mechanism is mainly EM related to field enhancement near the Ag nanoislands; however, some chemical enhancement models were also proposed. Enhancement of chemisorbed molecules is usually larger than that of physisorbed molecules. One of the reasons for that is the orientational effect associated with chemisorptions, which causes stronger dipole transitions for vibrations with a dipole perpendicular to the surface than parallel to it. Another reason for the enhancement is the coupling between the vibrational modes and the charge transfer between the adsorbed molecule and the metal surface. Recently[122] SEIRA was investigated on molecules near more sophisticated NPs such as nanoshell aggregates on substrates. SEIRA substrates were developed, including metal island films[123] and perforated metal films based on the EOT phenomenon.[124]

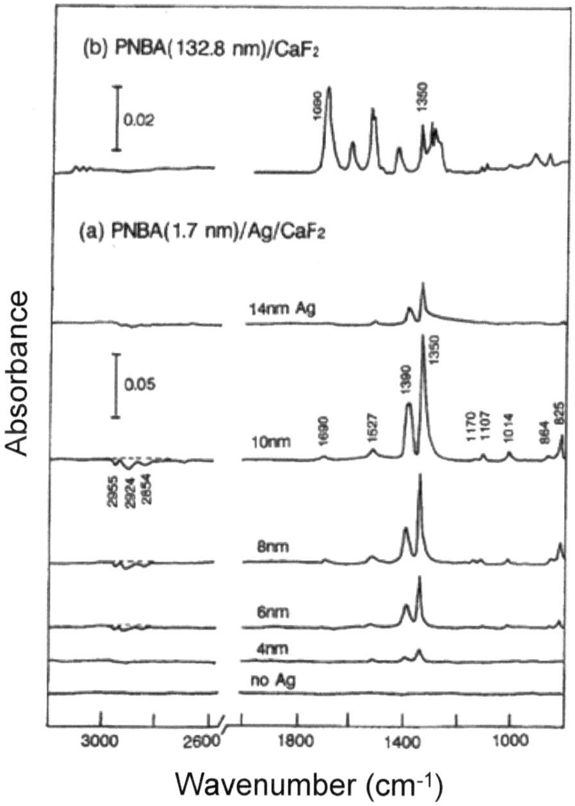

Figure 7.26 Absorbance spectra of thin films of PNBA on an Ag nanostructured thin film on a CaF_2 substrate (a) compared with the bare substrate (b). The thicknesses of the Ag films are shown, and the thicknesses of the organic film are (a) 1.7 nm and (b) 132.8 nm. (Adapted from Fig. 7.1 of Osawa.[122])

7.4 Nanosculptured Thin Films for Sensing

7.4.1 What are Sculptured Thin Films?

STFs are nanostructured inorganic materials with anisotropic properties. They show the potential to be designed in a controllable manner using physical vapor deposition (PVD) for potentially multifunctional materials and products. The glancing angle deposition (GLAD) is a sophisticated method of creating 3D chiral nanostructures with a tailored geometry. Here, the self-organized nanostructure growth is based on a concurrent growth mechanism because of geometrical shadowing in combination with a kinetic limitation for surface adatoms.[125] This technique requires a particle flux reaching the substrate under an extremely oblique angle of incidence (typically >85°, see Fig. 7.27a). The growth conditions support columnar growth, and the samples consist of 3D needles, which are slanted into the direction of the particle flux (Fig. 7.27b,c).

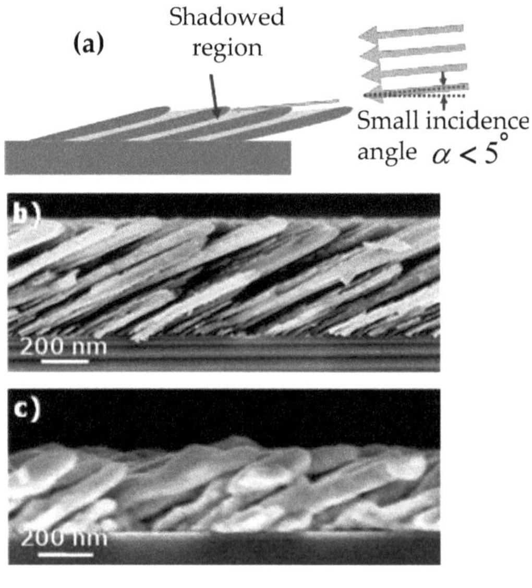

Figure 7.27 Material deposition under oblique particle incidence. (a) Basic principle: The shadowing between neighboring structures results in the self-organized formation of inclined nanocolumns. (b) Si and (c) Ag STFs. (Picture from the IOM, Leipzig, Germany.)

The instantaneous change of the growth direction due to a simple variation of the incident vapor flux (by substrate rotation) allows for the fabrication of 3D nanostructures with manifold morphologies. Since the growth process is mainly determined by the shadowing length and the surface diffusion length, materials ranging from insulators, metals, and semiconductors can be grown with 3D morphologies like spirals, chevrons, screws, pillars, needles, etc. With continuous substrate rotation vertical posts, screws or even spirals can be realized (Fig. 7.28). Shaped nanostructures with *n*-fold symmetry result from a step-wise substrate rotation by the angle 360/n degrees. The application of GLAD to fabricate metallic (Ag, Al, Co, Cr, Cu, Fe, Mg, Nb, Ni, Ru, Ta, Ti, W), semiconductor (Ge, Si), and compound nanostructures as oxides (SiO_2, TiO_2, ZnO) or chalcogenides (GeSbSe) has been reported.[126,127] Furthermore, as long as chemical interactions can be neglected, the deposition is independent of the used substrate. So also cheap lime glass or flexible polyimide foils may serve as an appropriate substrate material for GLAD.

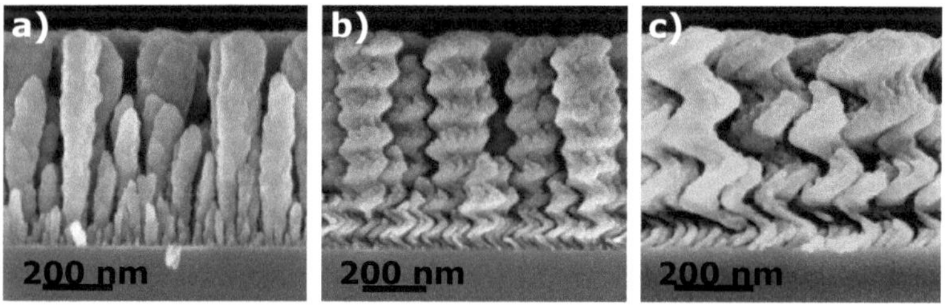

Figure 7.28 Si nanostructures fabricated by GLAD with continuous substrate rotation.[129] With respect to the angular velocity, different morphologies were obtained: (a) $\omega > 0.175$ rev/min: vertical posts; (b) $0.03 < \omega < 0.175$ rev/min: screws; and (c) $\omega < 0.03$ rev/min: spirals. (Picture from the IOM, Leipzig, Germany.)

To improve the arrangement and uniformity of GLAD-deposited nanostructures, patterned substrates with symmetrically ordered surface mounds were successfully applied (Fig. 7.29). The artificial definition of prepatterns can be performed by electron beam lithography or nanosphere lithography.[128] On the basis of the latter, honeycomb and hexagonal patterns of extremely uniform Si and Ge nanostructures of different shapes were demonstrated.

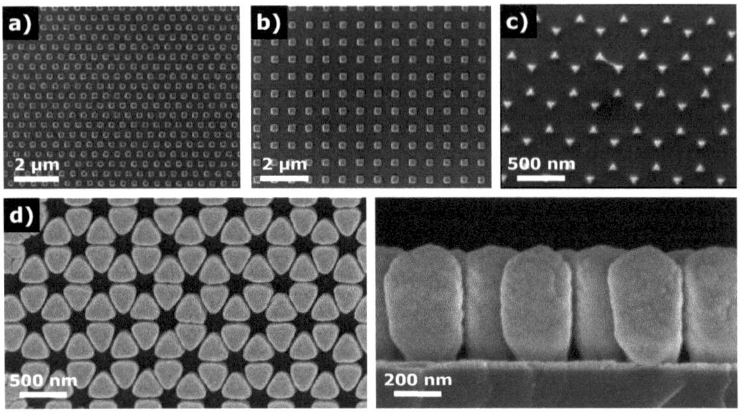

Figure 7.29 Fabrication of ordered Au prepatterns on an Si (100) substrate. (a, b) hcp and square arrangement by electron beam lithography, (c) honeycomb arrangement by nanosphere lithography, and (d) columnar GLAD nanostructures in honeycomb arrangement (left: top view; right: cross section). (Picture from the IOM, Leipzig, Germany.) *Abbreviation*: hcp, hexagonal close packing.

Metallic nanofeatures on different substrates are the most comfortable beds for building optical biosensors in a variety of analytes, such as water and blood. A promising option is to prepare nano-STFs using OAD techniques with a suitable substrate rotation, thus giving the advantageous possibility to grow nearly arbitrarily shaped, separated nano-STFs in a single deposition step without any pre- or postdeposition patterning of the film. The nanostructure and porosity both cause enhancement of the optical signature from biomolecules attached or in close proximity to the nanofeatures. Periodicity and birefringence yield polarization dependence of the optical response, a fact that can be used for multisensing action. The wide variety of material possibilities for engineering STFs helps to match special needs. Compared with existing methodologies of preparing metallic nanoparticles and porous silicon, STFs exhibit a wider range of possibilities, yet they cover the benefits that one expects from such biosensing systems. The porosity can be engineered to be within the 10%–90% range. Being porous, an STF can function as a nanoreactor. This capability can be harnessed for a variety of biosensing applications.

7.4.2 SPR Sensing Using Nano-STFs

The use of porous materials for the purpose of SPR biosensors has been under investigation by several groups. Columnar thin films

(CTFs), which can be considered porous materials, were investigated first by Yang *et al.*[129] They used a prism coupling of SPs in the Otto geometry at 3.391 μm to determine the surface optical anisotropy of several obliquely deposited nickel films. Yang *et al.* measured the SPR response from these CTFs of nickel and estimated by Bruggeman effective medium theory the optical properties and geometrical factors of the metallic films. Nanoporous gold films in the context of SPR were also investigated by Maaroof *et al.*,[130] In the pioneering study of Shalabney *et al.*,[131] the authors demonstrated that thin films of porous metals can be used as biosensors. They showed that using metallic CTFs facilitates the development of sensors with high sensitivity. For Ag and Au CTFs, the SPR dip sensitivity increases by about a factor of 2 with 30% porosity compared with nonporous films. As shown in Fig. 7.30, the metallic CTF was divided into two

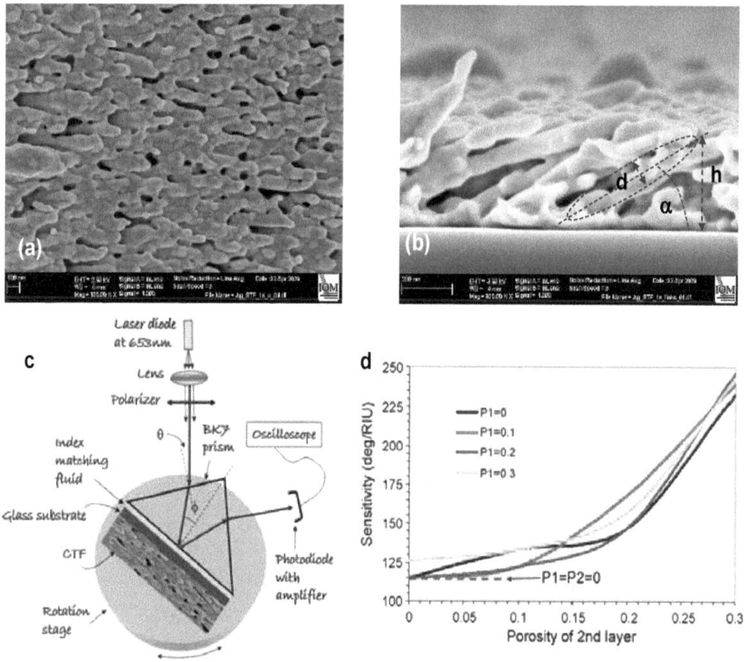

Figure 7.30 (a) Silver CTFs (top view), (b) silver CTF (cross section), (c) schematic of the Kretschmann configuration using a CTF as the metal film, and (d) simulation results to demonstrate sensitivity enhancement as porosity deceases. Calculated sensitivity when the porosity of the first layer (thinner layer) is kept fixed is indicated in the legend, while the porosity of the thicker layer is varied from 0 to 0.3. (Reproduced from Shalabney *et al.*[132] Parts (c) and (d) were taken from Figs. 7.2 and 7.10 in the original article.) See also Color Insert.

porous layers with different thicknesses and the porosity was varied to obtain an optimal fit with the experiment's results. The authors demonstrated theoretically that the sensitivity of a CTF-based SPR sensor is higher than the conventional SPR sensor, and furthermore the sensitivity can be tuned by varying the porosity of the CTF. As the porosity increases, the SPR dip widens and becomes asymmetric because of increasing scattering losses in the CTF that are due to the nonhomogeneous distribution of matter therein. As the porosity increases beyond 0.75, the SPR dip almost disappears, with a vestigial peak near the onset to the TIR regime, which can be also used as a *peak sensor*.

The appearance of a peak at the onset of the TIR regime when the SPR broadens can also be seen when the broadening is due to absorption loss rather than scattering loss inside a thin metal film. Al and Cr, for example, have bulk refractive indexes with high imaginary parts at wavelengths in the Vis region; therefore, absorption loss is strong in these metals, and the peak can be seen even with dense films. The appearance of the peak in a conventional SPR sensor with a closed metallic film indicates that a peak sensor can be produced also because of absorption losses (see Fig. 7.31).

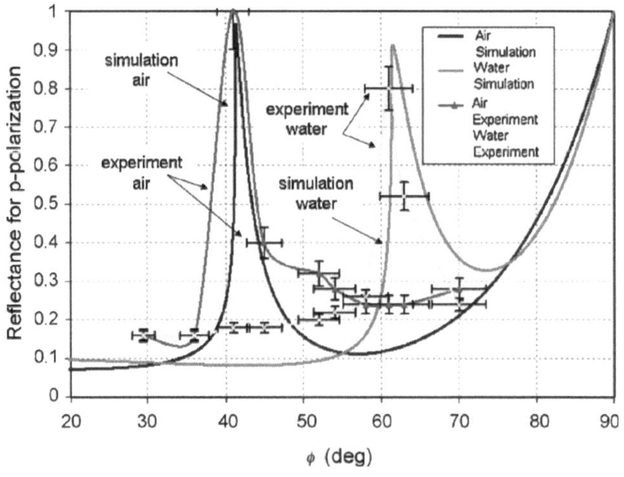

Figure 7.31 Measured/simulated data for p-polarization reflectance R vs. the angle ϕ when the dense/nonporous ($p = 0$) metal film is 8 nm thick and made of Cr. The curves exhibit a broad dip and a sharp and symmetric peak due to the absorption in bulk Cr. The analyte sensed is either air or water (refractive index = 1.33). (Graph reproduced from Fig. 7.4 of Shalabney *et al.*[132]) See also Color Insert.

7.4.3 Surface-Enhanced Spectroscopies from Nano-STFs

The interaction between light and NPs causes further optical signals to arise and be enhanced, such as SERS and SEF. SERS is a very sensitive method that allows the detection of organic molecules adsorbed on noble metal substrates (silver, gold, copper) at submicromolar concentration. Xu *et al.*[132] obtained vibrational Raman spectra from single hemoglobin molecules attached to 100-nm-sized immobilized Ag particles. Since SERS depends on the local field to the fourth power ($|E|^4$) in the vicinity of an adsorbate molecule,[133,134] one can significantly enhance the SERS intensity by enhancing the field near the scattering molecules. In the previous section we reviewed field-enhancing methods by using different shapes of nanostructures characterized by large field enhancement factors due to excitation with the resonance wavelengths corresponding to the specific structures. Félidj *et al.*[135] found that by engineering the NP structure size and shape to yield a resonance wavelength that is suitably located between the exciting laser and the Raman frequency, one can obtain large enhancement factors of Raman-scattering signals. The effects of the nanostructures' features (substrate material, NP material, NP size, NP shape, etc.) on SERS enhancement were intensively investigated during the past few years.[136–138] In the work of Kumar *et al.*[139] it was shown that an SP, propagating along a metal surface, embedded with regularly arranged NPs, undergoes SERS from molecules adsorbed over the particles. A significant scattered wave is detected if a relative relationship between the particles' arrangement period, Raman frequency, and the SP wave is satisfied. In Fig. 7.32, the effects of temperature and substrate rotation during the deposition on SERS spectra are demonstrated. SERS has been shown recently[140] to give significant and reproducible signals using ultrasmooth metal surfaces on a variety of Si-based nanostructures.

The potential of SEF was rediscovered during the past decade because of the emerging developments in the optics of metallic nanostructures.[141] SEF is a very useful phenomenon with significant applications in biotechnology and life sciences.[142] Localization of the EM field near nano tips, corners, holes, needles, etc., has been shown to produce large SEF by factors of up to a few hundreds in what is known as the lightening nanoantenna effect.[143] A recent review[144] article summarized fluorophore molecule interactions with the metal surface under SP wave excitation.

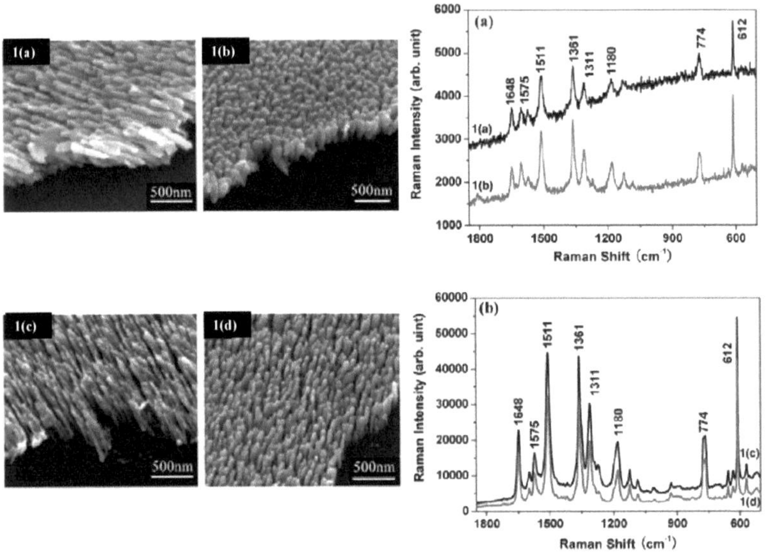

Figure 7.32 Growth morphology of Ag thin films by GLAD at various conditions (1a) at 120°C without substrate rotation, (1b) at 120°C with substrate rotation at 0.2 rpm, (1c) at −40°C without substrate rotation, (1d) at −40°C with substrate rotation at 0.2 rpm, (a) Raman spectra of R6G on Ag thin films consisting of (a) joined nanorods shown by 1(a) and 1(b), and (b) separated Ag nanorods shown by 1(c) and 1(d), respectively, at a concentration of 1×10^{-6} mol·L^{-1}. (Reproduced from Zhou *et al.*[137] The parts 1(a d) were taken from Fig. 7.1, while parts (a) and (b) were taken from Fig. 7.3 in the original article.)

SEF from porous, metallic STFs was demonstrated for sensing of bacteria in water by Abdulhalim *et al.*[145] Enhancement factors larger than 15 were observed using STFs made of silver, aluminum, gold, and copper with respect to their dense film counterparts. The STFs used are assemblies of tilted, shaped, parallel nanowires prepared with several variants of OAD. Abdulhalim *et al.*[146] examined the effect of the substrate material, constituent rod material, porosity, and the rod tilt angle on the enhancement factor of SEF. Enhancement of SEF from porous, metallic STFs was applied to biosensing in water. The main SEF mechanisms, according to the authors, are believed to be the lightening nanoantenna effect and the dipole-dipole interaction. This observation makes STFs potential candidates as

SEF nanobeds for biosensing and bioimaging. In Fig. 7.33, one can see the advantage of such sculptured films on their counterpart closed and flat films.

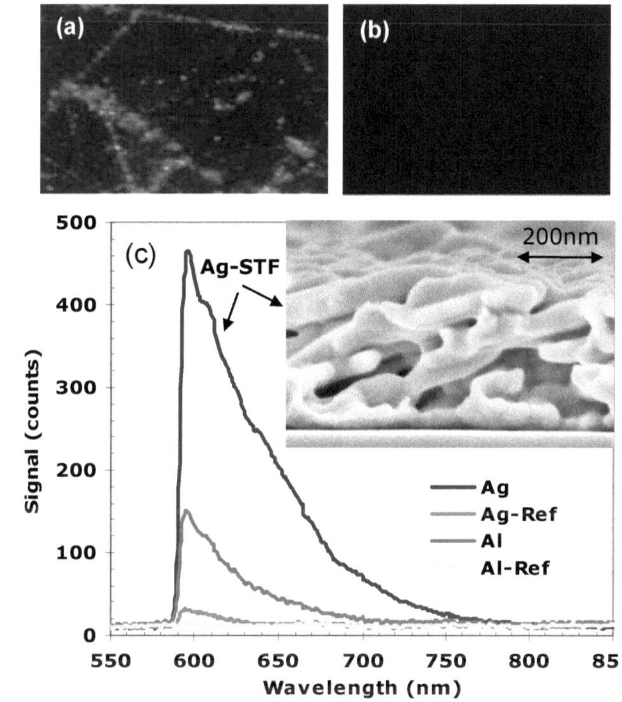

Figure 7.33 Fluorescence images from (a) Ag-nanorod STF and (b) a dense Ag film immersed in an aqueous solution of luminescent *E. coli*. (c) Typical SEF spectra from an Ag-nanorod STF and an Al-nanorod STF and from the corresponding reference films. Inset: SEM micrograph of an Ag-nanorod STF showing the highest enhancement factor. (Reproduced from Fig. 7.2 of Abdulhalim *et al.*[146]) *Abbreviation*: *E. coli*, *Escherichia coli*. See also Color Insert.

7.5 Concluding Remarks

Plasmonic nanostructured thin films for sensing were reviewed with emphasis both on the physics and on the device side. Extended SPs within the Kretschmann configuration were reviewed, and then enhanced transmission through perforated metallic films until LSPs on NPs and the associated surface-enhanced spectroscopies. Special attention was given to STFs as potential planar beds for biosensing using SPR, SEF, SERS, or SEIRA.

Planar optical sensing platforms have a great advantage over other geometrical configurations in their simplicity in fabrication, light coupling and reading, a large interaction region with the analyte, and the possibility of patterning them for biochip buildup and microfluidic sensing schemes. Planar sensors in the form of thin films can come in a variety of forms, such as arrays of nanofeatures (columns, prisms, holes, slits, etc.), waveguides, STFs, SP-based substrates using grating coupling, photonic crystals, porous materials (e.g., porous Si), and polymer hydrogels. Optical effects on which the sensing mechanism is based vary from refractive index influence on the evanescent field (such as with waveguides and SPR) to resonant absorption, Raman scattering, fluorescence, diffraction from the holographic structure, the thermo-optic effect, or interference effects in multilayered thin films and photonic crystals. The latest developments in nanophotonics and plasmonics have revealed a wide selection of nanostructures that enhance the sensitivity and lower the detection limit.[146] This fact, together with the technological development in miniaturized optoelectronic devices, such as highly sensitive low-voltage solid-state photomultipliers, miniature spectrometers, arrays of detectors, and complementary metal-oxide-semiconductor (CMOS) cameras, is expected to revolutionize the field of optical biosensing.

Acknowledgments

I am thankful to my students whose work occupied a large part of this chapter: Atef Shalabney, Olga Krasnykov, Amit Lahav, and Alina Karabchevsky. Discussions and collaborations with Profs. Mark Auslender, Akhlesh Lakhtakia, Robert Marks, and Bernd Rauschenbach were very fruitful all the way through many of our sensing activities during the past few years.

References

1. Maier, S. A. (2006). Plasmonics: metal nanostructures for subwavelength photonic devices, *IEEE J. Sel. Top. Quantum Electron.*, **12**, 1214–1220.

2. Maier, S. A., Brongersma, M. L., Kik, P. G., Meltzer, S., Requicha, A. A. G., and Atwater, H. A. (2001). Plasmonics—a route to nanoscale optical devices, *Adv. Mater.*, **13**, 1501–1505.

3. Kalele, S. A., Tiwari, N. R., Gosavi, S. W., and Kulkarni, S. K. (2007). Plasmon-assisted photonics at the nanoscale, *J. Nanophoton.*, **1**, 12501.

4. Zalevsky, Z., and Abdulhalim, I. (2010). *Integrated Nanophotonic Devices (Micro and Nano Technologies)*, William Andrew.

5. Maier, S. A. (2007). *Plasmonics: Fundamentals and Applications*, Springer.

6. Gai, H., Wang, J., and Tian, Q. (2007). Tuning the resonant wavelength of a nanometric bow-tie aperture by altering the relative permittivity of the dielectric substrate, *J. Nanophoton.*, **1**, 013555.

7. Abdulhalim, I., Zourob, M., and Lakhtakia, A. (2008). Surface plasmon resonance sensors—a mini review, *Invited* review to a special issue on the topic "Electromagnetic Surface Waves," *J. Electromagn.*, **28**(3), 213–242.

8. Homola, J., Koudela, I., and Yee, S. S. (1999). Surface plasmon resonance sensors based on diffraction gratings and prism couplers: sensitivity comparison, *Sens. Actuators A*, **54**, 16–24.

9. Kuang-Li, L., Chia-Wei, L., Way-Seen, W., and Pei-Kuen, W. (2007). Sensitive biosensor array using surface plasmon resonance on metallic nanoslits, *J. Biomed. Opt.*, **12**, 044023.

10. Kim, S., Jung, J.-M., Choi, D.-G., Jung, H.-T., and Yang, S.-M. (2006). Patterned arrays of Au rings for localized surface plasmon resonance, *Langmuir*, **22**, 7109–7112.

11. Hao, F., Nordlander, P., Burnett, M. T., and Maier, S. A. (2007). Enhanced tunability and linewidth sharpening of plasmon resonances in hybridized metallic ring/disk nanocavities, *Phys. Rev. B*, **76**, 245417–245422.

12. Zhang, X., Yonzon, C. R., Young, M.A., Stuart, D.A., and Van Duyne, R.P. (2005). Surface-enhanced Raman spectroscopy biosensors: excitation spectroscopy for optimisation of substrates fabricated by nanosphere lithography, *IEE Proc. Nanobiotechnol.*, **152**, 195–206.

13. Hatab, N. A., Hsueh, C.-H., Gaddis, A. L., Retterer, S. T., Li, J.-H., Eres, G., Zhang, Z., and Baohua, G. (2010). Free-standing optical gold bowtie nanoantenna with variable gap size for enhanced Raman spectroscopy, *Nano Lett.*, **10**, 4952–4955.

14. Chung, P.-Y., Lin, T.-H., Schultz, G., Batich, C., and Jiang, P. (2010). Nanopyramid surface plasmon resonance sensors, *Appl. Phys. Lett.*, **96**, 261108.
15. Evlyukhin, A. B., Bozhevolnyi, S. I., Pors, A., Nielsen, M. G., Radko, I. P., Willatzen, M., and Albrektsen, O. (2010). Detuned electrical dipoles for plasmonic sensing, *Nano Lett.*, **10**, 4571–4577.
16. Martín-Palma, R. J., Manso-Silván, M., Lakhtakia, A., and Pantano, C. G. (2009). Ordered arrays of nanocolumns grown by the oblique angle deposition technique on a self-assembled layer of polystyrene spheres, *Mater. Lett.*, **63**, 197–199.
17. Kubo, W., and Fujikawa, S. (2011). Au double nanopillars with nanogap for plasmonic sensor, *Nano Lett.*, **11**, 8–15.
18. Sinton, D., Gordon, R., and Brolo, A. G. (2008). Nanohole arrays in metal films as optofluidic elements: progress and potential, *Microfluid Nanofluid*, **4**, 107–116.
19. Wood, R. (1902). On a remarkable case of uneven distribution of light in a diffraction grating spectrum, *Phl os. Mag.*, **4**, 396–402.
20. Fano, U. (1941). The theory of anomalous diffraction gratings and of quasi-stationary waves on metallic surfaces (Sommerfeld's waves), *J. Opt. Soc. Am.*, **31**, 213–222.
21. Kogelnik, H. (1975). "Theory of dielectric waveguides," in *Integrated Optċ s* (ed. T. Tamir), Springer, Berlin.
22. Mashev, L., and Popov, E. (1985). Zero order anomaly of dielectric coated grating, *Opt. Commun.*, **55**, 377–380.
23. Ebbesen, T. W., Lezec, H. J., Ghaemi, H. F., Thio, T., and Wolff, P. A. (1998). Extraordinary optical transmission through sub-wavelength hole arrays, *Nature*, **391**, 667–669.
24. Bethe, H. A. (1944). Theory of diffraction by small holes, *Phys. Rev.*, **66**, 163.
25. Bouwkamp, C. J. (1950a). On Bethe's theory of diffraction by small holes, *Phl ṗ s Res. Rep.*, **5**, 321–332.
26. de Abajo, F. J. G. (2002). Light transmission through a single cylindrical hole in a metallic film, *Opt. Express*, **10**, 1475–1484.
27. Baida, F. I., and van Labeke, D. (2002). Light transmission by sub-wavelength annular aperture arrays in metallic films, *Opt. Commun.*, **209**, 17–22.
28. Xiaolei, S., Lambertus, H., and Thornton, R. L. (2003). Ultrahigh light transmission through a C-shaped nanoaperture, *Opt. Lett.*, **28**(15), 1320–1322.

29. Gordon, R. (2007). Bethe's aperture theory for arrays, *Phys. Rev. A*, **76**, 053806.

30. Popov, E., Nevière, M., Enoch, S., and Reinisch, R. (2000). Theory of light transmission through subwavelength periodic hole arrays, *Phys. Rev. B*, **62**, 16100.

31. Popov, E., Enoch, S., Tayeb, G., Neviere, M., Gralak, B., and Bonod, N. (2004). Enhanced transmission due to nonplasmon resonances in one- and two-dimensional gratings, *Appl. Opt.*, **43**, 999–1008.

32. Lalanne, P., Hugonin, J. P., Astilean, S., Palamaru, M., and Moller, K. D. (2000). One-mode model and airy-like formulae for one-dimensional metallic gratings, *J. Opt. A: Pure Appl. Opt.*, **2**, 48.

33. López-Rios, T., Mendoza, D., García-Vidal, F. J., Sánchez-Dehesa, J., and Pannetier, B. (1998). Surface shape resonances in lamellar metallic gratings, *Phys. Rev. Lett.*, **81**, 665.

34. Porto, J. A., García-Vidal, F. J., and Pendry, J. B. (1999). Transmission resonances on metallic gratings with very narrow slits, *Phys. Rev. Lett.*, **83**, 2845.

35. Wang, B., and Lalanne, P. (2010). Surface plasmon polaritons locally excited on the ridges of metallic gratings, *J. Opt. Soc. Am. A*, **27**, 1432–1441.

36. Catrysse, P. B., and Fan, S. (2008). Propagating plasmonic mode in nanoscale apertures and its implications for extraordinary transmission, *J. Nanophoton.*, **2**, 021790.

37. Kang, J. H., Choe, J.-H., Kim, D. S., Park, Q-H. (2009). Substrate effect on aperture resonances in a thin metal film, *Opt. Express*, **17**, 15652–15658.

38. Pang, Y., Genet, C., and Ebbesen, T. W. (2007). Optical transmission through subwavelength slit apertures in metallic films, *Opt. Commun.*, **280**, 10–15.

39. Hsu, S.-Y., Lee, K.-L., Lin, E.-H., Lee, M.-C., and Wei, P.-K. (2009). Giant birefringence induced by plasmonic nanoslit arrays, *Appl. Phys. Lett.*, **95**, 0131051–0131053.

40. Degiron, A., and Ebbesen, T. W. (2005). The role of localized surface plasmon modes in the enhanced transmission of periodic sub-wavelength apertures, *J. Opt. A: Pure. Appl. Opt.*, **7**, S90–S96.

41. Garcia-Vidal, F. J., Martin-Moreno, L., Ebbesen, T. W., and Kuipers, L. (2010). Light passing through subwavelength apertures, *Rev. Mod. Phys.*, **82**, 729–787.

42. Karabchevsky, A., Krasnykov, O., Abdulhalim, I., Hadad, B., Goldner, A., Auslender M., and Hava, S. (2009). Metal grating on a substrate nanostructure for sensor applications, *Photon Nanostruct.: Fundam. Appl.*, **7**, 170–175.

43. Karabchevsky, A., Krasnykov, O., Auslender, M., Hadad, B., Goldner, A., and Abdulhalim, I. (2009). Theoretical and experimental investigation of enhanced transmission through periodic metal nanoslits for sensing in water environment, *J. Plasmon.*, **4**, 281–292.

44. Krasnykov, O., Karabchevsky, A., Auslender, M., and Abdulhalim, I. (2011). Sensor with increased sensitivity based on enhanced optical transmission in the infrared, *Opt. Commun.*, **284**, 1435–1438.

45. Moharam, M. G., Grann, E. B., and Pommet, D. A. (1995). Formulation for stable and efficient implementation of the rigorous coupled-wave analysis of binary gratings, *J. Opt. Soc. Am. A*, **12**, 1068–1076.

46. Lalanne, P., and Morris, G. M. (1996). Highly improved convergence of the coupled wave method for TM polarization, *J. Opt. Soc. Am. A*, **13**, 779–784.

47. Granet, G., and Guizal, B. (1996). Really efficient implementation of the coupled-wave method for metallic lamellar gratings in TM polarization, *J. Opt. Soc. Am. A*, **13**, 1019–1023.

48. Lalanne, P. (1997). Improved formulation of the coupled-wave method for two-dimensional gratings, *J. Opt. Soc. Am. A*, **14**, 1592–1598.

49. Granet, G. (1999). Reformulation of the lamellar grating problem through the concept of adaptive spatial resolution, *J. Opt. Soc. Am. A*, **16**, 2510–2516.

50. Li, L., and Haggans, C. (1993). Convergence of the coupled-wave method for metallic lamellar diffraction gratings, *J. Opt. Soc. Am. A*, **10**, 1184–1189.

51. Li, L., Chandezon, J., Granet, G., and Plumey, J. P. (1999). Rigorous and efficient grating-analysis method made easy for optical engineers, *Appl. Opt.*, **38**, 304–313.

52. Li, L. (1996). Use of Fourier series in the analysis of the discontinuous structures, *J. Opt. Soc. Am. A*, **13**, 1870–1876.

53. Popov, E., and Neviere, M. (2000). Grating theory: new equations in Fourier space leading to fast converging results for TM polarization, *J. Opt. Soc. Am. A*, **17**, 1773–1784.

54. Chu, H. (2003). Finite difference approach to optical scattering of gratings, *Proc. SPIE*, **5188**, 358–370.

55. Li, L. (1993). A modal analysis of lamellar diffraction gratings in conical mountings, *J. Mod. Opt.*, **40**, 553–573.

56. Li, L. (1994). Multilayer modal method for diffraction gratings of arbitary profile, depth, and permittivity, *J. Opt. Soc. Am. A*, **10**, 2583–2501, Addendum *J. Opt. Soc. Am A*, **11**, 1685.

57. Li, L. (1996). Formulation and comparison of two recursive matrix algorithms for modeling layered diffraction gratings, *J. Opt. Soc. Am. A*, **13**, 1024–1035.

58. Auslender, M., and Hava, S. (1996). Scattering-matrix propagation algorithm in full-vectorial optics of multilayer grating structures, *Opt. Lett.*, **21**, 1765–1767.

59. Karabchevsky, A., Auslender, M., and Abdulhalim, I. (2011). Periodically limited surface plasmon excitation at the metallic nanoslits, *J. Nanophoton.*, **5**, 051821.

60. de Abajo, F. J. G., Sáenz, J. J., Campillo, I., and Dolado, J. S. (2006). Site and lattice resonances in metallic hole arrays, *Opt. Express*, **14**, 7–18.

61. Chang, S. H., Gray, S. K., and Schatz, G. C. (2005). Surface plasmon generation and light transmission by isolated nanoholes and arrays of nanoholes in thin metal films, *Opt. Express*, **13**, 3150–3165.

62. *Nobel Lectures in Chemistry 1901–1941* (1965), Elsevier, Amsterdam.

63. Schultz, S., Smith, D. R., Mock, J. J., and Schultz, D. A. (2000). Single-target molecule detection with nonbleaching multicolor optical immunolabels, *Proc. Natl. Acad. Sci. USA*, **97**, 996.

64. Sonnichsen, C., Franzl, T., Wilk, T., von Plessen, G., Feldmann, J., Wilson, O., and Mulvaney, P. (2002). Drastic reduction of plasmon damping in gold nanorods, *Phys. Rev. Lett.*, **88**, 077402.

65. Kuwata, H., Tamaru, H., Esumi, K., and Miyano, K. (2003). Resonant light scattering from metal nanoparticles: practical analysis beyond Rayleigh approximation, *Appl. Phys. Lett.*, **83**, 4625.

66. Jackson, J. D. (1998). *Classical Electrodynamics*, 3rd ed., Academic Press.

67. Kreibig, U., and Vollmer, M. (1995). *Optical Properties of Metal Clusters*, Springer, Berlin.

68. Khlebtsov, N. G. (2008). Optics and biophotonics of nanoparticles with a plasmon resonance, *Quantum Electron.*, **38**, 504–529.

69. Kelly, K. L., Coronado, E., Zhao, L. L., and Schatz, G. C. (2003). The optical properties of metal nanoparticles: the influence of size, shape, and dielectric environment, *J. Phys. Chem. B*, **107**, 668–677.

70. Van de Hulst, H. C. (1981). *Lġ ht Scatterṅ g by Small Partċ les*, Dover, New York.

71. Doyle, W. T. (1958). Absorption of light by colloids in alkali halide crystals, *Phys. Rev.*, **111**, 1067–1072.

72. Doremus, R. H. (1964). Opticalproperties of small gold particles, *J. Chem. Phys.*, **40**, 2389.

73. Kreibig, U., and Vollmer, M. (1995). *Optċ al Propertė s of Metal Clusters*, Springer-Verlag, Berlin.

74. Doyle, W. T., and Agarwal, A. (1965). Optical extinction of metal nanospheres, *J. Opt. Soc. Am.*, **55**, 305–308.

75. Doyle, W. T. (1989). Optical properties of a suspension of metal spheres, *Phys. Rev. B*, **39**, 9852–9858.

76. Pitarke, J. M., Silkin, V. M., Chulkov, E. V., and Echenique, P. M. (2007). Theory of surface plasmons and surface-plasmon polaritons, *Rep. Prog. Phys.*, **70**, 1–87.

77. Zayats, A. V., Smolyaninov, I. I., and Maradudin, A. A. (2005). Nano-optics of surface plasmon polaritons, *Phys. Rep.*, **408**, 131–314.

78. Montgomery, J. M., Lee, T. W., and Gray, S. K. (2008). Theory and modeling of light interactions with metallic nanostructures, *J. Phys.: Condens. Matter*, **20**, 323201–323212.

79. Fort, E., and Gresillon, S. (2008). Surface enhanced fluorescence, *J. Phys. D: Appl. Phys.*, **41**, 13001–13032.

80. Stoner, E. C. (1945). The demagnetizing factors for ellipsoids, *SER*, **7**, 36, 803–812.

81. Shuford, K. L., Ratner, M. A., and Schatz, G. C. (2005). Multipolar excitation in triangular nanoprisms, *J. Chem. Phys.*, **123**, 114713–114719.

82. Shalaev, V. M. (2000). *Nonlṅ ear Optċ s of Random Medï* , Springer, Berlin.

83. Genov, D. A., Sarychev, A. K., Shalaev, V. M., and Wei, A. (2004). Resonant field enhancements from metal nanoparticle arrays, *Nano Lett.*, **4**, 153–158.

84. Hao, E., and Schatz, G. C. (2004). Electromagnetic fields around silver anoparticles and dimmers, *J. Chem. Phys.*, **120**, 357–366.

85. Kall, M., Xu, H., and Johansson, P. (2005). Field enhancement and molecular response in surfaceenhanced Raman scattering and fluorescence spectroscopy, *J. Raman Spectrosc.*, **36**, 510–514.

86. Fleischmann, M., Hendra, P. J., and McQullan, A. J. (1974). Raman spectra of pyridine adsorbed at a silver electrode, *Chem. Phys. Lett.*, **26**, 163–166.

87. Jeanmaire, D. L., and Van Duyne, R. P. (1977). Surface Raman spectroelectrochemistry, part 1, heterocyclic, aromatic, and aliphatic amines adsorbed on the anodized silver electrode, *J. Electroanal. Chem.*, **84**, 1–20.

88. Cotton, T. M., Kim, J.-H., and Chumanov, G. D. (2005). Application of surface-enhanced Raman spectroscopy to biological systems, *J. Raman Spectrosc.*, **22**, 729–742.

89. Schatz, G. C., and Van Duyne, R. P. (2002). "Electromagnetic mechanism of surface-enhanced spectroscopy," in *Handbook of Vibrational Spectroscopy* (ed. M. J. Chalmers and P. R. Griffiths), Wiley, Chicester, UK.

90. Schatz, G. C. (1984). Theoretical studies of surface enhanced Raman scattering, *Acc. Chem. Res.*, **17**, 370–376.

91. Kerker, M., Wang, D.-S., and Chew, H. (1980). Surface enhanced Raman scattering (SERS). by molecules adsorbed at spherical particles, *Appl. Opt.*, **19**, 3373–3388.

92. Wokaun, W., Gordon, J. P., and Liao, P. F. (1982). Radiation damping in surface enhanced Raman scattering, *Phys. Rev. Lett.*, **48**, 957–960.

93. Kreibig, U. (1974). Electronic properties of small silver particles: the optical constants and their temperature dependence, *J. Phys. F: Metal Phys.*, **4**, 999–1014.

94. Kraus, W. A., and Schatz, G. C. (1983). Plasmon broadening in spheroidal metal particles, *Chem. Phys. Lett.*, **99**, 353–357.

95. Chance, R. R., Prock, A., and Sylbey, R. (1978). Molecular fluorescence and energy transfer near interfaces, *Adv. Chem. Phys.*, **37**, 1–65.

96. Aslan, K., Gryczynski, I., Malicka, J., Malveeva, E., Lakowics, J. R., Geddes, C. D. (2005). Metal-enhanced fluorescence: an emerging tool in biotechnology, *Curr. Opin. Biotechnol.*, **16**, 55–62.

97. Fort, E., and Gresillon, S. (2008). Surface enhanced fluorescence, *J. Phys. D: Appl. Phys.*, **41**, 013001.

98. Lakowicz, J. R., Shen, Y., D'Auria, S., Malicka, J., Fang, J., Gryczynski, Z., and Gryczynski, I. (2002). Radiative decay engineering 2. Effects of silver island films on fluorescence intensity, lifetimes, and resonance energy transfer, *Anal. Biochem.*, **301**, 261–277.

99. Goldys, E. M., Barnett, A., Xie, F., Drozdowicz-Tomsia, K., Gryczynski, I., Matveeva, E. G., Gryczynsky, Z., and Shtoyko, T. (2007). Plasmon-enhanced fluorescence near metallic nanostructures: biochemical applications, *Appl. Phys. A*, **89**, 265–271.

100. Zhang, Y., Aslan, K., Previte, M. J. R., and Geddes, C. D. (2007). Metal-enhanced fluorescence from copper substrates, *Appl. Phys. Lett.*, **90**, 173116.

101. Zhang, J., Malicka, J., Gryczynski, I., and Lakowicz, J. R. (2005). Surface-enhanced fluorescence of fluorescein-labeled oligonucleotides capped on silver nanoparticles, *J. Phys. Chem. B*, **109**, 7643–7648.

102. Xie, F., Baker, M. S., and Goldys, E. M. (2006). Homogeneous silver-coated nanoparticle substrates for enhanced fluorescence detection, *J. Phys. Chem. B*, **110**, 23085–23091.

103. Matveeva, E., Gryczynski, Z., Malicka, J., Gryczynski, I., and Lakowicz, J. R. (2004). Metal-enhanced fluorescence immunoassays using total internal reflection and silver island-coated surfaces, *Anal. Bio chem.*, **334**, 303–311.

104. Song, J. H., Atay, T., Shi, S., Urabe, H., and Nurmikko, A. V. (2005). Large enhancement of fluorescence efficiency from CdSe/ZnS quantum dots induced by resonant coupling to spatially controlled surface plasmons, *Nano Lett.*, **5**, 1557–1561.

105. Tam, F., Goodrich, G. P., Johnson, B. R., and Halas, N. J. (2007). Plasmonic enhancement of molecular fluorescence, *Nano Lett.*, **7**, 496–501.

106. Liebermann, T., and Knoll, W. (2000). Surface plasmon field-enhanced fluorescence spectroscopy, *Colloid Surf. A*, **171**, 115–130.

107. Moskovits, M. (1985). Surface-enhanced spectroscopy, *Rev. Mod. Phys.*, **57**, 783–826.

108. Lee, I.-Y. S., Suzuki, H., Ito, K., and Yasuda, Y. (2004). Surface-enhanced fluorescence and reverse saturable absorption on silver nanoparticles, *J. Phys. Chem. B*, **108**, 19368–19372.

109. Lakowicz, J. R. (2001). Radiative decay engineering: biophysical and biomedical applications, *Anal. Bio chem.*, **298**, 1–24.

110. Lakowicz, J. R., Shen, Y., D'Auria, S., Malicka, J., Fang, J., Gryczynski, Z., and Gryczynski, I. (2002). Radiative decay engineering 2. Effects of silver island films on fluorescence intensity, lifetimes, and resonance energy transfer, *Anal. Bio chem.*, **301**, 261–277.

111. Lakowicz, J. R., Shen, Y., Gryczynski, Z., D'Auria, S., and Gryczynski, I. (2001). Intrinsic fluorescence from DNA can be enhanced by metallic particles, *Bio chem. Bio phys. Res. Commun.*, **286**, 875–879.

112. Gryczynski, I., Malicka, J., Shen, Y., Gryczynski, Z., and Lakowicz, J. R. (2002). Multiphoton excitation of fluorescence near metallic particles: enhanced and localized excitation, *J. Phys. Chem. B*, **106**, 2191–2195.

113. Yguerabide, J., and Yguerabide, E. E. (1998). Light-scattering submicroscopic particles as highly fluorescent analogs and their use as tracer labels in clincal and biological applications, 1. Theory, *Anal. Biochem.*, **262**, 137–156.

114. Sokolov, K., Chumanov, G., and Cotton, T. M. (1998). Enhancement of molecular fluorescence near the surface of colloidal metal films, *Anal. Chem.*, **70**, 3898–3905.

115. Kummerlen, J., Leitner, A., Brunner, H., Aussenegg, F. R., and Wokaun, A. (1993). Enhanced dye fluorescence over silver island films: analysis of the distance dependence, *Mol. Phys.*, **80**, 1031–1046.

116. Malicka, J., Gryczynski, I., Geddes, C. D., and Lakowicz, J. R. (2003). Metal-enhanced emission from indocyanine green: a new approach to *in vivo* imaging, *J. Biomed. Opt.*, **8**, 472–478.

117. Dorfman, A., Kumar, N., and Hahm, J.-i. (2006). Highly sensitive biomolecular fluorescence detection using nanoscale ZnO platforms, *Langmuir*, **22**, 4890–4895.

118. Mohamed, M. B., Volkov, V. V., Link, S., and El-Sayed, M. B. (2000). The "lightening" gold nanorods: fluorescence enhancement of over a million compared to the gold metal, *Chem. Phys. Lett.*, **317**, 517–523.

119. Zhang, J., and Lakowicz, J. R. (2005). Enhanced luminescence of phenyl-phenalthridine dye on aggregated small silver nanoparticles, *J. Phys. Chem. B*, **109**, 8701–8706.

120. Harstein, A., Kirtley, J. R., and Tsang, J. C. (1980). Enhancement of the infrared absorption from molecular monolayers with thin metal overlayers, *Phys. Rev. Lett.*, **55**, 201–204.

121. Osawa, M. (1997). Dynamic processes in electrochemical reactions studied by surface enhanced infrared absorption spectroscopy (SEIRAS), *Bull. Chem. Soc. Jpn.*, **70**, 2861–2880.

122. Kundu, J., Le, F., Nordlander, P., and Halas, N. J. (2008). Surface enhanced infrared absorption (SEIRA). spectroscopy on nanoshell aggregate substrates, *Chem. Phys. Lett.*, **452**, 115–119.

123. Jensen, T. R., Duyne, R. P. V., Johnson, S. A., and Maroni, V. A. (2000). Surface- enhanced infrared spectroscopy: a comparison of metal island films with discrete and nondiscrete surface plasmons, *Appl. Spectrosc.*, **54**, 371–377.

124. Rodriguez, K. R., Shah, S., Williams, S. M., Peeters-Kenedy, S., and Coe, J. V. (2004). Enhanced infrared absorption spectra of self-assembled alkanethiol monolayers using the extraordinary infrared transmission

of metallic arrays of subwavelength apertures, *J. Chem. Phys.*, **121**, 8571–8575.

125. Lakhtakia, A., and Messier, R. (2005). *Sculptured Thin Films: Nano-engineered Morphology and Optics*, SPIE Press, Bellingham, WA, USA.

126. Hawkeye, M. M., and Brett, M. J. (2007). Glancing angle deposition: fabrication, properties, and applications of micro- and nanostructured thin films, *J. Vac. Sci. Technol. A*, **25**, 1317–1336.

127. Mukherjee, S., and Gall, D. (2010). Power law scaling during physical vapor deposition under extreme shadowing conditions, *J. Appl. Phys.*, **107**, 084301–084307.

128. Patzig, C., Rauschenbach, B., Erfurth, W., and Milenin, A. (2007). Ordered silicon nanostructures by ion beam induced glancing angle deposition, *J. Vac. Sci. Technol. B*, **25**, 833–838.

129. Yang, F., Bradberry, G. W., and Sambles, J. R. (1991). The study of the optical properties of obliquely evaporated nickel films using IR surface plasmons, *Thin Solid Films*, **196**, 35–46.

130. Maaroof, A. I., Gentle, A., Smith, G. B., and Cortie, M. B. (2007). Bulk and surface plasmons in highly nanoporous gold films, *J. Phys. D: Appl. Phys.*, **40**, 5675–5682.

131. Shalabney, A., Lakhtakia, A., Abdulhalim, I., Lahav, A., Patzig, C., Hazek, I., Karabchevsky, A., Rauschenbach, B., Zhang, F., and Xu, J. (2009). Surface plasmon resonance from metallic columnar thin films, Photon. Nanostruct.—Fundam. Appl., 7, 176–185.

132. Xu, H., Bjerneld, E. J., Käll, M., and Börjesson, L. (1999). Spectroscopy of single hemoglobin molecules by surface enhanced Raman scattering, *Phys. Rev. Lett.*, **83**, 4357–4360.

133. Kerker, M., Wang, D.-S., and Chew, H. (1980). Surface enhanced Raman scattering (SERS). by molecules adsorbed at spherical particles, *Appl. Opt.*, **19**, 3373–3388.

134. Schatz, G. C., and Van Duyne, R. P. (2002). "Electromagnetic mechanism of surface-enhanced spectroscopy," reproduced from *Handbook of Vibrational Spectroscopy* (ed. J. M. Chalmers and P. R. Griffiths), John Wiley & Sons, Chichester.

135. Félidj, N., Aubard, J., Lévi, G., Krenn, J. R., Hohenau, A., Schider, G., Leitner, A., and Aussenegg, F. R. (2003). Optimized surface-enhanced Raman scattering on gold nanoparticle arrays, *Appl. Phys. Lett.*, **82**, 3095–3097.

136. Zhou, Q., Li, Z., Yang, Y., and Zhang, Z. (2008). Arrays of aligned, single crystalline silver nanorods for trace amount detection, *J. Phys. D: Appl. Phys.*, **41**, 152007–152011.

137. Jackson, J. B., Westcott, S. L., Hirsch, L. R., West, J. L., and Halas, N. J. (2003). Controlling the surface enhanced Raman effect via the nanoshell geometry, *Appl. Phys. Lett.*, **82**, 257–259.

138. Wang, Y., Zou, X., Ren, W., Wang, W., and Wang, E. (2007). Effect of silver nanoplates on Raman spectra of p-aminothiophenol assembled on smooth macroscopic gold and silver surface, *J. Phys. Chem. C*, **111**, 3259–3265.

139. Kumar, G., Singh, D. B., and Tripathi, V. K. (2006). Surface enhanced Raman scattering of a surface plasma wave, *J. Phys. D: Appl. Phys.*, **39**, 4436–4439.

140. Nagpal, P., Lindquist, N. C., Oh, S.-H., and Norris, D. J. (2009). Ultrasmooth patterned metals for plasmonics and metamaterials, *Science*, **325**, 594–597.

141. Aslan, K., Gryczynski, I., Malicka, J., Matveeva, E., Lakowicz, J. R., and Geddes, C. D. (2005). Metal-enhanced fluorescence: an emerging tool in biotechnology, *Curr. Opin. Biotechnol.*, **16**, 55–62.

142. Fort, E., and Grésillon, S. (2008). Surface enhanced fluorescence, *J. Phys. D: Appl. Phys.*, **41**, 013001–013032.

143. Moskovits, M. (1985). Surface enhanced spectroscopy, *Rev. Mod. Phys.*, **57**, 783–826.

144. Fu, Y., and Lakowicz, J. R. (2009) Modification of single molecule fluorescence near metallic nanostructures, *Laser Photon. Rev.*, **3**, 221–232/DOI 10.1002/lpor.200810035

145. Abdulhalim, I., Karabchevsky, A., Patzig, C., Rauschenbach, B., Fuhrmann, B., Eltzov, E., Marks, R., Xu, J., Zhang, F., and Lakhtakia, A. (2009). Surface enhanced fluorescence from metal sculptured thin films with application to biosensing in water, *Appl. Phys. Lett.*, **94**, 063106–063200.

146. Shalabney, A., and Abdulhalim, I. (2011). Sensitivity enhancement methods for surface plasmon sensors, *J. Laser Photon. Rev.*, **5**, 571–606.

Chapter 8

Nanostructures for Plasmonics: Design, Fabrication, and Characterization

Andrea Toma,[a] Gobind Das,[a] Remo Proietti Zaccaria,[a] Manohar Chirumamilla,[a] Francesco Gentile,[a] Federico Mecarini,[a] Maria Laura Coluccio,[a] Marco Leoncini,[a] Carlo Liberale,[a] Marco Francardi,[b,c] Francesco De Angelis,[a] and Enzo Di Fabrizio[a,b]

[a]*Nanob tech Facl t y, Istt uto Itali no di Tecnologi , vi Morego 30, Genova, I-16163 Italy*

[b]*Lab BIONEM, Di artim ento di Medi n a S erim entale e Cli i a, Univ erst à "Magna Greci " di Catanzaro, vi le Euro a, Catanzaro, I-88100, Italy*

[c]*Internati nal School for Advanced Studi s (SISSA), Area Sci nce Park SS 14 Km 163.5 Edificio Q1, Trieste, I-34012, Italy*

andrea.toma@iit.it

8.1 Introduction

In the last couple of decades we have witness enormous advances in the field of microtechnology, to the extent that nowadays we talk about nanotechnology. Faster computers, LCD-based mobiles, and nanoparticles for UV absorption in suntan lotions are just few of many ways in which nanotechnology plays a fundamental role today. The merit of this is mainly in the advance of fabrication methods. Techniques such as focused ion beam (FIB) lithography guarantee

Nanoantenna: Plasmon-Enhanced Spectroscopies for Biotechnological Applications
Edited by Marc Lamy de la Chapelle and Annemarie Pucci

ISBN 978-981-4303-61-3 (Hardcover), 978-981-4303-62-0 (eBook)
www.panstanford.com

a resolution of less than 10 nanometers, which is about five times more precise than the technology was capable of ten years ago.

Dramatic improvements in fabrication have brought about a renaissance for some well-established topics, such as plasmonics. Plasmonics has an interesting history. At the beginning of the twentieth century, Robert Williams Wood observed unknown optical features in the reflection spectrum of metallic gratings.[1] In 1956 Ugo Fano introduced the term *polart on* to define the coupling between bound electrons and incoming photons,[2] and in 1968 the scientists Dr. Heinz Raether, Eric Kretschmann, and Andreas Otto presented methods for coupling photons on a flat metallic surface.[3,4] By that point the most important aspects related to plasmonics were well known, but fabrication technique limitations prevented their application at the nanoscale level. Luckily, we are now in a situation where this is possible, which clearly explains the vast interest in plasmonics more than a hundred years after the work of Robert Wood.

8.2 Light-Electron Coupling on a Flat Metallic Surface

In order to offer a panoramic view of the plasmonic world, we begin where Raether, Kretschmann, and Otto began in 1968, with an ideally flat metallic surface surrounded by a dielectric environment. It is well known that when light hits a metallic flat surface it is reflected and absorbed. However, under special circumstances, a surface-confined propagating electromagnetic wave can be generated. This is called *surface plasmon polart on* (SPP) and results from the coupling of incoming photons and free electrons in the metal. To understand this, we need to recall some equations that perfectly explain the phenomenon. The general equation relating frequency ω and wave vector k can be written as

$$\varepsilon\frac{\omega^2}{c^2} = k^2 = k_x^2 + k_y^2 + k_z^2 \tag{8.1}$$

where ε and c are the material dielectric function and light velocity, respectively. Equation 8.1, known as the dispersion relation, is a very important equation of electromagnetism because it contains two fundamental conservation laws: energy and momentum. In fact, in any physical optical phenomenon both energy and wave vector parallel to a given interface must always be conserved. Considering that our present goal is to analyze a flat metallic surface, we have to identify which dispersion relation describes such a system. For simplicity, but without any loss of generality, we shall consider a two-dimensional system where only the coordinates x and z are taken into account. A sketch of the problem is shown in Fig. 8.1.

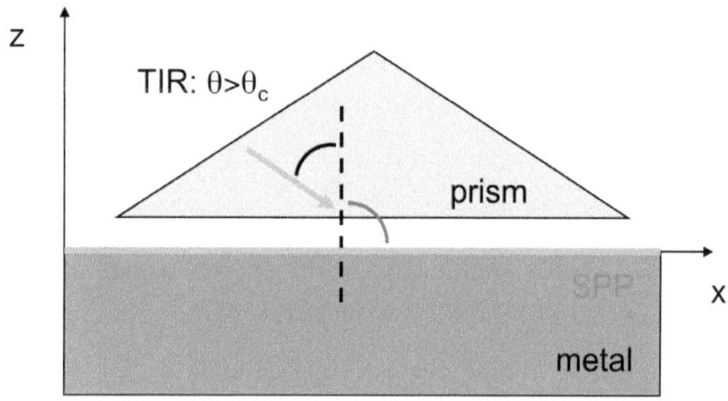

Figure 8.1 Two-dimensional representation of SPP generation. Both a prism and a flat metallic surface are shown. Light impinges on the metal travelling in the xz plane. The prism allows the light to assume the correct characteristics for the photons-electrons coupling to be successful.

For the moment let us neglect the prism at the top of the metallic slab and consider only some radiation impinging on the flat metallic surface. Here the key question is "Is it possible to excite an electromagnetic wave with the property of existing just on the interface between the flat metal and the surrounding medium?" If we look at Eq. 8.1, and we consider that the electric field would have the form $E \propto e^{ik_x x} e^{ik_z z}$, we must conclude that k_z has an imaginary component that would determine the decay away from the interface. If this happens, the SPP is created. In order to verify whether this is a feasible condition, we need to recall another equation valid for a flat and infinitely extended interface[5]:

$$k_x^2 = \frac{\omega^2}{c^2} \frac{\varepsilon_\mathrm{m} \varepsilon_\mathrm{d}}{\varepsilon_\mathrm{m} + \varepsilon_\mathrm{d}} \tag{8.2}$$

where $\varepsilon_m = \varepsilon_{m,real} + i\varepsilon_{m,imag}$ and ε_d stand for the metallic and dielectric functions, respectively. Equations 8.1 and 8.2 uniquely define the expression for k_z:

$$k_{z,j}^2 = \varepsilon_j \frac{\omega^2}{c^2} - k_x^2 = \frac{\omega^2}{c^2}\left(\varepsilon_j - \frac{\varepsilon_m \varepsilon_d}{\varepsilon_m + \varepsilon_d}\right) \tag{8.3}$$

where j represents either the dielectric or the metallic interface.

As already mentioned, the condition needed to realize a wave confined at the interface is to have an imaginary component for k_z, namely it must be valid in the relation

$$\varepsilon_j - \frac{\varepsilon_m \varepsilon_d}{\varepsilon_m + \varepsilon_d} < 0 \tag{8.4}$$

(this condition implies a purely imaginary k_z). This expression can be simplified at least in the visible IR region where $|\varepsilon_{m,real}| > \varepsilon_{m,imag}$. In this case the condition for realizing SPP is

$$\varepsilon_j < \frac{\varepsilon_{m,real} \varepsilon_d}{\varepsilon_{m,real} + \varepsilon_d} \tag{8.5}$$

namely $\varepsilon_{m,real} < 0$ and $|\varepsilon_{m,real}| > \varepsilon_d$.

By describing (for simplicity) ε_m with the Drude model,[6] it is easy to show that Eq. 8.5 is always verified, at least for the range of frequencies of interest (below the plasma frequency). Hence, under the Drude assumption, Eq. 8.2 becomes

$$k_x^2 = \frac{\omega^2}{c^2} \frac{(1 - \omega_p^2/\omega^2)\varepsilon_d}{1 - \omega_p^2/\omega^2 + \varepsilon_d} \tag{8.6}$$

with k_x always being a real quantity (the discussion can be generalized to the case of k_x being complex). The plot of the dispersion relation $\omega(k_x)$ from Eq. 8.6 takes the form indicated in Fig. 8.2a (black line), where $c = 1$ and the plasma frequency ω_p is equal to 9.01 eV (Ag).[6] The plateau is reached at $\omega = \omega_p/\sqrt{(\varepsilon_d + 1)}$. Hence, the X component of the wave vector of a radiation confined on any dielectric-metal interface must show a dispersion relation similar to the one in Fig. 8.2a. In turn this means that, in order to excite surface waves, the incoming light must satisfy the conservation of both ω and k_x; that is, it must *intersect* the curve in Fig. 8.2a. Is this possible? The answer is no; in fact any incoming radiation will always be associated with a light line higher than the surface wave curve of Fig. 8.2a.

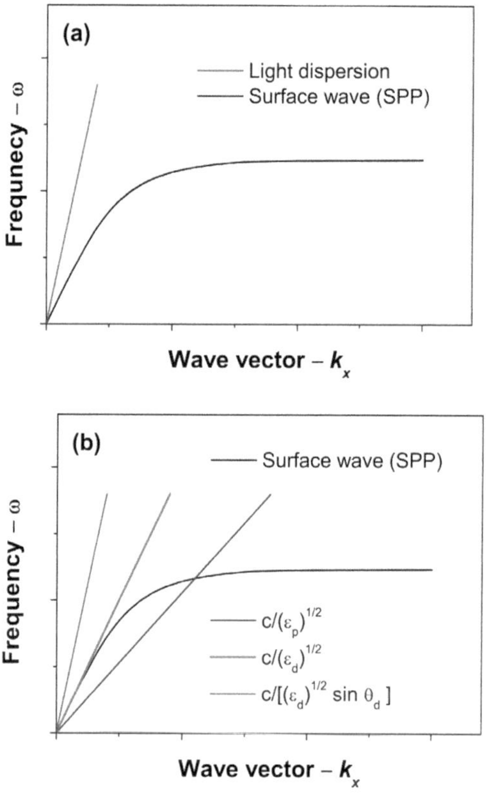

Figure 8.2 (a) SPP dispersion relation for an interface silver-air under the Drude approximation. The light dispersion and the SPP curve show no intersections. (b) In addition to the surface wave relation, three light dispersion relations are shown. They are related to the silver-glass (blue trace), silver-air (red trace), and glass-air-silver (cyan trace) interfaces. The silver-glass line intersects the SPP curve in one point. See also Color Insert.

Even the extreme and ideal case of incoming radiation tangent to the interface, which would correspond to the maximum possible k_x (and $k_z = 0$), would be simply represented by a light line tangent to the bottom part of the surface wave curve. No intersection points are possible; hence no SPP can be generated. This can be easily understood by putting together Eq. 8.1, 8.2, and 8.4 and Fig. 8.2a:

$$k_x^2 + k_z^2 = \varepsilon \frac{\omega^2}{c^2}, \quad k_x^2 = \frac{\omega^2}{c^2} \frac{\varepsilon_{\mathrm{m}} \varepsilon_{\mathrm{d}}}{\varepsilon_{\mathrm{m}} + \varepsilon_{\mathrm{d}}}, \quad \varepsilon_j < \frac{\varepsilon_{\mathrm{m,real}} \varepsilon_{\mathrm{d}}}{\varepsilon_{\mathrm{m,real}} + \varepsilon_{\mathrm{d}}} \tag{8.7}$$

In fact, in order to have an intersection between the light line and the SPP curve, the conservation of k_x implies

$$\varepsilon_j > \frac{\varepsilon_{\mathrm{m,real}} \varepsilon_{\mathrm{d}}}{\varepsilon_{\mathrm{m,real}} + \varepsilon_{\mathrm{d}}} \tag{8.8}$$

which is exactly the *oppost e* of the condition necessary for creating a confined (evanescent) surface wave (this is valid even in the limit case of $k_z = 0$). Hence, direct light (we have been referring to light transporting only a real wave vector, such as a plane wave) cannot excite SPP on an ideally flat and infinitely extended surface.

In 1968 Raether, Kretschmann, and Otto showed a solution to this problem. For the sake of space, we are going to deal only with the well-known Otto configuration, but the results are extendable to the Kretschmann configuration. The scheme showing the Otto configuration is given in Fig. 8.1 when the prism is also taken into account. In particular, the structure can be seen as a sequence of three materials defined by three different dielectric functions: ε_{p}, ε_{d}, and ε_{m}. These represent the prism, the dielectric, and the metal, respectively. Three conditions are needed to create SPP at the dielectric-metal interface: (1) $\varepsilon_{\mathrm{p}} > \varepsilon_{\mathrm{d}}$; (2) the incoming light goes through the prism above the critical angle θ_{c} for generating total internal reflection (TIR); and (3) the metal slab must be close enough to the prism to catch the evanescent field originating at the prism-dielectric interface.

The steps leading to the generation of SPP when the previous three points are fulfilled are as follows. Fig. 8.2b shows the dispersion relation for the interface between silver and air. Furthermore, two light lines are shown: One is associated with the propagation of free photons in air (ε_{d}), the next with glass (ε_{p}, material composing the prism). In both cases the limiting situation of $k_z = 0$ is assumed. The figure clearly shows the existence of a light line crossing the SPP curve at at least one point. Remember, this is the condition necessary for creating SPP. Hence Fig. 8.2b provides a very important hint for realizing a device that can couple incoming photons with local electrons to realize SPP. In fact, if it is possible to transfer the characteristics of the glass light line to the air-metal interface, then SPP can be generated.

The goal is easily reached thanks to the three conditions mentioned above. If the incoming light enters in a prism (ε_p) and it undergoes TIR, an evanescent field will be created at the glass-air interface, as shown in Fig. 8.1. This decaying field will carry the optical characteristic of the prism (glass). Finally, if the metallic slab is located in proximity of the prism then it will be able to feel the evanescent field and a situation such as that shown in Fig. 8.2b will be realized: SPP will be created thanks to the coupling of incoming light and local free electrons.

At this point the reader might wonder on the necessity of having a TIR for the purpose of transferring the optical information from the prism to the metal. Could just a standard diffracted beam be enough? This can be explained by means of the Snell relation. In fact, when light impinges on the glass-air interface at an angle θ, the diffracted beam is related to the incoming one by the expression $n_p \sin\theta_p = n_d \sin\theta_d$, where n_p and n_d are the refractive index of the prism and the surrounding dielectric, respectively. By adjusting this expression with the proper terms, we can obtain

$$k_{x,p} = \frac{\varepsilon}{c}\sqrt{\varepsilon_p}\sin\theta = \frac{\omega}{c}\sqrt{\varepsilon_d}\sin\theta_d = k_{x,d} \tag{8.9}$$

Finally, from Eq. 8.9 we can conclude that $\varepsilon_d > \varepsilon_d \sin^2\theta_d = \varepsilon_p \sin^2\theta$. This condition cannot produce a light line intersecting the SPP curve of the air-metal interface, as shown in Fig. 8.2b. Namely, no diffracted beam can create SPP. The solution, then, is to employ the total internal reflection effect. In fact, when the incident angle θ is equal to the critical angle θ_c, the resulting beam travels at an angle $\theta_d = \pi/2$. For $\theta > \theta_c$ the incoming $k_{x,p}$ will keep growing, determining an increase of $k_{x,d}$ over the limit $\varepsilon_d\omega^2/c^2 = k^2_{x,d}$ dictated by Eq. 8.1. This will determine the rise of an imaginary $k_{z,d}$, which will guarantee conservation of the energy. This imaginary quantity is also responsible for the evanescent behavior of the diffracted beam. At this point the system will present a $k_{x,d}$ much bigger than the one originating from a simple dielectric-silver interface, namely that the associated light line will be able to intersect the SPP curve.

Even though we have been focusing only on a flat metallic surface and on the possibility of creating SPP, the results we have obtained can in principle be extended to many more plasmonic systems. For example, metallic cylinders surrounded by dielectric or adiabatic cones can also support SPP. Because all these devices present their own dispersion relations, specific theoretical investigations are needed. However, the general idea of having a surface wave strongly confined at the metal-dielectric interface remains valid.

The situation is slightly different when a nanoparticle of dimensions much smaller than the wavelength of the incoming light is considered (quasi-static approximation). In this case no dispersion relation can be defined and the plasmonic effects rely mostly on the material characteristics. However, this is a regime which starts working, for the visible light, with particles less than 20 nm in diameter, so we will not consider it here. When the nanoparticles have larger dimensions, the situation is quite different. For example, surface-enhanced Raman scattering (SERS) works based on the signal originating by single metallic (silver or gold) scatterers diffused on a substrate. Each scatterer can be described as a nanoparticle that enhances the local electric field for better Raman measurements. Similarly, tip-enhanced Raman spectroscopy (TERS) relies on devices with a tip-like shape. This geometrical design, usually realized by a metal coating of STM or AFM tips, has the characteristic of enhancing the electric field, especially around the tip apex. The spatial resolution depends on the tip radius, which can be as small as 10 nm. Improvement of TERS devices is offered by surface plasmon polariton–enhanced Raman scattering (SPPERS) systems, where a decoupling between light source and detection spot allows us to neglect the background signal.[7]

8.3 Plasmonic Devices

One of the main advantages of plasmonic phenomena is the possibility of shortening the wavelength of the light without changing its frequency. This fundamental characteristic results in devices that work at wavelengths well below the diffraction limit.

Clearly, this can be used in a number of ways, such as transferring a signal along extremely limited spatial channels or confining light in pointlike spots.

This last aspect is key for any kind of nanoimaging or nanospectroscopy. For example, SERS needs a number of local hot spots on a wide surface, while TERS and SPPERS make use of elongated structures for obtaining a single but easily controlled hot spot. Also the fabrication procedure for SERS and TERS/SPPERS devices is quite different. In fact, while SERS simply requires a properly prepared surface, either electrochemically or through metal nanoparticles deposition (e.g., metal evaporation on flat surfaces or on self-organized free standing anodic porous alumina), TERS/SPPERS assumes more sophisticated fabrication techniques such as electron beam (e.g., nanoaggregate or nanoantenna arrays) and focused ion beam (e.g., adiabatic nanocone) lithography. These approaches, even though time- and cost-consuming, allow a fabrication precision down to few nanometers, guaranteeing the realization of extremely fine nanostructures.[8–10]

8.3.1 Gold-Coated Si Substrate

Since SERS was discovered,[11] it is well known that when a Raman scatterer is placed on or near to a roughened noble metal substrate, Raman intensity can be greatly enhanced. Strong electromagnetic fields are generated when a localized-surface plasmon resonance (LSPR) of nanoscale roughness on a silver, gold, or copper substrate is excited by visible light.

In the example in Fig. 8.3, a Si (001) substrate was gold coated to a thickness of around 50 nm. The samples were then utilized for chemical analysis of myoglobin proteins. Myoglobin, a heme protein, was deposited using the drop-coating deposition (DCD) technique,[12] in which the 2 μL volume substance was dropped over the metal surface and left until excess water evaporated. After the water evaporation, a coffee-ring shape formed over the substrate, with a high density of molecules on the ring periphery and fewer molecules as one moves from the periphery to the center. SERS measurements were performed on different points, from the periphery to the center of the coffee-ring structure.

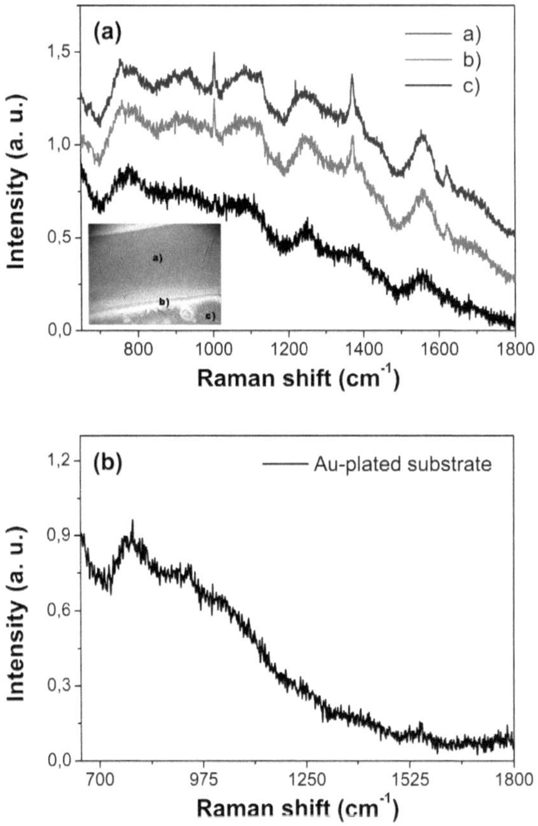

Figure 8.3 SERS spectra of myoglobin collected at different positions on a hydrated drop, from periphery (black line) to center (blue line), deposited on a flat gold surface. The associated point of measurement is indicated in the inset. The SERS background spectrum is reported in the second panel. See also Color Insert.

The SERS measurements were carried out using an 830 nm laser wavelength (power = 1.2 mW and accumulation time = 100 s). SERS spectra of hydrated myoglobin (60 μM) at different positions on the drop are shown in Fig. 8.3a. The characteristic peaks of myoglobin (1003, 1126, 1373, and 1560 cm^{-1}) are clearly observed and can be attributed to the phenylalanine C–N stretching,[13] an oxidation marker band of heme iron,[14] and a C–C vibrational band, respectively. The myoglobin spectral peak intensity decreases as the focus moves from periphery to the center, due to the fact that molecular density decreases closer to the center. The associated points of measurement are shown in the inset in Fig. 8.3a. The SERS background spectrum is illustrated in Fig. 8.3b. The evaluated enhancement factor for the gold flat surface with respect to the Si surface is of just one order.

8.3.2 Gold-Coated Honeycomb APA Structure

Recently, the use of anodic porous alumina (APA) has been proposed as a self-organized template for patterning over a large area of a noble metal structure, due to its high potential in the opto/plasmonic field.[15] The design of a honeycomb-like device was then implemented in order to theoretically simulate the gold structures supported by the APA template. The electric field distribution, when an X-polarized source is employed, is shown in Fig. 8.4. We can easily recognize that the maxima of the field tends to localize inside the hexagonal air holes (diameter = 70 nm) providing an enhancement factor around 10 (evaluated with respect to the light source).

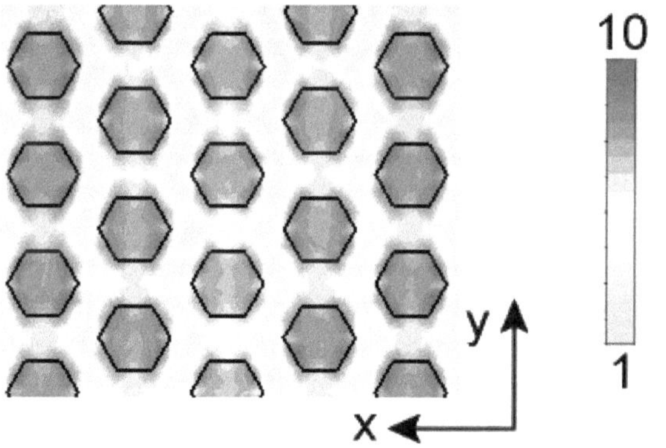

Figure 8.4 Distribution of the absolute value of the total electric field in the XY plane. The chosen wavelength is λ = 633 nm. The vertical bar shows a logarithmic scale.

Here a commercial anodic porous alumina (Whatman Anopore, purchased from SPI Supplies, Westchester, PA) with a pore diameter of around 200 nm was used as a starting substrate for the fabrication of a novel SERS device. This APA comes as self-standing discs, of which the oxide layer has been set free of the aluminum substrate.

The APA has a nominal pore density of $10^9/cm^2$ and a surface porosity of 25%–30%. In Fig. 8.5a, a SEM image of the typical surface of bare APA is presented. Large pores are visible, with some very thin oxide wires occasionally bridging the opposite walls. In order to endow the surface with plasmonic functionality, gold was thermally evaporated from a tungsten boat onto one side face of the APA, starting the deposition at a base chamber pressure of 2.0×10^{-6} mbar, and proceeding at a rate of 0.5 Å/s until a final total gold thickness of 25 nm was reached. The surface topography was once again inspected by SEM, Fig. 8.5b, in order to investigate the morphological modification induced by the gold coating.

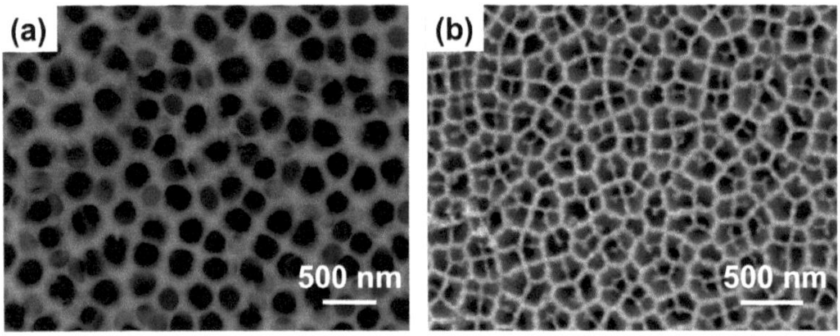

Figure 8.5 Representative SEM micrographs of APA surface (a) before and (b) after 25 nm gold coating.

The free-standing APA sheet was simply dipped in a solution containing cresyl violet (CV) at a concentration of 3.46 μM and incubated at room temperature for 15 minutes. The sheet was then removed from the solution and gently rinsed to eliminate excess dye molecules not chemisorbed to the surface, blown dry with nitrogen flow, and stored in a dessicator before Raman measurements.

SERS measurements were carried out by means of inVia (Renishaw, New Mills, UK) microspectroscopy. The samples were excited at 633 nm with a laser power of 0.12 mW for an accumulation time of 60 seconds by using a 50X objective. The measurements were performed on a gold-coated APA sheet (sample gold APA), CV chemisorbed on the gold-coated APA sheet (sample CV gold APA), and CV drop-coated over a Si substrate (sample CV Si). The results of large-scale SERS measurements for different types of substrates are plotted in Fig. 8.6a. The characteristic vibrational bands of CV are observed in the SERS spectrum.[16,17] Intense Raman bands centered at around 591, 882, 927, and 1189 cm^{-1} can be attributed to the N–H_2 rocking vibration, two benzene group bending, out-of-phase N–H_2 rocking vibration, and a combination of N–H_2 rocking and C–H_x rocking, respectively.[17]

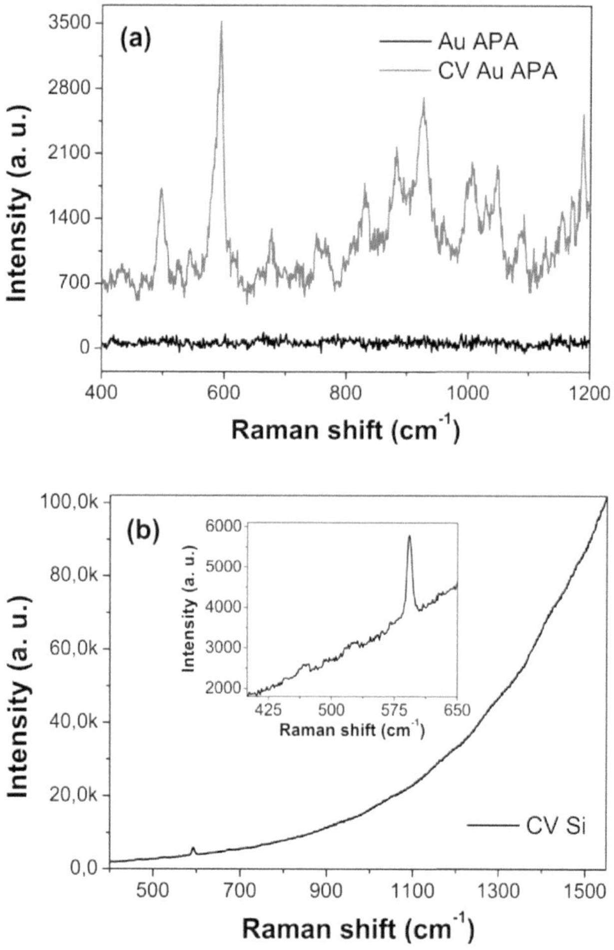

Figure 8.6 (a) Raman spectra of gold APA background and sample covered with CV molecules (concentration 3.46 μM). (b) Sample with CV, in the same concentration as (a), deposited on Si substrate. The inset shows a detailed view between 425 and 650 cm^{-1}. See also Color Insert.

From the Raman bands observed in Fig. 8.6a we can expect that the CV is oriented with the N–H_2 group closer to the gold film, leading to the strong Raman absorption and, consequently, a higher Raman signal. In order to compare the SERS capability of this device (CV gold APA sample), further measurements of CV drop deposited on a silicon wafer were also carried out (CV Si sample). The Raman spectrum of this structure is reported in Fig. 8.6b. We can clearly observe that a sharp band appears around 591 cm^{-1}, with an exponentially increasing fluorescence background. No other peaks related to CV are observed in this case. We should also point out that our SERS device is acting with a very strong fluorescence quenching, similar to the results observed in the past.[18] The evaluated enhancement factor is around 10^6.

8.3.3 Gold Square-Like Nanostructure

In order to increase local field enhancement, a single square-like nanoparticle (side = 80 nm and height = 20 nm) was investigated.

Figure 8.7 shows the absolute value of the electric field distribution with an X-polarized source. Four hotspots, presenting an enhancement around 75 compared to the field source, are observed. It is interesting to note that the hot spots are more spatially confined with respect to the honeycomb device thus explaining the higher enhancement factor of the square-like nanoparticle.

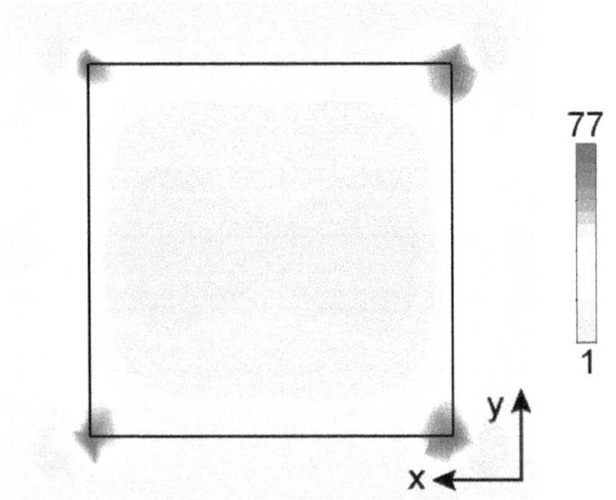

Figure 8.7 Distribution of the absolute value of the total electric field in the XY plane. Four hotspots corresponding to the square corners are observed. The chosen wavelength is λ = 633 nm. The vertical bar is on a logarithmic scale.

Similar devices have been fabricated by electron-beam lithography (Raith 150-Two) on Si substrates (Fig. 8.8). A layer of polymethyl-methacrylate (PMMA) resist with a thickness of around 85 nm was spin-coated on top of the cleaned Si surfaces. Thereafter, the resist film was dried in air at 180°C for 8 minutes. An e-beam direct writing system was used to define the pattern. After the resist was developed in a 1:3 solution of methylisobutyleketone (MIBK) and isopropanol (IPA) for 40 seconds, titanium and gold were deposited by means of e-beam evaporation technique (Kurt J. Lesker deposition system).

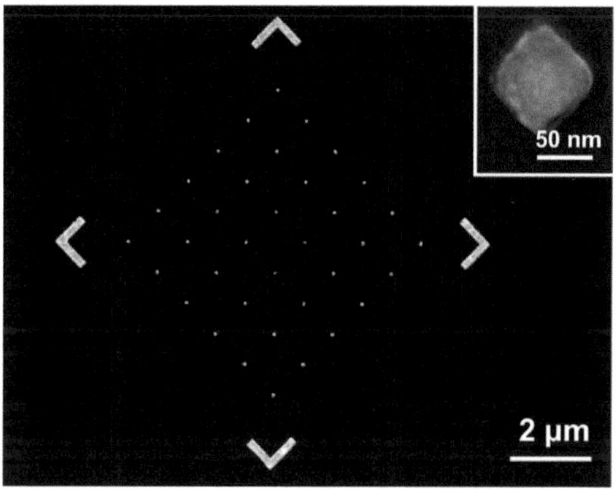

Figure 8.8 SEM image of the whole SERS device. Big markers facilitate the recognition of the Raman active region. The square-like particle dimensions are 80 × 80 nm^2 with a thickness of 20 nm. The inset emphasizes not only the good liftoff condition but also the well-controlled nanoparticle dimensions.

R6G with a concentration of 3.46 μM was deposited over a SERS substrate using chemisorption, as discussed in Section 8.2.2. SERS measurements were performed by means of InVia (Renishaw) microspectroscopy with a 100 X objective using a He–Ne laser (633 nm).

Figure 8.9 shows the SERS spectrum on the bare nanostructured device (black line) reporting no evidence of Raman band in the range of 1100 and 1900 cm^{-1}. After chemisorption deposition of R6G, characteristic peaks of the molecules are clearly evidenced at 1183 cm^{-1} (combination of C–H bending and N–H bending vibration of xanthene ring), 1361, 1509, and 1575 cm^{-1} (C–C stretching vibration of xanthene ring), and 1650 cm^{-1} (C=O stretching). The calculated SERS enhancement in this case is found to be 10^5.

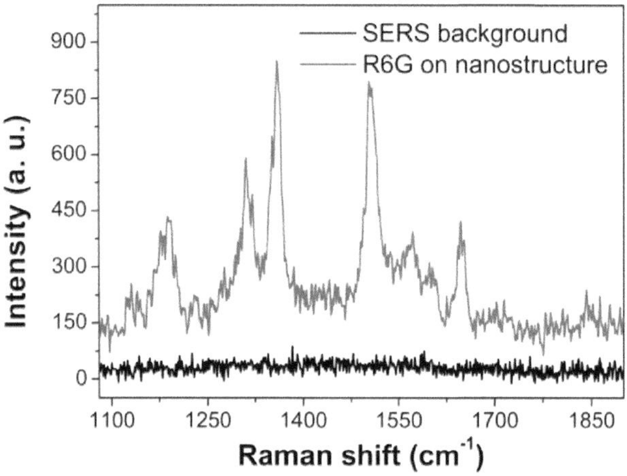

Figure 8.9 SERS spectra of bare nanostructured device (black line) and square-like nanoparticle covered with R6G molecules (red line). See also Color Insert.

8.3.4 Gold Nanoaggregate Array

A SERS device, consisting in array of gold nanoaggregates, was simulated as shown by the distribution of the absolute value of the electric field in Fig. 8.10c. Hotspots can be identified all over the structure. Their presence is motivated by the morphological irregularity of the device. The combination of sharp structure and gap variability among the nanoaggregates (2–10 nm) plays a fundamental role in the enhancement factor, that under a X-polarized source at λ = 830 nm is around 600 (evaluated with respect to the light source).

Similar SERS devices were fabricated on a Au–Cr base-plated Si wafer using a high resolution fabrication technique, electroplating, and e-beam lithography, as shown in Fig. 8.10. Several patterns of nanoholes were designed, using a Crestec CABL-9000C EBL system, on a layer of electronic-resist spun onto the wafer. Gold nanograins were grown inside the nanoholes by using gold-plating growing, and then the resist film was removed. In this way, nanostructures of gold were obtained. The nanograin size and the mutual distance between two nanoaggregates are 80–140 nm and 10–30 nm, respectively.

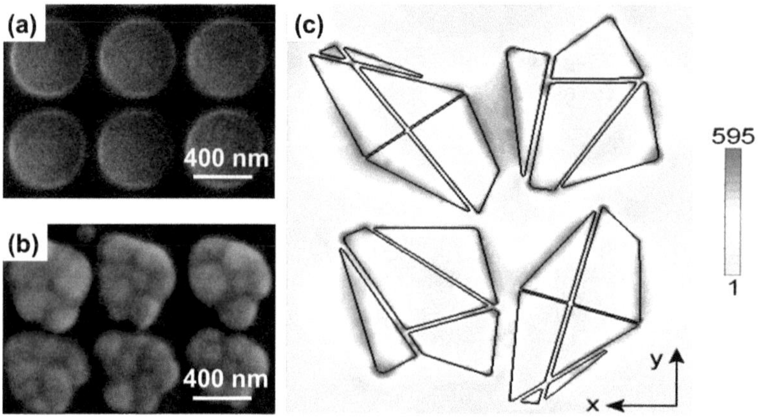

Figure 8.10 SEM micrographs of (a) an array of empty holes fabricated by EBL technique and (b) gold nanoaggregates selectively grown inside the nanohole pattern. (c) Distribution of the absolute value of the total electric field in the XY plane. The chosen wavelength for the illuminating source is λ = 830 nm. The vertical bar is on a logarithmic scale.

Micro-confocal Raman measurements (Renishaw) were excited by means of an 830 nm laser wavelength (power = 6 mW and accumulation time = 100 s) in backscattering geometry with a spectral resolution of about 1.3 cm^{-1}. Various proteins (myoglobin, bovine serum albumin (BSA), ribonuclease-B (RNase-B), and lysozyme) dissolved in 2 μL water was deposited, following the DCDR method, on the nanostructures.[12] It should be noted that Zhang, *et al.*[12] carried out many measurements for the aqueous samples and their corresponding DCDR samples, and observed that the spectra resemble each other. Moreover, after each deposition, it was ensured through optical image that the ring (inner diameter 3.5 mm and width 0.08 mm) of adsorbed protein remains on the nanograin-aggregates array.

Figure 8.11a shows the SERS spectra of myoglobin protein with different concentration in the range of 500–2000 cm^{-1}, deposited over the SERS device. Myoglobin (~75% α-helix) contains one planar Fe protoporphyrin prosthetic heme group implanted in polypeptide chains. The vibrational bands for ring π-bindings fall in the 1000–1700 cm^{-1} region. Raman spectrum shows well-known Mb marker vibrational bands of protein centered at around 1125, 1373, and 1560 cm^{-1}, which are attributed to the C–N stretching, an oxidation marker band of heme group, and the C–C vibrational band, respectively.[13,14]

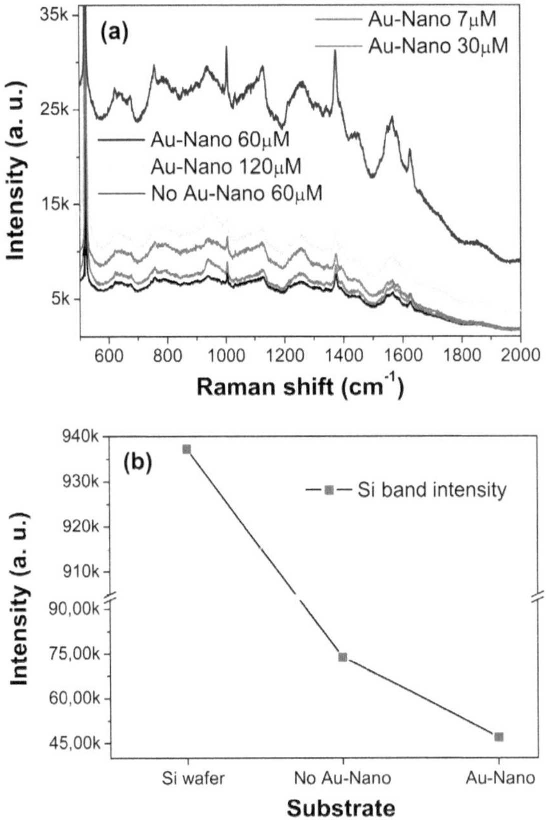

Figure 8.11 (a) SERS spectra of myoglobin at different concentrations deposited over gold nanoaggregates. Panel (b) shows the Raman intensity of the reference Si band for Si, no-gold nanostructures, and gold nano-samples. See also Color Insert.

In addition, the bands at around 760, 1013, 1365, and 1555 cm^{-1} in the same figure are related to the tryptophan, whereas the bands around 1005 and 1033 cm^{-1} are related to the phenylalanine residues. The sharp band centered at 520 cm^{-1} in the Raman spectrum is from the Si substrate. Raman spectra, performed at both positions, of myoglobin with various concentrations on the gold nanoaggregates (Au-Nano) structure and myoglobin on the no-gold nanostructures (No Au-Nano) are illustrated in Fig. 8.11a. The figure shows that the myoglobin SERS intensity reduces with the concentration and the Si substrate contribution increases at position Au-Nano in comparison to position No Au-Nano. The measurements were also performed for myoglobin deposited on Si wafer. A faint peak of Phenylalanine residue is observed, whereas rest of the protein Raman is covered by strong Si Raman band (figure not shown here). The intensity of Si band (520 cm^{-1}) for various measurements carried out for myoglobin on a Si wafer, nanograin aggregates, and on a base plate without nanostructures is illustrated in Fig. 8.11b. If we assume the size of one myoglobin molecule to be 20 nm^2, the minimum amount of myoglobin within the focal spot of laser (radius = 1 μm) could be, therefore, estimated at around 10–240 attomole.[19]

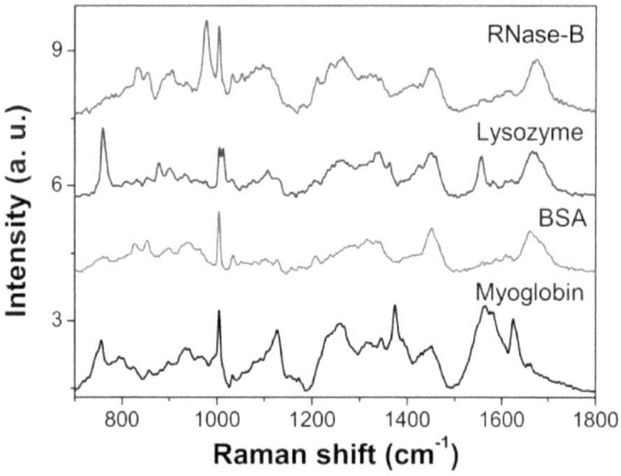

Figure 8.12 SERS spectra of different proteins deposited over a gold nanoaggregate device: myoglobin, bovine serum albumin, lysozyme, and ribonuclease-B refer to black, red, blue, and cyan traces respectively. See also Color Insert.

The SERS spectra of proteins (myoglobin, BSA, RNase-B and lysozyme) were illustrated in Fig. 8.12. The SERS spectrum of

BSA (red line) is dominated by bands due to the amide I (around 1650 cm^{-1}, mainly by ν–C O vibration), C–H_3 and C–H_2 deformation vibration from side chains of different amino acids (in the range of 1440–1480 cm^{-1}), the symmetry stretching of a –COO^- group (around 1420 cm^{-1}),[20] the amide III band from the in-plane δ–NH group and ν–C–N (1230–1300 cm^{-1}), and the contribution from different amino acids: Phe (~620, 1005, and 1033 cm^{-1}),[21] Trp (1011 and 1560 cm^{-1}),[22] Tyr (645, 825, 855, 1210, and 1620 cm^{-1}), and cysteine (655, 672, and 720 cm^{-1}).[21,23]

The SERS spectrum for lysozyme, Fig. 8.12 (blue line), shows its characteristic bands. The amide I band with a major contribution of C=O stretching vibration can clearly be observed ranging between 1630 and 1700 cm^{-1}, whereas amide II (N–H bending coupled with C–N stretching) and amide III (C–N stretching mixed with N–H bending vibration) are observed in the range between 1530–1580 cm^{-1} and 1220–1300 cm^{-1}, respectively. The Raman spectrum shows a high signal-to-noise ratio with respect to the lysozyme Raman spectrum, carried out on the gold-coated Teflon substrate (concentration = 100 μM, integration time = 100 s)[12] even though the lysozyme, in our case, has much less concentration, 7 μM, and less integration time, 50 seconds. The intense peak centered at 760 cm^{-1} is clearly observed for the lysozyme protein, irrespective of other proteins. This band, sensitive to the hydrophobic environment of Trp residues, is attributed to the indole ring breathing vibration of Trp residues (W18).

The SERS spectrum of RNase–B (cyan trace) shows various vibrational bands, attributed to different amino acids, especially from Phe residues at around 1005, 1033, and 1210 cm^{-1}, Tyr residues at around 830 and 855 cm^{-1}, different scissor and wagging vibration related to the C–H_x ($x = 1 - 3$) at about 1450 cm^{-1} and 1320 cm^{-1}, and C–S vibrational bands centered at 625 and 645 cm^{-1}, respectively. One peak centered at around 980 cm^{-1} can be observed only for RNase–B among the other proteins described above. This band is attributed to the symmetric stretching vibration of the sulphate ion, which acts as an inhibitor of enzymatic action. It has been found[24] that the reactive sites for the sulphate ion of RNase–B are histidine residues. Daly, *et al.*[25] discusses the loss of symmetry of sulphate ion by protonation causes the disappearance of band at around 980 cm^{-1}.

The SERS spectra for various proteins by varying the temperature in the range between –65°C and 90°C are discussed by Das, *et al.*[26]

8.3.5 Gold Nanoantenna Array

The influence of the geometrical shape on the optical properties of metal nanoparticles has been extensively demonstrated in Sections 8.2.2, 8.2.3, and 8.2.4. However, while a symmetrical geometry allows polarization independent plasmon excitation, a dichroic absorption is well expected for nanoantenna (NA) fabrication.

The schematic diagram summarized in Fig. 8.13 elucidates the main steps involved in the fabrication process. A 120 nm–thick layer of PMMA (950 K) was spin-coated on a CaF_2 (100) substrate. To prevent the charging of the sample during electron exposure, a thin aluminum layer (10 nm in thickness) was thermally evaporated on the PMMA surface. Electron beam direct-write of nanoantenna patterns were carried out with a high-resolution Raith150-Two e-beam writer at 10 keV beam energy and 180 pA beam current. After the aluminum was removed in a KOH solution, the e-beam exposed resist was developed in a conventional solution of MIBK: IPA (1:3) for 30 seconds, resulting in a mask for the subsequent metal evaporation process. Then e-beam evaporation was employed to produce a 5 nm adhesion layer of titanium and a 60 nm gold film with a 0.3 Å/s deposition rate in a low 10^{-7} mbar vacuum chamber (Kurt J. Lesker PVD75). Finally, the unexposed resist was removed with acetone and rinsed out in IPA.

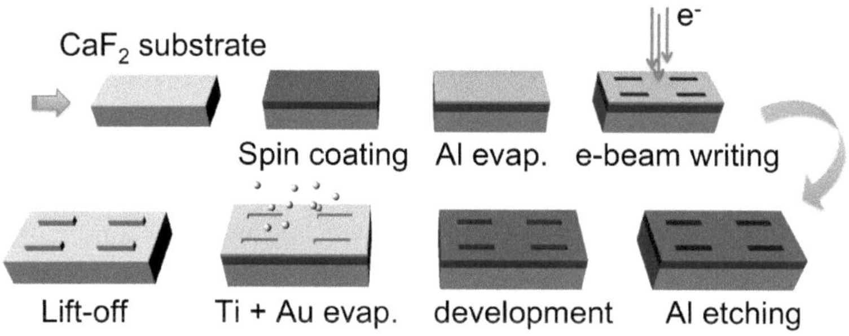

Figure 8.13 Schematic block diagram of the fabrication process adopted in nanoantenna production.

In the following procedure, we fabricated large arrays of nanoantennas achieving a high degree of reproducibility at the sub-10 nanometer scale. Large scan overviews of different nanoantenna designs are reported in Fig. 8.14a and b. The sample topography has been characterized recurring to scanning electron microscopy (SEM) and atomic force microscopy (AFM, Veeco MultiMode with NanoScope V controller) equipped with ultra-sharp Si probes (ACLA–SS, AppNano) and operated in tapping mode. Fig. 8.14c and d show a representative SEM image of an individual structure and its polycrystalline nature. The nanoantenna top surface is not ideally smooth but has an RMS roughness value around 1 nm.

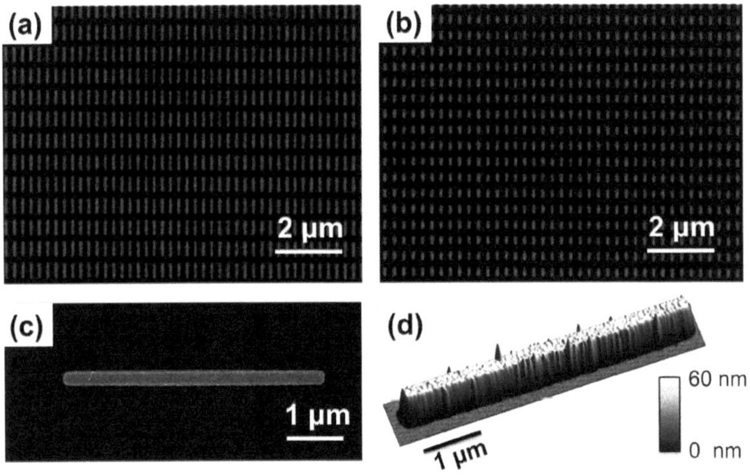

Figure 8.14 Representative SEM images of two different nanoantenna arrays fabricated using the EBL technique (a,b). The dimensions of the NAs are of length 410 nm (a) and 200 nm (b), while width and height both are 60 nm for all NA. SEM micrograph of individual NA is shown in (c). (d) AFM 3D image.

The optical properties of the nanoantenna array, shown in Fig. 8.15a, were investigated by means of spectroscopic transmission of polarized light in the 400–900 nm range. We used a fiber-optic spectrometer (AvaSpec-256, Avantes) and the light source was a combined deuterium-halogen lamp (AvaLight–DHc, Avantes).

The polarization of the incident light was varied from transverse magnetic TM (electric field perpendicular to the short

NA axis) to transverse electric TE (electric field parallel to the long NA axis), as reported in Fig. 8.15a. The optical transmittance spectra present evidence of a clear anisotropic behavior; for TM polarization a localized minimum around 620 nm is found. This is the typical behavior exhibited by subwavelength metal nanoparticles sustaining a localized surface plasmon resonance.[27] For TE polarization, no particle plasmon excitation is observed, while the weak maximum in the transmittance spectrum can be associated with the grating periodicity of the nanoantenna array.[28]

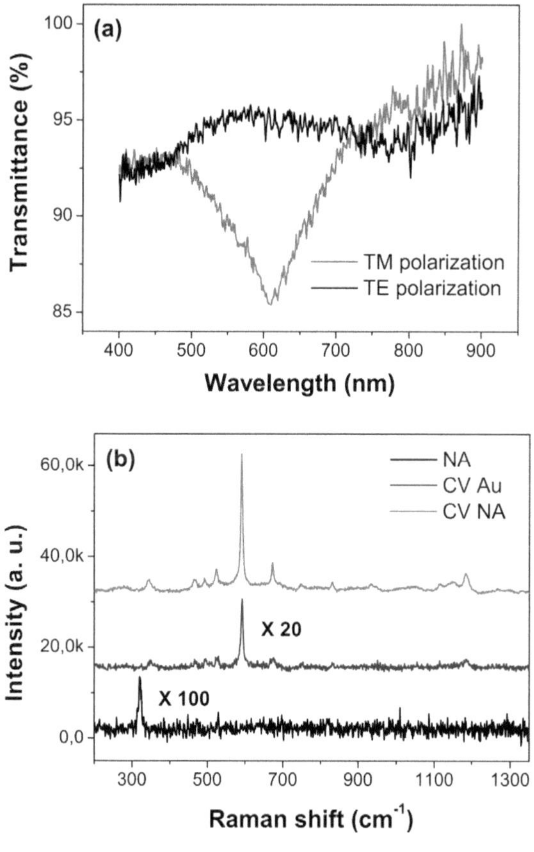

Figure 8.15 (a) Transmission optical spectra of the NA array of Fig. 8.12a. Red and black lines are for TM and TE polarization, respectively. (b) SERS spectra of cresyl violet molecules deposited on NA sample (red line) and flat gold film (blue line). Background measurement performed on bare NA structure is also reported (black line). See also Color Insert.

In order to probe the enhancement behavior in Raman spectroscopy related to plasmon excitation, the NA sample was immersed in a CV solution (3.46 μM in H_2O) for 15 minutes. The sample was gently rinsed in DI water to remove any molecules not chemisorbed on the metal surface and then dried in nitrogen flow. SERS measurements were carried out by means of inVia microspectroscopy (Renishaw). The samples were excited by 633 nm laser wavelength (laser power = 0.14 mW and accumulation time = 50 s) through a 150 X objective. The measurements were performed on bare nanoantenna arrays (sample NA), on cresyl violet molecules chemisorbed on flat gold film (sample CV Au), and on CV deposited on the nanoantenna arrays (sample CV NA).

To ensure the quality of the sample surface, the background measurements were performed at different positions of the nanoantenna SERS device. As shown in Fig. 8.15b (NA line) no characteristic Raman band, except at around 320 cm^{-1} related to the Ca–F vibration from the substrate, is observed. As discussed in Section 8.2.2, the characteristic vibrational bands of CV molecule, centered at 591, 882, 927, and 1189 cm^{-1}, are observed in the SERS spectrum (CV NA trace).[16,17] CV molecules chemisorbed over flat gold film are also shown in Fig. 8.13b. We can clearly observe that there is a giant enhancement in Raman signal for the CV NA sample compared to the CV Au sample. The evaluated SERS enhancement for the fabricated nanoantenna device is 10^7.

8.3.6 Silver-Coated Adiabatic Nanocone

Nanostructures with sharp features can produce highly localized electromagnetic fields, exploiting the strong combination between plasmon resonances and geometrical effects (the so-called lightening rod effect).[10,29–31]

A new, simulated device based on the adiabatic propagation of surface plasmon polaritons on a conical tip is shown in Fig. 8.16. The structure presents an extremely high electromagnetic field corresponding to the tip apex, a result that can be explained, in a rough approximation, through a simple electrostatic model or, more precisely, through the decrease in wavelength associated to the field.

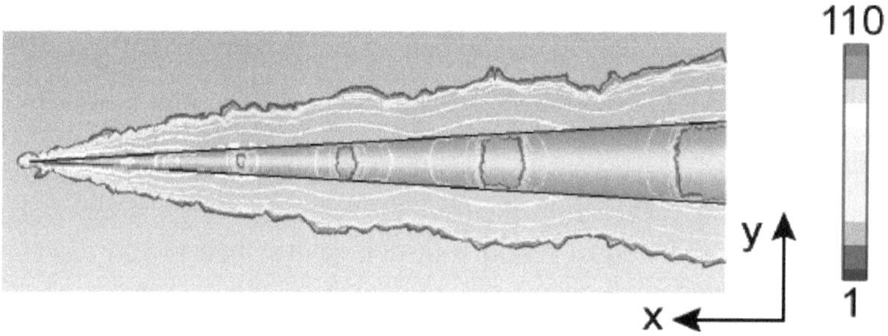

Figure 8.16 Distribution of the absolute value of the total electric field in the *yz* plane. The chosen wavelength is λ = 514 nm. The dimensions of the silver cone are length 2500 nm and base diameter 300 nm, corresponding to an angle of 0.06 rad. Only the tip apex (the last 250 nm) is shown in the figure.

Figure 8.16 shows a three-dimensional numerical calculation of the field distribution around the silver nanostructure (height = 2500 nm and base diameter = 300 nm), which was analyzed recurring to a commercial software (CST Studio Suite 2010). The silver optical dispersion was considered in a Drude-Lorentz scheme, while no spatial dispersion was taken into account. The results confirm both the adiabatic compression and the associated field enhancement.

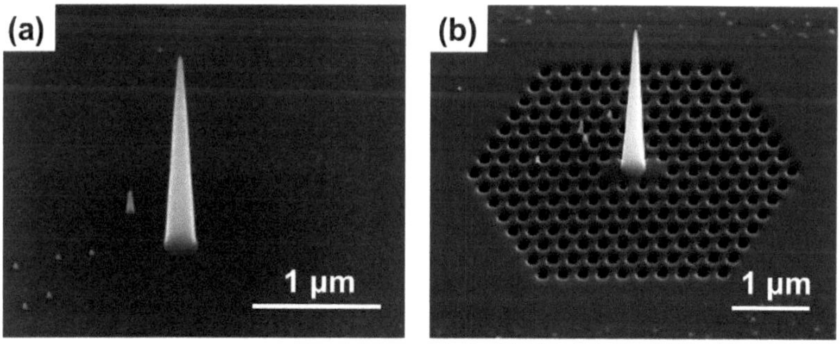

Figure 8.17 (a) SEM image of a nanocone fabricated on a Si_3N_4 membrane. The nanocone height is about 2 μm, the base is 280 nm, and the radius of curvature of the tip is below 5 nm. (b) SEM image of the whole device including the photonic cavity and the adiabatic cone.

A bottom-up nanofabrication approach has been exploited in order to fabricate a cone with identical geometrical properties as the simulated one, combined with a photonic crystal (PC) cavity acting as an efficient coupler between the external optical source and the adiabatic resonator.[32] The device was fabricated in two different steps: (a) First, focused ion beam milling (FEI Novalab 600) was employed to produce the PC cavity, then (b) an electron beam-induced CVD was used to fabricate the nanocone at the center of the PC cavity. During the ion milling process, the beam current and the acceleration voltage were set at 50 pA and 30 kV, respectively. The adiabatic cone has been grown in the centre of the cavity, exploiting a gas precursor which contains platinum-carbon polymer, $(CH_3)3Pt(C_pCH_3)$. The final device is reported in Fig. 8.17a and b.

In order to access the SPPERS capabilities of our device, a monolayer of benzenethiol was deposited, via chemisorption technique, over silver-coated nanocones with various radius of curvature (RoC from 30 nm to 10 nm). In all the experiments a 150 X objective lens with a power source equal to 0.09 mW, an accumulation time of 100 seconds, and an excitation wavelength of 514 nm was used. The SPPERS spectra associated with the benzenethiol monolayer deposited on the nanocones with different RoC are shown in Fig. 8.18.

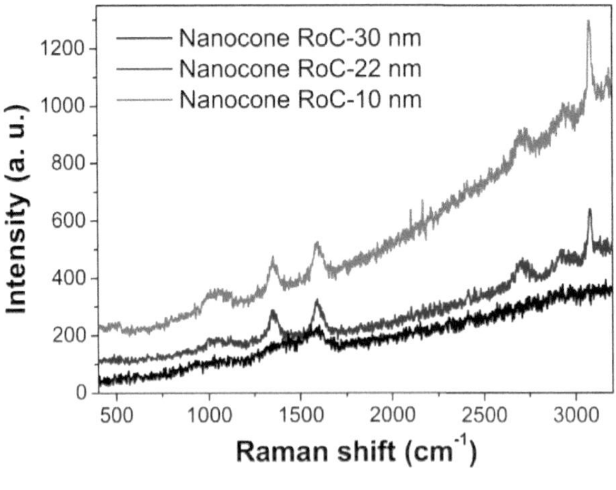

Figure 8.18 SPPERS spectra coming from a layer of benzenethiol deposited on conical silver structures with three different RoC. See also Color Insert.

We found that for the nanocone with RoC 30 nm the chemical information of the substance under investigation is not revealed because of the low field enhancement. However, when the nanocone RoC is sharpened, the vibrational bands of the molecule start to become visible, showing the molecular chemical information.

8.4 Conclusion

Various plasmonic-based devices, starting from the simplest technique to the highly specialized and precise geometry for SERS/SPPERS application, were created by combining numerical simulation and nanofabrication tools. Today, advanced nanofabrication methods provide strong control over the geometry, architecture, and functionality of the resulting nanostructures. The combination of nanostructures with the plasmon generation allows us to achieve an enhanced optical signal by means of spectroscopic techniques that increase detection sensitivity down to the single-molecule level. Some nanostructures (nanocone) are compatible with the scanning probe system, which makes it possible to study single molecules on cell surfaces and tissues.

We hope that, in the future, a new chip array based on the nanoantenna concept will emerge as an element for a highly sensitive sensor.

References

1. Wood, R. W. (1902). On a remarkable case of uneven distribution of light in a diffraction grating spectrum, *Phl . Mag.*, **4**, 396–402.
2. Fano, U. (1956). Atomic theory of electromagnetic interactions in dense materials, *Phys. Rev.*, **103**, 1202–1218.
3. Otto, A. (1968). Excitation of nonradiative surface plasma waves in silver by the method of frustrated total reflection, *Z. Phys.*, **216**, 398–410.
4. Kretschmann E., and Raether H. (1968). Radiative decay of non-radiative surface plasmons excited by light, *Z. Naturf.*, **23A**, 2135–2136.
5. Sakoda, K. (2001). *Optical Properties of Photonic Crystals*, Springer-Verlag, Berlin.

6. Rakić, A. D., Djurišić, A. B., Elazar, J. M., and Majewski, M. L. (1998). Optical properties of metallic films for vertical-cavity optoelectronic devices, *Appl. Opt.*, **37**, 5271–5283.
7. De Angelis, F., Patrini, M., Das, G., Maksymov, I., Galli, M., Businaro, L., Andreani, L. C., and Di Fabrizio, E. (2008). A hybrid plasmonic–photonic nanodevice for label-free detection of a few molecules, *Nano Lett.*, **8**, 2321–2327.
8. Tormen, M., Businaro, L., Altissimo, M., Romanato, F., Cabrini, S., Perennes, F., Proietti, R., Sun, H.-B., Kawata, S., and Di Fabrizio, E. (2004). 3D patterning by means of nanoimprinting, X-ray and two-photon lithography, *Microelectron. Eng.*, **73–74**, 535–541.
9. Cabrini, S., Carpentiero, A., Kumar, R., Businaro, L., Candeloro, P., Prasciolu, M., Gosparini, A., Andreani, C., De Vittorio, M., Stomeo, T., and Di Fabrizio, E. (2005). Focused ion beam lithography for two-dimensional array structures for photonic applications, *Microelectron. Eng.*, **78–79**, 11–15.
10. Romanato, F., Businaro, L., Vaccari, L., Cabrini, S., Candeloro, P., De Vittorio, M., Passaseo, A., Todaro, M. T., Cingolani, R., Cattaruzza, E., Galli, M., Andreani, C., and Di Fabrizio, E. (2003). Fabrication of 3D metallic photonic crystals by X-ray lithography, *Microelectron. Eng.*, **67–68**, 479–486.
11. Jeanmaire, D. L., and Van Duyne, R. P. (1977). Surface Raman spectrochemistry: Part I. heterocyclic, aromatic, and aliphatic amines adsorbed on anodized silver electrode, *J. Electroanal. Chem. Interfac. Electrochem.*, **84**, 1–20.
12. Zhang, D., Xie, Y., Mrozek, M. F., Ortiz, C., Jo Davisoon, V., and Ben-Amotz, D. (2003). Raman detection of proteomic analytes, *Anal. Chem.*, **75**, 5703–5709.
13. Spiro, T. G. (1985). Resonance Raman spectroscopy as a probe of heme protein structure and dynamics, *Adv. Prot. Chem.*, **37**, 111–159.
14. Sato, H., Chiba, H., Tashiro, H., and Ozaki, Y. (2001). Excitation wavelength-dependent changes in Raman spectra of whole blood and hemoglobin: comparison of the spectra with 514.5-, 720-, and 1064-nm excitation, *J. Biomed. Opt.*, **6**, 366–370.
15. Nuntawong, N., Horprathum, M., Eiamchai, P., Wong-ek, K., Patthanasettakul, V., and Chindaudom, P. (2010). Surface-enhanced Raman scattering substrate of silver nanoparticles depositing on AAO template fabricated by magnetron sputtering, *Vacuum*, **84**, 1415–1418.

16. Vogel, E., Gbureck, A., and Kiefer, W. (2000). Vibrational spectroscopic studies on the dyes cresyl violet and coumarin 152, *J. Mol. Struct.*, **550**, 177–190.

17. Sackmann, M., Bom, S., Balster, T., and Materny, A. (2007). Nanostructured gold surfaces as reproducible substrates for surface-enhanced Raman spectroscopy, *J. Raman Spect.*, **38**, 277–282.

18. Dulkeith, E., Morteani, A. C., Niedereichholz, T., Klar, T. A., Feldmann, J., Levi, S. A., Van Veggel, F. C. J. M., Reinhoudt, D. N., Moller, M., and Gittins, D. I. (2002). Fluorescence quenching of dye molecules near gold nanoparticles: radiative and nonradiative effects, *Phys. Rev. Lett.*, **89**, 203002_1–4.

19. Das, G., Mecarini, F., De Angelis, F., Prasciolu, M., Liberale, C., Patrini, M., and Di Fabrizio, E., (2008). Attomole (amol) myoglobin Raman detection from plasmonic nanostructures, *Microelectron. Eng.*, **85**, 1282–1285.

20. Gilmanshin, R., Beek, V. B., and Callender, R. (1996). Study of the ribonuclease S- peptide/S-protein complex by means of Raman difference spectroscopy, *J. Phys. Chem.*, **100**, 16754–16760.

21. Li, T., Chen, Z., Johnson, J. E., and Thomas Jr., G. J. (1990). Structural studies of bean pod mottle virus, capsid, and RNA in crystal and solution states by laser Raman spectroscopy, *Biochemistry*, **29**, 5018–5026.

22. Miura, T., Takeuchi, H., and Harada I., (1991). Raman spectroscopic characterization of tryptophan side chains in lysozyme bound to inhibitors: role of the hydrophobic box in the enzymic function, *Biochemistry*, **30**, 6074–6080.

23. Tuma, R., Russell, M., Rosendahl, M., and Thomas Jr., G. J. (1995). Solution conformation of the extracellular domain of the human tumor necrosis factor receptor probed by Raman and UV-resonance Raman spectroscopy: structural effects of an engineered PEG linker, *Biochemistry*, **34**, 15150–15156.

24. Wyckoff, H. W., Tsernoglou, D., Hanson, A. W., Knox, J. R., Lee, B., and Richards, M. F. (1970). The three-dimensional structure of ribonuclease-S: interpretation of an electron density map at a nominal resolution of 2 A, *J. Biol. Chem.*, **245**, 305–328.

25. Daly, F. P., Brown, C. W., and Kester, D. R. (1972). Sodium and magnesium sulfate ion pairing: evidence from Raman spectroscopy, *J. Phys. Chem.*, **76**, 3664–3668.

26. Das, G., Mecarini, F., Gentile, F., De Angelis, F., Kumar M. H. G., Candeloro, P., Liberale, C., Cuda, G., and Di Fabrizio, E. (2009). Nano-patterned SERS substrate: application for protein analysis vs. temperature, *Bio sen. Bio elect.*, **24**, 1693–1699.

27. Bohren, C. F., and Huffman, D. R. (1998). *Abs orptio n and Scatterin g of Lig ht by Small Partic les*, Wiley, New York.

28. Schider, G., Krenn, J. R., Gotschy, W., Lamprecht, B., Ditlbacher, H., Leitner, A., and Aussenegg, F. R. (2001). Optical properties of Ag and Au nanowire gratings, *J. Appl. Phys.*, **90**, 3825–3830.

29. Stockman, M. I. (2004). Nanofocusing of optical energy in tapered plasmonic waveguides, *Phys. Rev. Lett.*, **93**, 137404-1–137404-4.

30. De Angelis, F., Das, G., Candeloro, P., Patrini, M., Galli, M., Bek, A., Lazzarino, M., Maksymov, I., Liberale, C., Andreani, L. C., and Di Fabrizio, E. (2010). Nanoscale chemical mapping using three-dimensional adiabatic compression of surface plasmon polaritons, *Nature Nanotech.*, **5**, 67–72.

31. Babadjanyan, A. J., Margaryan, N. L., and Nerkararyan, Kh. V. (2000). Superfocusing of surface polaritons in the conical structure, *J. Appl. Phys.*, **87**, 3785–3788.

32. Galli, M., Agio, M., Andreani, L. C., Atzeni, L., Bajoni, D., Guizzetti, G., Businaro, L., Di Fabrizio, E., Romanato, F., and Passaseo, A. Optical properties and photonic bands of GaAs photonic crystal waveguides with tilted square lattice, *Eur. Phys. J. B*, **27**, 79–87.

Chapter 9

Surface-Enhanced Infrared Spectroscopy

Frank Neubrech and Annemarie Pucci
Kirchhoff-Institute for Physics, University of Heidelberg, Im Neuenheimer Feld 227, 69120 Heidelberg, Germany
neubrech@kip.uni-heidelberg.de

In studies of small amounts of molecules, conventional infrared (IR) spectroscopy suffers from relative low molecular IR absorption signals compared to background noise. One way to increase sensitivity is to use field-enhanced spectroscopy techniques as surface-enhanced infrared spectroscopy (SEIRS). In this chapter, we introduce two approaches to the SEIRS method: surface-enhanced infrared absorption (SEIRA) using metal nanoparticle films and SEIRS using well-defined resonant metal nanoparticles (nano antenna-assisted SEIRS with individual nanoantennas, multimers, or antenna arrays, respectively).

9.1 SEIRA with Metal Island Films

In 1980, Hartstein, *et al.*[1] found that the IR absorption of molecules on metal films can be considerably enhanced. The enhancement

Nanoantenna: Plasmon-Enhanced Spectroscopies for Biotechnological Applications
Edited by Marc Lamy de la Chapelle and Annemarie Pucci

ISBN 978-981-4303-61-3 (Hardcover), 978-981-4303-62-0 (eBook)
www.panstanford.com

factor can reach the order of 1000, if the molecules are adsorbed on gold or silver island films.[2–5] The effect was called surface-enhanced infrared absorption (SEIRA). The strength of the enhancement depends on the metal type according to the specific dielectric function in the infrared and, because of the relation to the plasmonic excitation spectrum in the IR range, on the metal-film morphology. For molecules directly adsorbed on the metal, interaction with electron-hole pair excitations may also contribute to SEIRA if the lowest unoccupied molecular orbit (LUMO) of a molecule is energetically overlapping with the Fermi energy of the metal,[6,7] In sensing applications with functionalization layers directly on the metal, this first-layer effect is certainly not important. For metal island films, maximum plasmonic enhancement is found at the percolation threshold.[7,8] This maximum enhancement can be understood well, since at percolation the number of adsorbate sites in the smallest possible grooves between neighbor islands becomes a maximum enhancement. In the smallest nanogaps, the local field enhancement is expected to be particularly strong. As the simplest approximation, the field-enhancement scale is the ratio of particle size to gap size. More precisely, the resulting SEIRA signal enhancement (or the enhancement factor, respectively) depends on the electromagnetic resonance spectrum of the near field. Since the islands are usually rather flat and broad and since they have a strong electromagnetic coupling to neighbor islands, such resonance spectra in the IR are extremely broadened. The oscillator strength of these broad resonances depends on the size, shape, and kind of array on the one hand, and, on the other hand, on the metallic properties of the particles; for example, on the relaxation rate of the free charge carriers and on their plasma frequency. In this way the enhancement depends strongly on the metal properties.[9] In reality this means that too-small particles give smaller enhancement because of the strong effect of surface scattering of free-charge carriers. SEIRA studies of extended areas of metal-nanoparticle films always give only information on the average enhancement.[9] The best average enhancement factors amount to three orders of magnitude,[4,5] which includes that higher local enhancement is balanced with lower ones at adsorption sites in wider grooves or at smaller islands.

Model calculations based on effective dielectric functions are able to reasonably describe the experimental trends in IR transmittance spectroscopy. Such model calculations revealed enhanced adsorbate-vibration signals but also a change in the adsorbate line shape with the metal-filling factor of the film,[10–12] in accordance to the measurements. The strongest asymmetry occurs at the percolation threshold.[7,8] The asymmetry is a consequence of a Fano-type effect[12] from the dipolar interaction of the vibrational excitation with the very broad plasmonic excitation. However, unlike a Fano-type interaction between the vibrational excitation and the rather narrow plasmonic excitation of an individual nanoparticle, the full antiphase situation (like an antiabsorption) is not reached in SEIRA from metal island films because of the very broad metal-particle film spectrum that is the result of multiple coupled oscillators.

9.2 Nanoantenna-Assisted Surface-Enhanced Infrared Spectroscopy

Surface-enhanced infrared spectroscopy with well-defined nanoantennas is expected to become well established in life sciences for detecting molecules and proteins. Considerable well-defined signal enhancement can be reached with specially designed nanoparticles, called nanoantennas. When this SEIRS approach is followed, enhancement factors up to 500,000, which correspond to an attomolar sensitivity, have been found.[13] Since enhanced electromagnetic near fields caused by collective charge carrier oscillations (plasmons) are the main contributor to SEIRS signal enhancement,[7] this method is not restricted to enhancing molecular vibrations. SEIRS can also be used to enhance other kinds of excitations, such phonon-polaritons as described in Section 9.2.2.[14]

9.2.1 Surface-Enhanced Infrared Spectroscopy Using Gold Nanoantennas

Plasmon polaritons on metal nanoparticles can be resonantly excited with light. These excitations accompany enhanced electromagnetic near fields in their vicinity which mainly determine the signal enhancement in SEIRS (as well as in surface-enhanced Raman scattering (SERS)). Maximum SEIRS enhancement can be reached with resonant excitation and a good match of plasmonic and vibrational excitation of interest.[13]

We first consider the application of individual gold nanorods (these can also be gold nanorods arranged in well-defined configurations with large distances) to SEIRS. For an individual nanorod with resonance in the IR, the length L scales linearly with the resonant photon wavelength λ_{res} (or inverse circular frequency ω_{res}) of the fundamental excitation (see Chapter 3).[15–19] With the help of the linear relationship we can reliably predict resonance positions and thus design the geometrical properties of nano-antennas to match plasmonic resonances to the vibrational band of interest. It may be important to note that the strength of the plasmonic signal increases with L.

For a proof of nanoantenna-assisted SEIRS, we used nearly perfect crystalline gold nanocylinders[20] supported by IR transparent CaF_2 substrates.[20,21] As a well-defined model adsorbate monolayer, we employed the long alkyl-chained octadecanethiol (CH_3–$(CH_2)_{17}$–SH, ODT), which contains the functional thiol (SH) group that forms a strong bond to the gold[22] but not to the CaF_2, which ensures well-defined sample preparation. Moreover, only one monolayer of ODT with a defined thickness (d = 2.7 nm) is formed.[23,24]

ODT features several sufficiently strong IR absorption bands.[25] In our studies, we focus on the most intense ones, the antisymmetric and symmetric CH_2 stretching vibrations at 2927 cm^{-1} and 2855 cm^{-1} (see Fig. 9.1b), denominated as ODT vibrations in the following. The monolayer preparation is rather common: Nanoantennas deposited on CaF_2 are exposed to a one millimolar solution of ODT in ethanol for at least eight hours. During this time, a selective adsorption of ODT takes place, which ensures a perfect prepared sample. To remove physisorbed ODT molecules from the substrate, the sample is rinsed with ethanol afterward.

In order to obtain absorption bands close to the fundamental plasmonic resonance and therefore exploit the maximum near field enhancement, nanoantennas with specific lengths L were selected, according to the linear relation discussed above (see also Chapter 6. After localizing the nanoantennas with white light microscopy, we performed normal IR transmittance measurements from the selected focal spot with the nanoantenna (see Fig. 9.1c) using microscopic infrared spectroscopy under similar conditions, as described in Chapter 6.

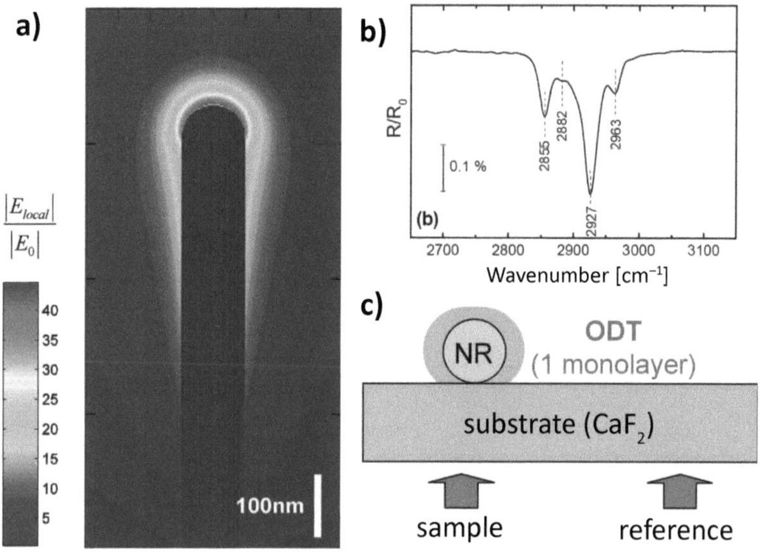

Figure 9.1 (a) Local electrical field E related to the incident electrical field E_0 in the vicinity of a gold nanorod ($L = 1.31$ μm, $D = 100$ nm) at its resonance frequency. The calculation was done using the boundary element method (see Aizpurua, *et al.*[26]). Because it is symmetrical, only one half of the antenna is shown. (b) IR spectrum[4] (acquired by IRRAS) of ODT molecules in the range of interest: The antisymmetric (2927 cm^{-1}) and symmetric (2855 cm^{-1}) CH_2 as well as the antisymmetric (2888 cm^{-1}) and symmetric (2963 cm^{-1}) CH_3 vibrations can be seen. (c) Schematic drawing (not to scale) of the applied measurement principle.

Relative transmittance spectra (transmittance of the nano-antenna covered with a monolayer ODT divided by the transmittance of the bare substrate without ODT) of selected individual nanoantennas reveal fundamental antenna resonances for parallel polarization, as seen in Fig. 9.2. Absorption bands at 2855 cm^{-1} and 2927 cm^{-1}, which can be identified as symmetric and antisymmetric CH_2 stretching vibrations of the ODT molecules,[25] can be observed on the antenna background. Those vibrational bands only occur if the plasmonic resonance matches the vibrational resonance of the ODT molecules, which is a clear indication of local field enhancement due to the antenna resonance. Thus, if no resonance can be excited, as for example for light polarized perpendicular to the long rod axis, no enhanced fields occur and therefore no SEIRS signal of the ODT molecules is detected (upper panel in Fig. 9.2).

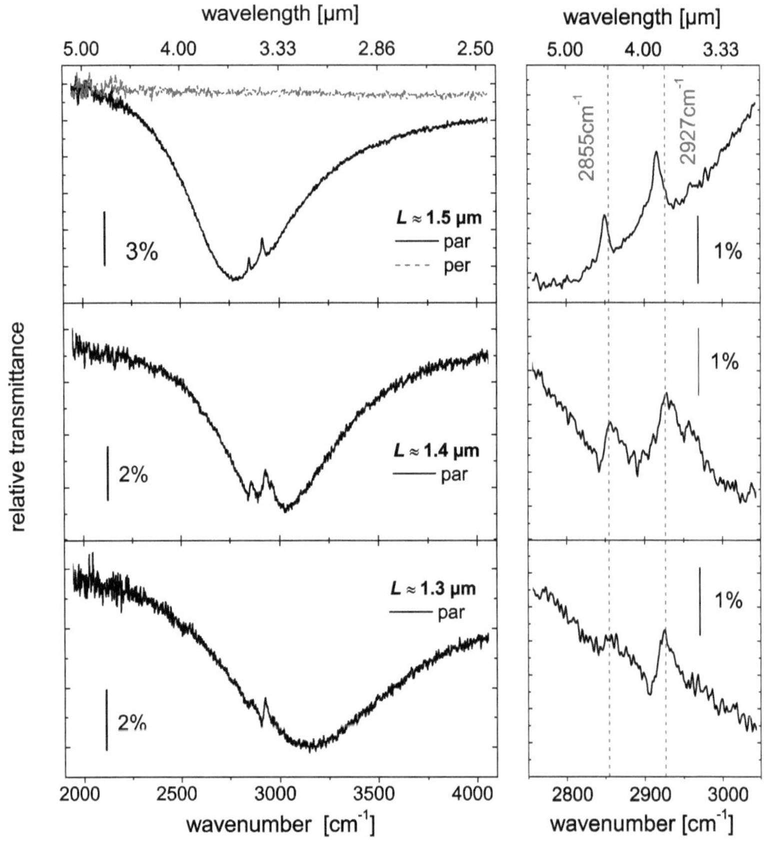

Figure 9.2 Relative transmittance of three ODT-coated nanorods with different L as indicated and D = 100 nm on CaF_2 measured with IR radiation polarized parallel (par) and perpendicular (per) to the long rod axis. The right panel shows the vibrational signals of ODT magnified. The asymmetric line shapes in dependence of the tuning can be clearly seen. The dotted red lines indicate ODT vibrational frequencies. See also Color Insert.

For perpendicular polarization the antenna is completely detuned (the plasmonic resonance is far away from the vibrational resonance), because the transverse resonance is shifted toward the visible spectral range.[26] A decrease of the SEIRS signal is also found for slightly detuned antennas. This is illustrated in Fig. 9.3a, where the intensity of the SEIRS signal is plotted against the ratio of resonance frequency ω_{res} divided by the respective vibrational frequency of ODT (ω_{ODT}) as a measure for detuning.

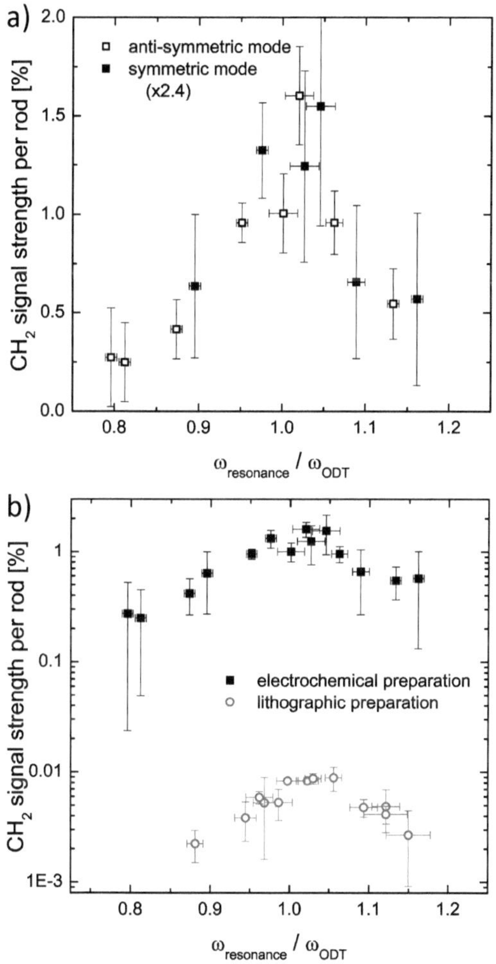

Figure 9.3 (a) Enhanced CH_2 signal strength (difference between minimum and maximum of the respective ODT vibration) in dependence of $\omega_{res}/\omega_{ODT}$. Signal strengths of antisymmetric stretching vibrations are multiplied by a factor of 2.4, which takes into account the different oscillator strengths of symmetric and antisymmetric vibration (see also Fig. 9.1b). It is assumed that both absorption bands are enhanced comparably. (b) Normalized CH_2 signal strength: nanorods prepared by electrochemistry (black symbols) and EBL (red circles). The differences in intensity are due to residue on nanoantenna surfaces. See also Color Insert.

As expected, the highest signal enhancement is observed for best matching ($\omega_{res}/\omega_{ODT} \approx 1$) and decreases to both sides. Theoretically, the resonance maximum of the near-field spectrum (and thus the maximum near-field intensity) is slightly red, shifted with respect to the resonance curve in the far field.[26] For SEIRS this means that a small shift between maximum far-field extinction and the SEIRS maximum could improve signal enhancement.

The line shape of the enhanced vibrational signal clearly differs from a typical Lorentzian absorption band (for example, see Fig. 9.1c) and depends on the position of the resonance frequency with respect to the vibrational lines. As seen in Fig. 9.2, asymmetric line shapes occur for detuned nanoantennas, whereas antiabsorption peaks occur for the best match. The line shape of the adsorbate signal results from the interaction of the broader plasmonic excitation with the spectrally narrower vibrational dipoles.[13] This effect resembles Fano profiles in wave and quantum mechanics, where the asymmetric line shape results from coupling between a discrete state and a continuum.[12] In this case, dipolar coupling is relevant and governed by the phase difference of the polarization fields of antenna and vibration and, therefore, changes for different $\omega_{res}/\omega_{ODT}$. Theoretical calculations using the boundary element method[27] show agreement with the experimental line shapes; this is a proof for the electromagnetic nature of the resonance enhancement and the Fano-type dipolar interaction.[13]

In order to evaluate the sensitivity of nanoantenna-assisted SEIRS, we estimated enhancement factors by comparing the resonantly enhanced absorption bands to the nonenhanced absorption bands from IR reflection-absorption spectroscopy (IRRAS) of ODT on smooth gold films.[4] Taking into account the different measurement setups and the different areas (focal spots on the sample) that contribute to the enhanced signal, we can estimate enhancement factors. For the best match of plasmonic and vibrational excitations (at $\omega_{res}/\omega_{ODT} \approx 1$), we achieved enhancement factors up to 500,000, corresponding to an attomolar sensitivity or the detection of about 100,000 ODT molecules. Until now there has been a discrepancy between measured and calculated enhancement of up to two orders of magnitude, with the smaller enhancement in the calculation giving the accurate line shape. The problem on the theoretical side is mainly the use of a continuum description of the molecular layer and the assumption of an abrupt gold surface.

The high experimental enhancement factors obtained with well-designed nanoantennas exceed those reached by common SEIRA techniques, such as rough film substrates,[3-5,28] by at least two orders of magnitude and those of homogeneous artificial gold particle films[29] by more than one order of magnitude.

Resonance frequencies of nanoparticles in ensembles depend in a complex way on their mutual arrangement (see Chapter 6),[19] which can be beneficial for SEIRS. In the experiment with the homogeneous gold particle films,[29] the authors took advantage of nm-separated interacting nanoparticles, which yields an additional near field enhancement[26,30,31] and therefore higher signal enhancements, since the field can be confined in small gaps between the particles. For nanoantennas, a further huge field enhancement was theoretically expected for a decreasing tip-to-tip distance down to few nanometers.[26] For two sets of interacting nanoantennas that effect was studied experimentally, as explained in the following. By electron beam lithography (EBL) and lift-off processes, cuboid-shaped gold nanorods (width and height of about 60 nm, see also Chapter 6) were produced in a well-defined arrangement of nanoantennas on ZnS substrates.[16] One set consisted of nanorods arranged in arrays with tip-to-tip gaps of 60 nm and side-by-side distances of 5 μm (see also Chapter 6). In the second set the gaps in tip-to-tip direction were decreased down to approximately 20 nm. Adsorption of ODT molecules and SEIRS measurements in relative transmittance geometry were done in the same manner as for individual nanorods. In order to make clear the influence of decreasing gap sizes on the signal enhancement, we calculated the extinction cross sections related to the geometric ones (see Chapter 6) for one nanorod. Besides plasmonic resonances, which are slightly different due to length deviations from EBL, enhanced ODT vibrational bands are observed at 2855 cm^{-1} and 2927 cm^{-1} (see Fig. 9.4).

Since both resonances are only slightly detuned from the vibrational bands, Fano-type asymmetric line shapes can be observed. It is important to note that the enhanced signal of the 20 nm separated nanorods is larger than that of the 60 nm separated ones (see inset in Fig. 9.4) even though the 20 nm separated antennas are more detuned. This clearly proves the additional near field enhancement due to interaction. Estimates of enhancement factors are done in the same way as for individual nanoantennas.

For 20 nm separated nanorods, we estimated a SEIRS factor of about 20,000 and for 60 nm separated ones a factor of 7,000.[32] Notice that enhancement factors would become about ten times larger for a better tuning. Unfortunately, because of substance residues from the EBL preparation process, the SEIRS signals are offen smaller than the ones from differently prepared individual nanocylinders (see Fig. 9.3b), even though the signal should be more enhanced due to interaction. Residues may remain on the gold nanorods' surfaces and therefore prevent ODT from adsorption, which leads to a decreased SEIRS intensity of ODT vibrations. Residues can be detected by the appearance of their vibrational bands that may be also enhanced (not shown). An additional influence of different crystalline qualities due to different preparation techniques on the near field enhancement is unlikely, since both show similar behavior in the far field (see Chapter 6 and Neubrech, *et al.*[16]). Despite of impurity-related effects, the dependence of Fano-type lines and signal enhancements on the tuning ratio $\omega_{res}/\omega_{ODT}$ clearly demonstrates the SEIRS activity of lithographically prepared nanorods.

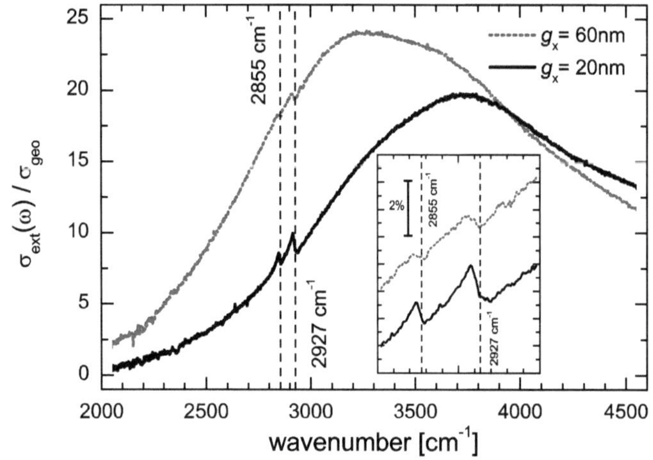

Figure 9.4 Extinction cross sections $\sigma(\omega)_{ext}$ (related to geometric ones σ_{geo}) normalized to one ODT-coated nanoantenna. The geometric dimensions of 20 nm separated nanorods (solid line) are L = 500 nm, $h \approx w \approx$ 60 nm, and L = 555 nm, $h \approx w \approx$ 60 nm for 60 nm separated ones (dotted line), respectively. The inset depicts an enlargement of the CH_2 stretching vibrations of ODT, illustrating the additional increase of enhancement due to near field interaction.

9.2.2 Antenna Sensing of Thin-Film Phonon-Polaritons

The precondition for nanoantenna-assisted surface-enhanced infrared spectroscopy is a good match between the dipolar vibrational excitation and the plasmonic excitation of the antenna (see Section 9.2.1). Thus, SEIRS is not only restricted to the enhancement of adsorbate vibrations but can also be used to detect other IR-active vibrations, such as phonon-polaritons.[33] In the following, we employ this technique to enhance polaritonic signals of a SiO_2 layer with nanometer thickness (3 nm) supported by a silicon substrate. On top of this natural SiO_2 layer, cuboid-shaped gold nanorods (width and height of about 100 nm) arranged in arrays were deposited by EBL and lift-off techniques.[14,19] First, for a clear experiment without mutual antenna interaction, we chose side-by-side and tip-to-tip distances of about five microns and performed microscopic IR spectroscopy at normal incidence of light (see Chapter 6 and Weber, *et al.*[19]). As in SEIRS studies with ODT, we observed an enhanced polaritonic signal of the SiO_2 layer if the polaritonic and the plasmonic excitation matched each other (e.g., Fig. 9.3 or Neubrech, *et al.*[14]). As proven by the polarization dependence and the Fano-type line shape, the appearance of the SiO_2 signal is due to enhanced electromagnetic near fields. Figure 9.5 illustrates the situation.

Figure 9.5 plots the baseline-corrected SiO_2 signal measured for different antenna lengths L and therefore for different resonance frequencies (ratios $\omega_{res}/\omega_{SiO_2}$, respectively). For the best match ($\omega_{res}/\omega_{SiO_2} \approx 1$), maximum signal strength and inverse absorption (antiresonance) of the SiO_2 signal is observed, since the coupling between both kinds of excitations is maximum. Decreasing ($\omega_{res}/\omega_{SiO_2} < 1$) or increasing ($\omega_{res}/\omega_{SiO_2} > 1$) the nanorod lengths slightly changes the rod's resonance frequencies and thus modifies the coupling strength. As a result, the line shape becomes asymmetric and the SiO_2 signal decreases. For larger mismatches, the signal diminishes further and its line shape becomes more and more a Lorentzian. In the approach to infinitely short rods ($L \to 0$ or $\omega_{res}/\omega_{SiO_2} \to \infty$) clearly no SiO_2 signal is detected, whereas the result for the approach to infinitely long rods ($L \to \infty$ or $\omega_{res}/\omega_{SiO_2} \to 0$) is not straightforward, as explained here. Besides the strong fundamental ($l = 1$) resonance, nanoantennas feature multipolar ($l > 1$) resonances at $\omega_{res} \cdot l$ (well below the onset of interband transitions and only in the linear range of the plasmon-polariton dispersion), which is also accompanied by enhanced electromagnetic near fields. Thus, higher order modes enhance the signal if the condition $\omega_{res}/\omega_{SiO_2} = 1/l$ (l being an odd number) is fulfilled. This is the case for very long rods. The effect should work also for adsorbate vibrations.

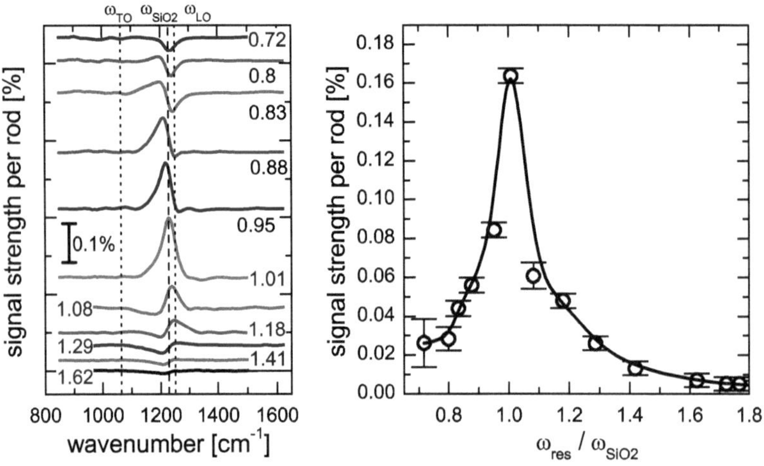

Figure 9.5 (a) Enhanced SiO_2 signals normalized to one nanorod for different $\omega_{res}/\omega_{SiO_2}$ as indicated. The SiO_2 signals are baseline corrected using polynomial baseline correction in order to fit the line shape of the antenna resonance The Fano-type behaviour can be seen clearly. Besides $\omega_{SiO_2} = 1230\ cm^{-1}$, the transversal ($\omega_{TO} = 1065\ cm^{-1}$) and the longitudinal ($\omega_{LO} = 1251\ cm^{-1}$) optical frequencies of the SiO_2 stretching vibration are indicated. (b) Intensity of the enhanced polaritonic signal normalized to one nanorod.

The situation for SiO_2 signal frequency is more complex than for less infrared-active vibrations like the signal frequency of ODT, because materials with ionic bonds (like SiO_2) feature very strong oscillators with negative values of the real part of the dielectric function. Under suitable conditions, such values give rise to an excitation of phonon-polaritons on surfaces or interfaces with frequencies between the transverse optical ω_{TO} and the longitudinal optical frequency ω_{LO} (reststrahlen region) of the respective vibration.[33] For the stretching vibration of SiO_2, values of $\omega_{TO} = 1065\ cm^{-1}$ and $\omega_{LO} = 1251\ cm^{-1}$ were found,[34] as indicated in Fig. 9.5. In our case, we have to regard the nanometer thickness of the SiO_2 layer as well, which leads to Fuchs-Kliewer surface phonon-polaritons[35,36] with frequencies between 1161 cm^{-1} and ω_{LO} = 1251 cm^{-1}. The frequency of the polaritonic feature observed in Fig. 9.5 was determined to be ω_{SiO_2} = 1230 cm^{-1} and thus attributed to the Fuchs-Kliewer surface phonon-polariton of the SiO_2 stretching mode of the thin oxide layer at the SiO_2/air interface, as we reported recently.[14] The wave vector component parallel to the surface, which is necessary to excite phonon-polaritons,[33] certainly is provided by near field scattering processes of nanoantennas, which is in agreement with scattering near field infrared microscopy measurements.[28,37]

References

1. Hartstein, A., Kirtley, J. R., and Tsang, J. C. (1980). Enhancement of the infrared absorption from molecular monolayers with thin metal overlayers, *Phys. Rev. Lett.*, **45**, 210–204.
2. Osawa, M. (2001). *Near-Field Optics and Surface Plasmon Polaritons*, Springer, Berlin/Heidelberg, *Topics in Applied Physics*, vol. 81.
3. Aroca, R. (2006). *Surface-Enhanced Vibrational Spectroscopy*, Wiley, Chichester, England.
4. Enders, D., Nagao, T., Pucci, A., Nakayama, T., and Aono, M. (2011). Surface-enhanced ATR-IR spectroscopy with interface-grown plasmonic gold-island films near the percolation threshold, *Phys. Chem. Chem. Phys.*, **13**, 4935–4941.
5. Enders, D., and Pucci, A. (2006). Surface-enhanced infrared absorption of octadecanethiol on wet-chemically prepared Au nanoparticle films, *Appl. Phys. Lett.*, **88**(18), 184104.
6. Priebe, A., Pucci, A., and Otto, A. (2006). Infrared reflection-absorption spectra of C_2H_4 and C_2H_6 on Cu: effect of surface roughness, *J. Phys. Chem. B*, **110**(4), 1673–1679.
7. Priebe A., Sinther M., Fahsold, G., and Pucci, A. (2003). The correlation between film thickness and adsorbate line shape in surface-enhanced infrared absorption, *J. Chem. Phys.*, **119**(9), 4887–4890.
8. Krauth, O., Fahsold, G., Magg, N., and Pucci, A. (2000). Anomalous infrared transmission of adsorbates on ultrathin metal films: Fano effect near the percolation threshold, *J. Chem. Phys.*, **113**(15), 6330–6333.

9. Pucci, A. (2005). IR spectroscopy of adsorbates on ultrathin metal films, *Phys. Stat. Sol. B*, **242**(13), 2704–2713.
10. Bjerke A. E., Griffiths, P. R., and Theiss W. (1999). Surface-enhanced infrared absorption of CO on platinized platinum, *Anal. Chem.*, **71**, 1967–1974.
11. Priebe, A., Fahsold, G., and Pucci, A. (2004). Strong pyramidal growth of metal films studied with IR transmittance and surface-enhanced IR absorption of CO, *J. Phys. Chem. B*, **108**(47), 18174–18178.
12. Fano, U. (1961). Effects of configuration interaction on intensities and phase shifts, *Phys. Rev.*, **124**(6), 1866–1878.
13. Neubrech, F., Pucci, A., Cornelius, T. W., Karim, S., García-Etxarri, A., and Aizpurua, J. (2008). Resonant plasmonic and vibrational coupling in a tailored nanoantenna for infrared detection, *Phys. Rev. Lett.*, **101**(15), 157403.
14. Neubrech, F., Weber, D., Enders, D., Nagao, T., and Pucci, A. (2010). Antenna sensing of surface phonon polaritons, *J. Phys. Chem. C*, **114**, 7299–7301.
15. Neubrech, F., Kolb, T., Lovrincic, R., Fahsold, G., Pucci, A., Aizpurua, J., *et al.* (2006). Resonances of individual metal nanowires in the infrared, *Appl. Phys. Lett.*, **89**(25), 253104.
16. Neubrech, F., Weber, D., Lovrincic, R., Pucci, A., Lopes, M., Toury, T., and Lamy de La Chapelle, M. (2008) Resonances of individual lithographic gold nanowires in the infrared, *Appl. Phys. Lett.*, **96**(16), 163105.
17. Neubrech, F., Garcia-Etxarri, A., Weber, D., Bochterle, J., Shen, H., Lamy de la Chapelle, M., *et al.* (2010). Defect-induced activation of symmetry forbidden infrared resonances in individual metallic nanorods, *Appl. Phys. Lett.*, **96**(21), 213111.
18. Novotny, L. (2007). Effective wavelength scaling for optical antennas, *Phys. Rev. Lett.*, **98**(26), 266802.
19. Weber, D., Albella, P., Alonso-Gonzalez, P., Neubrech, F., Gui, H., Nagao, T., Hillenbrand, R., Aizpurua, J., and Pucci, A. (2011). Longitudinal and transverse coupling in infrared gold nanoantenna arrays: Long range versus short range interaction regimes, *Opt. Express*, **19** (16), 15047–15061.
20. Liu, J., Karim, S., Duan, J. L., Toimil-Molares, M. E., Cornelius, T. W., Dobrev, D., *et al.* (2006). Electrochemical fabrication of single-crystalline and polycrystalline Au nanowires: the influence of deposition parameters, *Nanotechnology*, **17**, 1922–1926.

21. Pucci, A., Neubrech, F., Aizpurua, J., Cornelius, T., and Lamy de la Chapelle, M. (2008). Electromagnetic nanowire resonances for field-enhanced spectroscopy, in *One-Dimensional Nanostructures*, (ed. Wang, Z. M., Salamo, G., Waag, A., and Kishimoto, N.), Springer, New York, pp. 175–213.

22. Schreiber, F. (2000). Structure and growth of self-assembling monolayers, *Prog. Surf. Sci.*, **65**, 151–256.

23. Seo, E. K., Lee, J. W., Sung-Suh, H. M., and Sung, M. M. (2004). Atomic layer deposition of titanium oxide on self-assembled monolayer-coated gold, *Chem. Mat.*, **16**(10), 1878–1883.

24. Ulman, A. (1996). Formation and structure of self-assembled monolayers, *Chem. Rev.*, **96**, 1533–1554.

25. Lin-Vien, D., Colthup, N. B., Fateley, W. G., and Grasselli, J. G. (1991). *The Handbook of Infrared and Raman Frequencies of Organic Molecules*, Academic Press, San Diego.

26. Aizpurua, J., Bryant, G. W., Richter, L. J., and García de Abajo, F. J. (2005). Optical properties of coupled metallic nanorods for field-enhanced spectroscopy, *Phys. Rev. B*, **71**(23), 235420.

27. García de Abajo, F. J., and Howie, A. (1998). Relativistic electron energy loss and electron-induced photon emission in inhomogeneous dielectrics, *Phys. Rev. Lett.*, **80**, 5180–5138.

28. Ocelic, N., and Hillenbrand, R. (2004). Subwavelength-scale tailoring of surface phonon polaritons by focused ion-beam implantation, *Nat. Mater.*, **3**, 606–609.

29. Kundu, J., Le, F., Nordlander, P., and Halas, N. J. (2008). Surface-enhanced infrared absorption (SEIRA) spectroscopy on nanoshell aggregate substrates, *Chem. Phys. Lett.*, **452**, 115–119.

30. Romero, I., Aizpurua, J., Bryant G. W., and Garcia de Abajo, F. J. (2006). Plasmons in nearly touching metallic nanoparticles: singular response in the limit of touching dimmers, *Opt. Express*, **14**(21), 9988–9999.

31. Yu, N., Cubukcu, E., Diehl, L., Belkin, M. A., Crozier, K. B., Capasso, F., *et al.* (2007). Plasmonic quantum cascade laser antenna, *Appl. Phys. Lett.*, **91**(17), 173113.

32. Pucci, A., Neubrech, F., Weber, D., Hong, S., Toury, T., and Lamy de la Chapelle, M. (2010). Surface-enhanced infrared spectroscopy using gold nanoantennas, *Phys. Stat. Sol. B*, **247**(8), 2071–2074.

33. Cottam, M. G., and Tilley, D. R. (1989). *Introduction to Surface and Superlattice Excitations*, Cambridge University Press.

34. Gunde, M. K. (2000). Vibrational modes in amorphous silicon dioxide, *Physc a B*, **292**, 286–295.

35. Fuchs, R., and Kliewer, K. L. (1966). Optical modes of vibration in an ionic crystal slab including retardation. II. radiative region, *Phys. Rev.*, **150**, 573–588.

36. Kliewer, K. L., and Fuchs, R. (1966). Collective electronic motion in a metallic slab, *Phys. Rev.*, **144**, 498–512.

37. Hillenbrand, R., Taubner, T., and Keilmann, F. (2002). Phonon-enhanced light-matter interaction at the nanometre scale, *Nature*, **418**, 159–162.

Chapter 10

Surface-Enhanced Raman Scattering: Application to Molecular Sensing

N. Félidj[a,*] **and M. Lamy de la Chapelle**[b,**]

[a]*University Paris Diderot, Department of Chemistry, 15 street Jean de Baïf, Paris 75013, France*

[b]*University Paris 13, Laboratoire CSPBAT, 74 rue Marcel Cachin, Bobigny 93017, France*

[*]nordin.felidj@univ-paris-diderot.fr

[**]marc.lamydelachapelle@univ-paris13.fr

In this chapter, the basic principles of surface-enhanced Raman scattering (SERS) are introduced. After a brief description of Raman spectroscopy and an introduction of the development of SERS in the past four decades, we will mainly focus on the mechanisms leading to the SERS effect.

10.1 Introduction

The Raman effect was discovered by Sir C. V. Raman in 1928.[1] He was awarded the Nobel Prize in 1930 for this discovery. In principle, when light is scattered from a molecule, most photons are elastically scattered (Rayleigh scattering). The scattered photons have the same energy (or frequency) as the incident photons. Nevertheless,

Nanoantenna: Plasmon-Enhanced Spectroscopies for Biotechnological Applications
Edited by Marc Lamy de la Chapelle and Annemarie Pucci

ISBN 978-981-4303-61-3 (Hardcover), 978-981-4303-62-0 (eBook)
www.panstanford.com

only a small fraction of the scattered light (around 1 in 10 million photons) is scattered with a frequency different from the frequency of the incident photons. This shift in frequency originates from vibrational motion of the molecule, leading to identification of its chemical and structural information. Note that the Raman-shifted photons can be higher or lower in energy, depending upon the vibrational state of the molecule under study. Therefore, Stokes radiation occurs at lower energy compared with Rayleigh scattering, while anti-Stokes radiation occurs at higher energy.

The Raman effect is a powerful vibrational spectroscopic technique for analyzing the composition of materials (solids, liquids, and gases).[2] Its versatility makes it very useful in different fields and applications, such as the analysis of organic compounds and minerals and the detection of explosives. However, the weak Raman signal is one of the biggest inconveniences of Raman spectroscopy. For this reason, it was not widely used until it was possible to combine metallic structures with molecular compounds, leading to the SERS effect, in the beginning of the seventies.

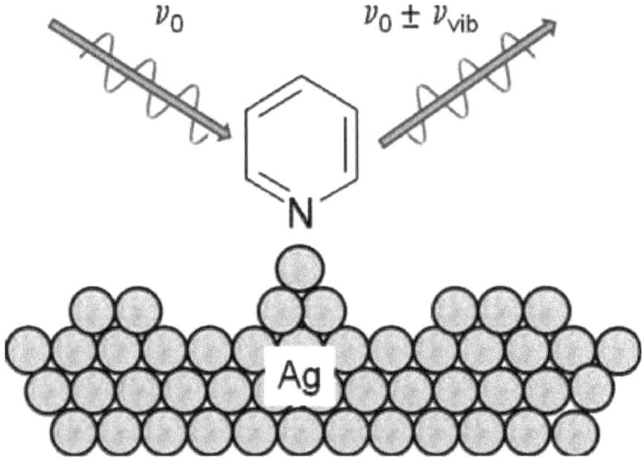

Figure 10.1 A schematic representation of a SERS experiment with pyridine adsorbed on silver, showing the incident laser and Raman scattered light, the intensities of which are both influenced by the enhanced field at the silver surface resulting from surface plasmon excitations (from Ref. 4).

The SERS effect was accidentally discovered by Fleishman *et al.* (University of Southampton, UK) in 1974.[3,4] The authors observed a huge enhancement of the Raman signal of adsorbed pyridines molecules with an unexpected high signal-to-noise ratio on a roughened silver substrate obtained by the electrochemical route. Initially, the main idea was to use a roughened metal surface in order to increase the surface area and thus the Raman signal of the pyridine molecules (Fig. 10.1). In 1977, Van Duyne and Jeanmaire (Northwestern University, USA) and Albrecht and Creighton (University of Kent, UK) reported similar results using silver colloids as SERS substrates.[5,6] Both groups demonstrated that the very high Raman signal of adsorbed molecules was generated by a real enhancement of the Raman scattering efficiency. The effect was named SERS.

The first SERS spectra were obtained from an electrochemical system (roughened silver electrode), but there exist many sorts of metallic substrates. The most commonly used substrates for obtaining SERS are disorganized media, including aggregated metal colloids, metal island films, and gold and silver colloidal monolayers.[7] Figure 10.2 illustrates the strong potential of SERS in detecting molecular compounds at very low concentrations in colloidal silver particles.

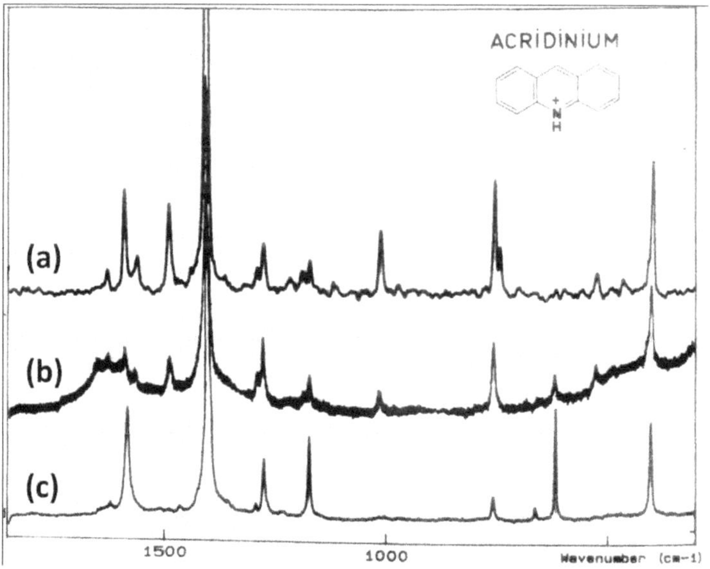

Figure 10.2 (a) Near-infrared Fourier transform Raman spectrum of acridinium (powder); (b) Raman spectrum of acridinium in aqueous solution (concentration 10^{-2} M); (c) SERS of acridinium (10^{-6} M) in a silver colloidal solution.

Figure 10.2a displays the Raman spectrum of acridinium in solid state (powder) with characteristic Raman bands at ca. 1,400 and 1,570 cm^{-1} from ring stretch vibrations. In a 10^{-2} M aqueous solution of acridinium, the signal/noise ratio is much lower as expected. However, when the molecular probe is introduced in a silver colloidal solution, with a concentration of 10^{-6} M, this ratio is much higher, although the concentration of the molecular probe is four orders of magnitude less.

Since its discovery, SERS has been considered one of the most powerful and sensitive tools for the chemical analysis of molecules adsorbed onto metallic nanoparticles (NPs).[8] This technique has a large potential in analytical chemistry and biological and optical applications.[9] Many aspects of the mechanisms involved in SERS have been described and resolved to a reasonable degree over the past two decades.[10] However, four decades later, the SERS community is still concerned by new challenges. One of the most spectacular is the observation by SERS of single molecules (SMs), or SM-SERS. SM-SERS was mentioned for the first time in 1997 by Kneipp *et al.* It has been demonstrated recently by Le Ru *et al.* using the so-called ƀ *analyte* method.[11,12]

The next paragraph mainly focuses on mechanisms leading to the SERS effect. It is now well established that the mechanisms can be divided into electromagnetic and chemical contributions: (i) The electromagnetic contribution arises from enhanced optical fields close to the metallic surface, because of the excitation of the localized surface plasmon (LSP) with a typical Raman enhancement factor (REF) of $|E|^4$ ca. 10^6–10^7 (where E is the local electric field); and (ii) a further chemical enhancement (CE) can be observed for molecules adsorbed onto specific metallic sites when a resonant charge transfer occurs.

10.2 Mechanisms of SERS

There are two mechanisms leading to the SERS phenomenona— n electromagnetic enhancement (EE) and a CE mechanism. The electromagnetic effect is mainly at the origin of the SERS effect. In a classical experiment, the Raman gain (RG) is typically 10^5–10^6. This EE originates from the excitation of the LSP sustained by the metallic

NPs.[13] However, a chemical effect might contribute but on the order of one or two orders of magnitude. The CE involves changes to the adsorbate electronic states because of chemisorption of the analyte through the enhancement in polarizability due to chemical effects such as charge-transfer excited states.[14] We now describe both SERS mechanisms.

The SERS effect is mainly associated with three metals: silver, gold, and copper.[15] Other metals have been explored since the discovery of SERS, but they present less efficiency. For instance, there is evidence that SERS can be observed for metals like platinum that is interesting in the context of catalysis.[15] The two main features for obtaining the SERS effect are summarized hereunder:

- SERS takes place for molecules close to the surface of metallic particles (mainly silver, gold, and copper), provided that the laser line wavelength is close to the maximum of LSP resonance. In particular, molecules adsorbed in the first surface layer display the largest REFs. The REFs also have a long-range effect until a few tens of nanometers.
- The REF depends on the metallic particle size and shape, the typical range of the NP size being between 10 nm and 200 nm.

10.2.1 Electromagnetic Mechanism

SERS mainly originates from an EE mechanism consecutive to the excitation of the LSP, occurring in metal NPs. As described previously, LSP excitation corresponds to collective oscillations of conductive electrons at the NP surface. This property results in a high degree of optical field confinement.[16] The LSP resonance wavelength can be changed with the NPs' size and shape, along their spacing, which is of main importance in the context of SERS applications.[17] Therefore, the scattered electric field due to LSP excitation, close to the NP surface, and experienced by the adsorbed molecule, is increased.

If we consider a molecule close to the NP, both the incident field due to LSP excitation and the scattered fields by the molecule (Raman scattered fields) will be enhanced.

Taking into account both enhanced fields, the average RG, $\langle G \rangle$, can be expressed as follows:

$$<G> = |<A(\nu_{exc})>|^2 \cdot |<A(\nu_R)>|^2$$

where $A(\nu_{exc})$ is the local electric-field EF at the incident frequency ν_{exc} and $A(\nu_R)$ is the corresponding factor at the Raman frequency ν_R. In general, $\langle A \rangle$ is averaged over the surface area of the particles where molecules are, while in SM-SERS, only the maximum enhancement A_{max} has to be considered.[18] Most of the time, to estimate the RG, $\langle G \rangle$ is approximated by assuming that $A(\nu_{exc})$ and $A(\nu_R)$ are identical, and hence $\langle G \rangle$ can be rewritten as $<G> \approx |A(\nu_{exc})|^4$.[12] This approximation takes advantage of the fact that the LSP width is often large compared with the Stokes shift, except for calibrated samples like lithographic structures,[17] where the LSP is often sharp (width lower than 100 nm.

In Fig. 10.3, we estimate the RG under this resonant condition in a simple case by considering a metal sphere, smaller than the incident wavelength. In this model, one can consider the quasistatic approximation, for which the incident electric field is uniform. This simple model will allow us to introduce the intrinsic expression of the RG, give some orders of magnitude, and point out the dependence of the RG versus the distance particle/molecule.[19]

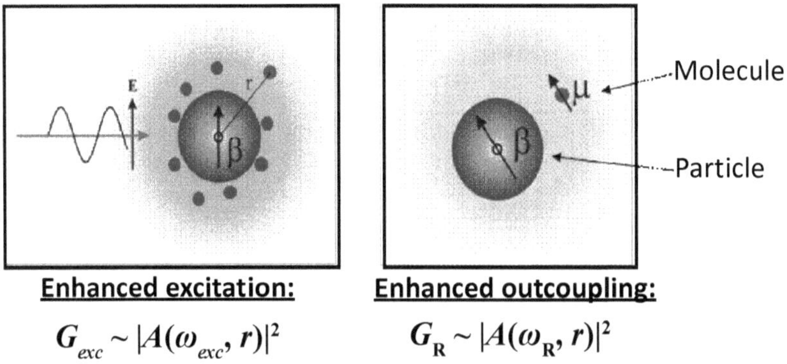

Figure 10.3 The RG is the product of two enhanced fields, (i) the scattered field by the NP at the excitation frequency ν_{exc} and (ii) the Raman scattered field by the molecule at the frequency ν_R.

We thus consider a small metal sphere, of radius r, illuminated with a plane monochromatic wave characterized by an electric field

E_0, at the laser frequency ν. The electric field E_M, at distance d from the NP surface, is the sum of E_0 and the field E_{sp} induced by excitation of the LSP of the NP. The scattered field E_M corresponds, within the quasistatic approximation, to the field of a point dipole, located at the center of the sphere and oscillating with the same frequency as the incident field. Its expression is as follows:

$$\vec{E}_{sp} = \left(\frac{\varepsilon(\nu) - \varepsilon_0}{\varepsilon(\nu) + 2\varepsilon_0}\right)\left(\frac{r}{r+d}\right)^3 \vec{E}_0$$

where $\varepsilon(\nu)$ represents the dielectric function of the metal and ε_0 the dielectric function of the surrounding medium. Since $E_0 << E_{sp}$, the electromagnetic EF $A(\nu)$ can be defined as the following ratio:

$$A(\nu_{ex}) = \frac{|\vec{E}_M(\nu_{ex})|}{|\vec{E}_0(\nu_{ex})|} \approx \frac{|\vec{E}_{sp}(\nu_{ex})|}{|\vec{E}_0(\nu_{ex})|} = \left(\frac{\varepsilon(\nu_{ex}) - \varepsilon_0}{\varepsilon(\nu_{ex}) + 2\varepsilon_0}\right)\left(\frac{r}{r+d}\right)^3$$

This expression points out the fact that $A(\nu)$ is maximum when the laser frequency coincides with the LSP frequency, that is, the real part of the dielectric function of the metal is Re $\varepsilon(\nu) = -2\varepsilon_0$. Note that for gold and silver, the imaginary part of the metal dielectric function is very weak. Therefore, those two metals are widely used in SERS experiments because of a weak damping of LSP resonances.

The RG can be rewritten:

$$G_{SERS}^{EM} \approx |A(\nu_{ex})|^4 = \left|\frac{\varepsilon(\nu_{ex}) - \varepsilon_0}{\varepsilon(\nu_{ex}) + 2\varepsilon_0}\right|^4 \left(\frac{r}{r+d}\right)^{12}$$

This formula shows that the RG falls rapidly with the growing distance from the surface. Practically, the RG, for a silver sphere, under resonant condition, leads to an RG of approximately 10^2–10^3. In the case of spheroids, the RG can be also estimated in the quasistatic approximation, with higher values of the RG, typically 10^4. The shape of the NP is thus crucial. Therefore, samples presenting very sharp structures (triangles, pyramids, rods) lead to much higher RGs due to a tip effect.[20]

Regular metallic gratings with a nanometer-scale structure appear to be suitable substrates for illustrating and understanding the electromagnetic mechanism involved in the SERS effect. Such arrays can be fabricated by electron beam lithography using a scanning electron microscope (SEM) equipped with a lithographic system, which enables the fabrication of highly reproducible patterns with a great variety of geometries and uniformly sized and shaped particles (Fig. 10.4). Consequently, the resonance frequency of the LSP, which depends on the size and shape of the particles, the substrate and surrounding medium dielectric functions, and the particle spacing grating constant, can be matched to almost any desired wavelength from the visible to the NIR region, as displayed in Fig.10.5a. This enables us to show how the SERS amplification is related to the resonance frequency of the LSP and thus directly connected to the particle geometrical features and organization.[7]

Figure 10.4 SEM image of a gold particle square grating fabricated by electron beam lithography. The diameter of the particles parallel to the substrate is 110 nm, and the particle height is 60 nm. The grating constant is 300 nm (from Ref. 7).

As shown in Fig. 10.5b, the arrays exhibit a significant SERS signal of *trans*-1,2-bis (4-pyridil) ethylene (BPE) (concentration 10^{-4} M). This molecular probe has intense and characteristic Raman bands located at about 1,200 cm^{-1}, 1,340 cm^{-1}, 1,610 cm^{-1}, and 1,640 cm^{-1} because of in-plane ring modes.

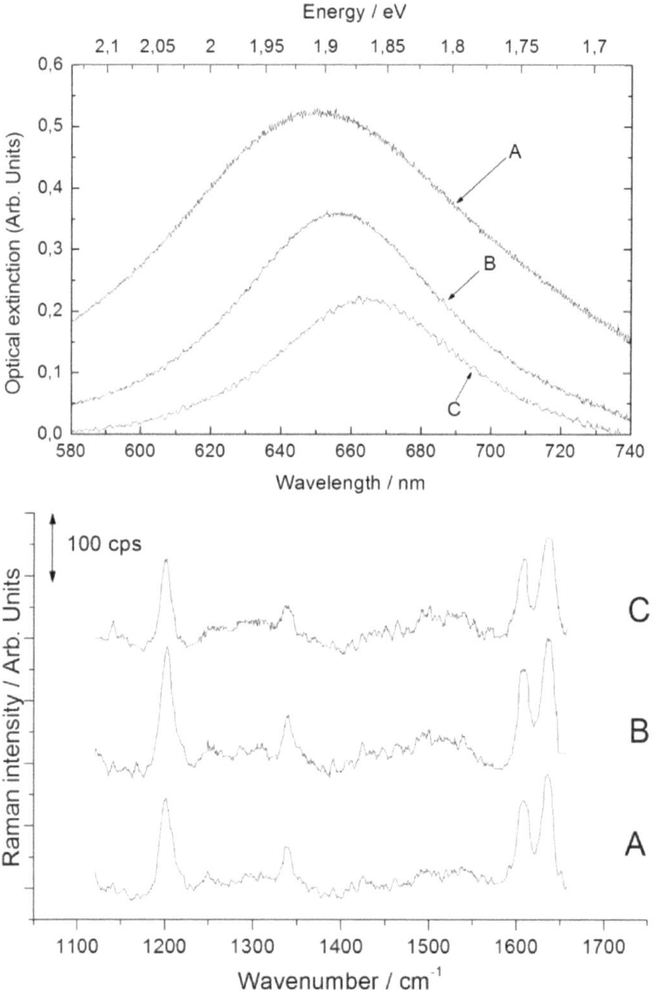

Figure 10.5 (a) Extinction spectra of gold gratings, acquired in aqueous solution, as a function of the grating constant: 200 nm (A); 250 nm (B); 300 nm (C). Particle height: 60 nm; diameter: 110 nm. (b) SERS spectra of 10^{-5} M BPE aqueous solution for different gratings: A, B, and C. All spectra were recorded under identical experimental conditions: laser excitation 632.8 nm; power 3 mW at the sample; slit width 300 mm; and three accumulations of 180 s counting time each. The spectra are vertically shifted for clarity (from Ref. 7).

Now, we try to estimate the EF per molecule for these bands. The intensities of the 1,610 and 1,640 cm^{-1} bands have been estimated together since they are close to each other and thus are easier to be treated as a single band. The EF per molecule is defined as $G = (N_2/N_1)\ (I_{SERS}/I_{ref})$, where I_{SERS} is the integrated intensity of the BPE band under consideration, recorded on the grating, I_{ref} the integrated intensity of the same Raman band, obtained by focusing the 632.8 nm laser line on an indium tin oxide (ITO)-coated glass immersed in a 0.5 M BPE solution, N_1 the number of molecules constituting the first monolayer adsorbed on the grating under the laser spot area, and N_2 the number of molecules excited within the volume of the laser waist for the 0.5 M BPE solution. Using the (×50) microscope objective, the area of the laser spot is about 5 μm^2. Thus, the value for N_1 is estimated from the approximative area of a single BPE molecule (~30 $Å^2$) to $1.6^7 \times 10^7$ molecules (i.e., 3.3×10^{14} molecules/cm^2). In these experiments, the Raman enhancement arises mainly from an electromagnetic amplification of the local field that exponentially decreases as moving away from the surface. In this case, one can question whether molecular layers other than the first one are involved in the SERS spectrum. However, it was presently observed that the molecular concentration (10^{-5} M) was that which yields the greatest gain. Increasing or decreasing this concentration leads to a weakening of the Raman signal. This suggests that layers other than the first one produce a screening effect, thus diminishing the Raman spectrum rather than increasing it. It is, therefore, justified that under the 10^{-5} M molecular concentration, there is only a single monolayer of absorbed molecules. Therefore, when estimating the RG, only the first layer of adsorbed molecules on the array is taken into account to determine the RG. The value of N_2 is estimated from the volume of the laser waist, which is assumed to be that of a cone of the apex angle given by the numerical aperture (NA) of the microscope objective and the height of the focusing scope; for the 503 objective used with NA ≈ 0.55, the apex angle is 33.4° and the focusing scope is approximately 20–25 μm, thus giving the volume of the waist lying between 3,000 and 7,000 μm^3. We took the volume of the laser waist to an average of 5,000 μm^3, thus leading to an N_2 value of about 1.5×10^{12} molecules, and we assumed that all the molecules excited in this volume are equally involved in the spontaneous Raman intensity. The EFs, thus measured for the bands located at 1,200, 1,340, and 1,610–1,640 cm^{-1} are gathered in Table 10.1. The standard deviation of the gain being estimated to about 10%, the relative variation of the gain from one array to the other is significant. For each band, two estimations for the Raman enhancement are given. In the first one, all the molecules adsorbed as the first layer under the 5 μm^2 laser spot are involved; this gain is merely proportional to the integrated intensity of the band within a fixed spectral range. In the second one, we try to get a physically more significant gain by accounting for the fact that only the fraction of molecules adsorbed onto the gold particles benefit from an electric field enhancement (effect of particle coverage q). This latter gain per molecule ranges between 105 and 6 × 105, which are values very often referred to in the literature for gold substrates. Greater gain values were obtained for experiments performed on silver island films because of the fact that silver yields higher enhancement than gold.

Table 10.1 SERS gain per molecule determined experimentally from bands of the BPE molecule adsorbed on various particle arrays

Grating		Particle	$G_{\alpha p}$		$G_{\alpha p}$		$G_{\alpha p}$	
#	$\lambda_{\alpha c}$	coverage	1200 cm^{-1}	(659 nm)	1340 cm^{-1}	(662 nm)	1610–1640 cm^{-1}	(669 nm)
A	650 nm	0.24	6.2×10^4 (†)	2.6×10^5 (‡)	4.1×10^4 (†)	1.7×10^5 (‡)	2.8×10^4 (†)	1.2×10^5 (‡)
B	656 nm	0.14	8×10^4 (†)	5.7×10^5 (‡)	4.6×10^4 (†)	3.3×10^5 (‡)	3.1×10^4 (†)	2.2×10^5 (‡)
C	664 nm	0.11	4×10^4 (†)	3.8×10^5 (‡)	3.1×10^4 (†)	2.9×10^5 (‡)	2.6×10^4 (†)	2.5×10^5 (‡)

†This gain is estimated by assuming that all molecules adsorbed on the array surface under the laser spot participate in the SERS effect.

‡This gain is estimated by assuming that only the molecules adsorbed onto the gold particles give rise to the SERS effect. The standard deviation of the gain is estimated to be about 10%.

This example shows that it is possible to estimate an RG from gold NP regular arrays, with a gain per molecule of about 105, depending on the position of the Raman band maximum relatively to the plasmon resonance wavelength.

It seems difficult to achieve stronger EFs with such organized systems. On the other hand, disordered systems, such as semi-continuous films, manifest hot spots strongly localized as a result of the fractal character of the system, leading to enormous local fields.[21] To emphasize the role of the coupling between particles, it is possible to design new organized metallic arrays consisting of gold ellipsoids deposited onto silver and gold substrates. The analysis of the optical properties and SERS spectra of such substrates was described by Felidj *et al.* [22]

10.2.2 Chemical Mechanism

Several experimental results indicate that another additional enhancement mechanism also operates, referring to a chemical effect or charge transfer mechanism. The molecule adsorbed onto the surface necessarily interacts with the surface. This chemical effect exists because of this interaction. The chemical effect is due to a charge transfer mechanism, which is due to the electronic coupling between the molecule and the metal.[23–25] The wave functions of the electron in the metal and adsorbate molecule overlap, which results in the tunneling of electrons between both the metal and the molecule.

The charge transfer process explains the experimental observations that the EE does not. For example, the dependence on the electrode potential, which accounts for the maximum SERS intensity, can also be explained in a straightforward manner by the charge transfer mechanism.[26] Variation of the electrode potential leads to changes of the relative energy between the metal Fermi energy level and the adsorbate energy levels. As a result, the resonance condition for charge transfer slightly differs and the incoming photon should match this altered condition. The charge transfer process can be summarized as follows:

(i) The absorption of a photon leads to the excitation of an electron in the metal.
(ii) The electron transfers to an excited electronic state level of the adsorbate.
(iii) The electron from the excited level of the adsorbate transfers back to the metal.
(iv) A photon is emitted.

The CE mechanism provides one or two orders of magnitude enhancement to the Raman signal intensity. In contrast to the EE mechanism, where no direct contact between the molecule and the metal is necessary, this is a primary condition for the charge transfer mechanism. For this reason, it is also known as the first layer effect in the literature.

If some difficulties in understanding SERS are in the way to be resolved, the current goal in theory is the quantification of the enhancements (EE) and mechanisms (CE) in the SERS effect. Therefore, if both physics and chemistry play a role in the enhancement mechanism, one question arises: what is the exact amount for both contributions? However, SERS provides a powerful technique and a promising way for the study of SM events, provided that the SERS substrates display strong field enhancements.

10.3 SERS Biosensor

Raman spectroscopy is already well known as a powerful tool for characterizing organic materials since the vibrational modes are actual fingerprints of the studied species (information on molecule conformation, chemical groups, and bonding).[27]

Several groups have shown the abilities of Raman spectroscopy and SERS to observe biological molecules[28–35] and characterized several specific biomolecule characterizations: hemoglobin,[36] myoglobin,[37] porphyrins,[38] amino acids,[39–41] nucleic acids,[42] and genes.[43–51] They have also been used for biomolecule detection as immunoassay methods.[52–56]

Several biosensors have yet been proposed using SERS as detection methods. Two main strategies have given some results using NPs in solution or NPs deposited on a surface.

In the first case, a SERS label molecule with a high cross section is deposited at the surface of gold or silver NPs. These NPs are then functionalized (often done by monoclonal antibodies) to bind selectively to targeted biomarkers. The detection of the SERS label molecule by Raman spectroscopy gives an indication of the presence of biomarkers. Such techniques have been employed to elaborate some immunoassays (microbead sandwich assay)[56]—for example, the detection of the P38 mitogen-activiated protein kinase.[57] This immunoassay reaches a low detection limit of 9.5×10^{-12} mol/L, but to get a result, a procedure with three steps of one hour is necessary, which means that it is not possible to obtain the effective result before several hours. This technique can also be effective for the detection of biomolecules in body fluids. The detection of folic acid in human serum has notably been demonstrated with a detection limit of 1.8×10^{-8} mol/L.[58] Some groups have also used this strategy to detect some tumors *ṅ vi̇ o*.[59]

Unfortunately, even if the concept of SERS biosensors has been demonstrated, several drawbacks can compromise such systems. Indeed, the NPs often aggregate together, which induces a change in SERS enhancement and could have some influence on the reliability of the detection. Such techniques are also restricted to water-soluble NPs. Furthermore, for *ṅ vi̇ o* measurements, the SERS agents should not be toxic to the body, which has not been clearly stated for NPs.

These drawbacks can be exceeded by using some NPs deposited on a substrate. Van Duyne *et al.* demonstrated the detection of glucose *ṅ vi̇ o* using a SERS substrate (a silver film under a nanosphere substrate) on an animal model. They were able to measure some variation of the glucose concentration and make some quantitative measurements between 10^{-3} and 300×10^{-3} mol/L.[60-61] They also demonstrated the detection of *Bacl lus* spores with a detection limit of 10^{-14} mol/L.[62]

Otherwise, the application of SERS substrates done by e-beam nanolithography for the detection of proteins has been demonstrated,[63-64] especially their conformation change with temperature.[64] It has notably been shown that with such a substrate, it could be possible to detect myoglobin with concentrations down to attomoles (10^{-18} mol).

The detection limit reached by SERS sensors is around 10^{-9} mol/L for NPs in solution and can be lower than 10^{-15} mol/L for NPs deposited on a substrate. For comparison these detection thresholds are largely higher than the ones obtained with classical techniquese– nzyme-linked immunosorbent assay (ELISA) (10^{-6}–10^{-8} mol/L) and high-performance liquid chromatography (HPLC) fluorescence (10^{-7} mol/L). Thus, using SERS techniques and, more especially, substrates, some very low detection limits could then be reached, paving the way for the development of highly sensitive nanobiosensors.

References

1. Singh, R. (2002). C. V. Raman and the discovery of the Raman effect, *Phys. Perspect.*, **4**, 399–420.
2. McCreery, R. I. (2000). *Raman Spectroscopy for Chemical Analysis*, John Wiley, New York.
3. Fleischmann, M., Hendra, P. J., and McQuillan (1974). RAMAN-spectra of pyridine adsorbed at a silver electrode, *J. Chem. Phys. Lett.*, **26**, 163–166.
4. McQuillan, J. (2009). The discovery of surface-enhanced Raman scattering, *A J Notes Rec. R. Soc.*, **63**, 105–109.
5. Jeanmaire, D. L., and Van Dunye, R. P. (1977). Surface Raman spectroelectrochemistry part I. Heterocyclic, aromatic, and aliphatic amines adsorbed on the anodized silver electrode, *J. Electroanal. Chem.*, **84**, 1–20.
6. Albrecht, M., Grant, J., and Creighton, A. (1977). Anomalously intense Raman spectra of pyridine at a silver electrode, *J. Am. Chem. Soc.*, **99**, 5215–5219.
7. Felidj, N., Aubard, J., Levi, G., Krenn, J. R., Salerno, M., Schider, G., Lamprecht, B., Leitner, A., and Aussenegg, F. R. (2002). Controlling the optical response of regular arrays of gold particles for surface-enhanced Raman scattering, *Phys. Rev. B*, **65**, 075419.
8. Moskovitz, M. (1985). Surface-enhanced spectroscopy, *Rev. Mod. Phys.*, **57**, 783–826.
9. Anker, J. N., Hall, W. P., Lyandres, O., Shah, N. C., Zhao, J., and Van Duyne, R. P. (2008). Biosensing with plasmonic nanosensors, *Nat. Mater.*, **7**, 442–453.
10. Stiles, P., Dieringer, J., Shah, N. C., and Van Duyne, R. P. (2008). Surface-Enhanced Raman Spectroscopy, *Ann. Rev. Anal. Chem.*, **1**, 601–626.
11. Kneipp, K., Wang, Y., Kneipp, H., Perelman, L., Itzkan, I., Dasari, R., Feld, M. (1997). Single molecule detection using surface-enhanced Raman scattering, *Phys. Rev. Lett.*, **78**, 1667.
12. Le Ru E. C., and Etchegoin, P. G. (2009). *Principles of Surface-Enhanced Raman Spectroscopy and Related Plasmonic Effects*, Elsevier, Amsterdam.
13. Kerker, M., Wang, D. S., and Chew, H. (1980). Surface enhanced Raman scattering (SERS). by molecules adsorbed at spherical particles, *Appl. Opt.*, **19**, 4159–4174.

14. Otto, A. J. (1991). Surface-enhanced Raman scattering of adsorbates, *Raman Spectrosc.*, **22**, 743–752.
15. Moskovitz, M. (1985). Surface-enhanced spectroscopy, *Rev. Mod. Phys.*, **57**, 783.
16. Hao, E., and Schatz, G. C. (2004). Electromagnetic fields around silver nanoparticles and dimers, *J. Chem. Phys.*, **120**, 357–366.
17. Felidj, N., Aubard, J., Levi, G., Krenn, J. R., Hohenau, A., Schider, G., Leitner, A., and Aussenegg, F. R. (2003). Optimized surface-enhanced Raman scattering on gold nanoparticle arrays, *Appl. Phys. Lett.*, **82**, 3095.
18. Etchegoin, P. G., and Le Ru, E. C. (2008). A perspective on single molecule SERS: current status and future challenges, *Phys. Chem. Chem. Phys.*, **10**, 6079–6089.
19. Weitz, D. A., Garoof, S., Gersten, J. J., and Nitzan, A. (1983). The enhancement of Raman scattering, resonance Raman scattering, and fluorescence from molecules adsorbed on a rough silver surface, *J. Chem. Phys.*, **78**, 5324–5338.
20. Lorenzo, L., Alvarez-Puebla, R., Pastoriza-Santos, I., Mazzucco, S., Stéphan, O., Kociak, M., Liz-Marzan, L. M., and Garcia de Abajo, J. (2009). Zeptomol detection through controlled ultrasensitive surface-enhanced Raman scattering, *J. Am. Chem. Soc.*, **131**, 4616–4618.
21. Stockmann, M. I., Shalaev, V. M., Moskovits, M., Botet, R., and George, T. F. (1992). Enhanced Raman scattering by fractal clusters: scale-invariant theory, *Phys. Rev. B*, **46**, 2821.
22. Felidj, N., Aubard, J., Levi, G., Krenn, J. R., Schider, G., Leitner, A., and Aussenegg, F. R. (2002). Enhanced substrate-induced coupling in two-dimensional gold nanoparticle arrays, *Phys. Rev. B*, **66**, 245407.
23. Otto, A. (1991). Surface-enhanced Raman scattering of adsorbates, *J. Raman Spectrosc.*, **22**, 743.
24. Aussenegg, F. R., and Lippitsch, M. E. (1978). On Raman scattering in molecular complexes involving charge transfer, *Chem. Phys. Lett.*, **59**, 214.
25. Furtak, T. E., and Macomber, S. H. (1983). Voltage-induced shifting of charge-transfer excitations and their role in surface enhanced Raman scattering, *Chem. Phys. Lett.*, **95**, 328.
26. Albrecht, M. A., and Creighton, J. A. (1977). Anomalously intense Raman spectra of pyridine at a silver electrode, *J. Am. Chem. Soc.*, **99**, 5215.
27. Mantsch, H. H., Choo-Smith, L.-P., and Show, R. A. (2002). Vibrational spectroscopy and medicine: an alliance in the making, *Vb . Spectrosc.*, **30**, 31.

28. Xu, H., Aizpurua, J., Käll, M., and Apell, P. (2000). Electromagnetic contributions to single-molecule sensitivity in surface-enhanced Raman scattering, *Phys. Rev. E*, **62**, 4318; Fabian, H., and Anzenbacher, P. (1993), New developments in Raman spectroscopy of biological systems, *Vib. Spectrosc.*, **4**, 125.

29. Lawson, E. E., Barry, B. W., Williams, A. C., and Edwards, H. G. M. (1998). Biomedical applications of Raman spectroscopy, *J. Raman Spectrosc.*, **28**, 111.

30. Carter, E. A., and Edwards, H. G. (2002). *Infrared and Raman Spectroscopy of Biological Materials* (ed. H.-U. Gremlich and B. Yan), Marcel Dekker, New York, pp. 421–475.

31. Koglin, E., and Sequaris, J. M. (1986). Surface-enhanced Raman scattering of biomolecules, *Top. Curr. Chem.*, **134**, 1.

32. Paisley, R. F., and Morris, M. D. (1988). Surface enhanced Raman spectroscopy of small biological molecules, *Prog. Anal. Spectrosc.*, **11**, 111.

33. Dou, X. M., and Ozaki, Y. (1999). Surface-enhanced Raman scattering of biological molecules on metal colloids: basic studies and applications to quantitative assay, *Rev. Anal. Chem.*, **18**, 285.

34. Kneipp, K., Kneipp, H., Itzkan, I., Dasari, R. R., and Feld, M. S. (2002). Surface-enhanced Raman scattering and biophysics, *J. Phys.: Condens. Matter*, **14**, R597.

35. Sokolov, K., Khodorchenko, P., Petukhov, A., Nabiev, I., Chumanov, G., and Cotton, T. M. (1993). Contributions of short-range and classical electromagnetic mechanisms to surface-enhanced Raman scattering from several types of biomolecules adsorbed on cold-deposited island films, *Appl. Spectrosc.*, **47**, 515.

36. Xu, H., Bjerneld, E. J., Käll, M., and Börjesson, L. (1999). Spectroscopy of single hemoglobin molecules by surface enhanced Raman scattering, *Phys. Rev. Lett.*, **83**, 4357.

37. Cotton, M. T., Schultz, S. G., and Van Duyne, R. P. (1980). Surface-enhanced resonance Raman scattering from cytochrome c and myoglobin adsorbed on a silver electrode, *J. Am. Chem. Soc.*, **102**, 7960.

38. Cotton, T. M., Schultz, S. G., and Van Duyne, R. P. (1982). Surface-enhanced resonance Raman scattering from water-soluble porphyrins adsorbed on a silver electrode, *J. Am. Chem. Soc.*, **104**, 6528.

39. Nabiev, I. R., Savchenko, V. A., and Efremov, E. S. (1983). Surface-enhanced Raman spectra of aromatic amino acids and proteins adsorbed by silver hydrosols, *J. Raman Spectrosc.*, **14**, 375.

40. Curley, D., and Siiman, O. (1988). Conformation and orientation of the haptens, 2,4-dinitrophenyl amino acids, on colloidal silver from surface-enhanced Raman scattering, *Langmuir*, **4**, 1021.

41. Chumanov, G. D., Efremov, R. G., and Nabiev, I. R. (1990). Surface-enhanced Raman spectroscopy of biomolecules. Part I. water-soluble proteins, dipeptides and amino acids, *J. Raman Spectrosc.*, **21**, 43.

42. Faulds, K., Jarvis, R., Smith, W. E., Graham, D., and Goodacre, R. (2008). Multiplexed detection of six labelled oligonucleotides using surface enhanced resonance Raman scattering (SERRS), *Analyst*, **133**, 1505.

43. Brabec, V. V., and Niki, K. K. (1985). Raman scattering from nucleic acids adsorbed at a silver electrode, *Biophys. Chem.*, **23**, 63.

44. Kneipp, K., Pohle, W., and Fabian, H. (1991). Surface enhanced Raman spectroscopy on nucleic acids and related compounds adsorbed on colloidal silver particles, *J. Mol. Struct.*, **244**, 183.

45. Graham, D., Smith, W. E., Linacre, A. M. T., Munro, C. H., Watson, N. D., and White, P. C. (1997). Selective detection of deoxyribonucleic acid at ultralow concentrations by SERRS, *Anal. Chem.*, **69**, 4703.

46. Dou, X., Takama, T., Yamaguchi, Y., Hirai, K., Yamamoto, H., Doi, S., and Ozaki, Y. (1998). Quantitative analysis of double-stranded DNA amplified by a polymerase chain reaction employing surface-enhanced Raman spectroscopy, *Appl. Opt.*, **37**, 759.

47. Deckert, V., Zeisel, D., Zenobi, R., and Vo-Dinh, T. (1998). Near-field surface-enhanced Raman imaging of dye-labeled DNA with 100-nm resolution, *Anal. Chem.*, **70**, 2646.

48. Graham, D., Mallinder B. J., and Smith, W. E. (2000). Surface-enhanced resonance Raman scattering as a novel method of DNA discrimination, *Angew. Chem.*, **39**, 1061.

49. Cao, Y. W. C., Jin, R., and Mirkin, C. A. (2002). Nanoparticles with Raman spectroscopic fingerprints for DNA and RNA detection, *Science*, **297**, 1536.

50. Vo-Dinh, T., Houck, K., and Stokes, D. L. (1994). Surface-enhanced Raman gene probes, *Anal. Chem.*, **66**, 3379.

51. Isola, N. R., Stokes, D. L., and Vo-Dinh, T. (1998). Surface-enhanced Raman gene probe for HIV detection, *Anal. Chem.*, **70**, 1352.

52. Rohr, T. E., Cotto, T., Fan, N., and Tarcha, P. J. (1989). Immunoassay employing surface-enhanced Raman spectroscopy, *Anal. Biochem.*, **182**, 388.

53. Dou, X., Takama, T., Yamaguchi, Y., Yamamoto, H., and Ozaki, Y. (1997). Enzyme immunoassay utilizing surface-enhanced Raman scattering of the enzyme reaction product, *Anal. Chem.*, **69**, 1492.

54. Ni, J., Lipert, R. J., Dawson, G. B., and Porter, M. D. (1999). Immunoassay readout method using extrinsic Raman labels adsorbed on immunogold colloids, *Anal. Chem.*, **71**, 4903.

55. Grubisha, D. S., Lipert, R. J., Park, H. Y., Driskell, J., and Porter, M. D. (2003). Femtomolar detection of prostate-specific antigen: an immunoassay based on surface-enhanced Raman scattering and immunogold labels, *Anal. Chem.*, **75**, 5936.

56. Xu, S., Ji, X., Xu, W. Zhao, B., Duo, X., Bai, Y., and Ozaki, Y. (2005). Surface-enhanced Raman scattering studies on immunoassay, *J. Biomed. Opt.*, **10**(3), 031112.

57. Douglas, P., Stokes, R. J., Graham, D., and Smith, W. E. (2008). Immunoassay for P38 MAPK using surface enhanced resonance Raman spectroscopy (SERRS), *Analyst*, **133**, 791.

58. Stokes, R. J., McBride, E., Wilson, C. G., Girkin, J. M., Smith, W. E., and Graham, D. (2008). Surface-enhanced Raman scattering spectroscopy as a sensitive and selective technique for the detection of folic acid in water and human serum, *Appl. Spectrosc.*, **62**(4), 371.

59. Qian, X., Peng, X.-H., Ansari, D., Yin-Goen, Q., Chen, G. Z., Shin, D. M., Yang, L., Young, A. N., Wang, M. D., and Nie, S. (2008). In vivo tumor targeting and spectroscopic detection with surface-enhanced Raman nanoparticle tags, *Nat. Biotechnol.*, **26**, 83–90.

60. Shafer-Peltier, K. E., Haynes, C. L., Glucksberg, M. R., and Van Duyne, R. P. (2003). Toward a glucose biosensor based on surface-enhanced Raman scattering, *J. Am. Chem. Soc.*, **125**, 588.

61. Stuart, D. A., Yuen, J. M., Shah, N., Lyandres, O., Yonzon, C. R., Glucksberg, M. R., Walsh, J. T., and Van Duyne, R. P. (2006). In vivo glucose measurement by surface-enhanced Raman spectroscopy, *Anal. Chem.*, **78**, 7211.

62. Zhang, X., Shah, N. C., and Van Duyne, R. P. (2006). Sensitive and selective chem/bio sensing based on surface-enhanced Raman spectroscopy (SERS), *Vib. Spectrosc.*, **42**, 2.

63. David, C., Guillot, N., Shen, H., Toury, T., Lamy de la Chapelle, M. (2010). SERS detection of biomolecules using lithographed nanoparticles towards a reproducible SERS biosensor, *Nanotechnology*, **21**, 475501.

64. Das, G., Mecarini, F., Gentille, F., De Angelis, F., Kumar, M. H., Candeloro, P., Liberale, C., Cuda, G., and Di Fabrizio, E. (2009). Nano-patterned SERS substrate: application for protein analysis vs. temperature, *Biosen. Bioelect.*, **24**, 1693–1699.

Chapter 11

Surface Plasmon Resonance Biosensors: Model and Optimization

Dominique Barchiesi

Project Group for Automatic Mesh Generation and Advanced Methods (GAMMA3UTT/INRIA), University of Technology of Troyes, 12 rue Marie Curie, CS 42060, FR-10004, Troyes Cedex, France

dominique.barchiesi@utt.fr

11.1 Introduction

Among the various biosensors, plasmonic biosensors have an increasing expansion since the success in the engineering control of nanostructure fabrication, especially the nanostructuration of gold. The most popular plasmonic biosensors are undoubtedly surface plasmon resonance (SPR) biosensors since they have become one of the most successful, label-free, and commercially developed optical biosensors. The SPR technique is currently employed in biomolecular engineering, drug design, monoclonal antibody characterization, virus-protein interaction, and environmental pollutants detection, among other interesting problems. A review of the various experimental setups and principles was made by Otto[27] and Hoaa *et al.*[17] The common feature of all these devices relies

Nanoantenna: Plasmon-Enhanced Spectroscopies for Biotechnological Applications
Edited by Marc Lamy de la Chapelle and Annemarie Pucci

ISBN 978-981-4303-61-3 (Hardcover), 978-981-4303-62-0 (eBook)
www.panstanford.com

on the shift of the plasmonic resonance caused by the surrounding dielectric environmental change in a binding event. Therefore, the sensing part of the setup is a gold layer, which is probed by an illuminating laser. Actually, since the works by Kretschman, Otto, and Raether, the metallic plate is known to be a resonator feature for incoming (p) polarized light.[21,27,29] This resonant behavior yields to a sudden absorption of light by the thin metallic layer, when the angle of incidence of the illumination varies, leading to a high sensitive device.[13,17,20,22,26] Kolomenskii *et al.* proposed models and experiments that demonstrate the linear relation between the measurement of the SPR shift and the concentration of NaCl and ethanol. They also discussed the limit of detection according to the specifications of the detector and the source.[20]

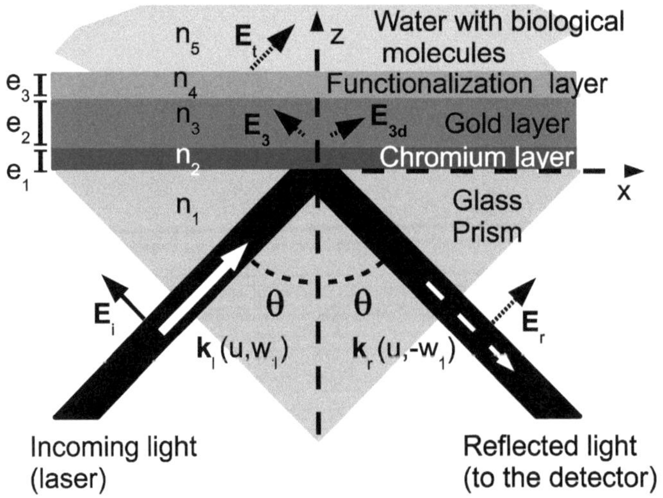

Figure 11.1 Schematic of the biosensor used to measure variations of the optical index around that of water n_5, as a function of biomaterials concentration. The index of the glass prism is n_1. The index of the gold layer n_3, as well as that of the chromium layer, is complex and depends on the incoming wavelength λ_0, which will be fixed in this study, as well as that of the chromium layer, $n_2(\lambda_0)$. The functionalization layer is used to fasten the biological molecules of interest in the vicinity of the gold layer. Its optical index is n_4. The thicknesses of the layers are e_1, e_2, and e_3 for the chromium, the gold, and the functionalization layer, respectively. The angle of incidence of the illumination is θ.

Basically, out of the resonance conditions, the intensity of the reflected light is that of a mirror, and around the plasmon resonance, the reflected light vanishes. For this, we consider the nanosensor the multilayered system shown in Fig. 11.1.

The active plasmonic part is the gold layer, and it has been demonstrated that the plasmon launch could be explained through the extinction theorem.[1] That paper focused on the dispersion relations for surface plasmon/polaritons and how the associated fields behave. In particular, the plasmon is considered a mode that the metallic medium may sustain as a result of the presence of the surface, without an incoming field. In that reference, the distinction of absorbing (dispersive) and nondispersive media illustrates the fundamental distinction between the Brewster angle and the plasmon effect.

Despite this apparent complexity, the surface plasmon/polariton or plasmon surface polariton in thin films is well known since decades,[32] and pedagogical papers and even educational papers related to this problem have allowed generations of students to face exercises at a degree level in electromagnetism.[31]

The model of laser illumination is a monochromatic, (p) polarized, plane wave whose electric field is, therefore, in the (x, z) plane and whose magnetic field lays along the y-axis.[10,29] Unless otherwise stated, the numerical values of the parameters are given in Table 11.1.

Table 11.1 Parameters used in the model, unless otherwise stated. The optical indexes of metals at the indicated wavelength λ_0 are taken from Palik.[28] i is a pure, imaginary number ($i^2 = -1$)

n_1	n_2	n_3	n_4	n_5	e_1	e_2	e_3	λ_0
1.55	$3.08 + 3.35i$	$0.163 + 3.47i$	1.50	1.33	2 nm	50 nm	5 nm	670 nm

The present chapter is devoted to theoretical considerations on the SPR biosensor, with applications to its optimization. In the following, the metallic layers are supposed to be infinite along x and y directions and therefore the problem reduces to one dimension and can be described, without loss of generality, in the plane (x, z). In the next section, we will recall the concept of plasmons on a single interface.

11.2 The Plasmon on a Single Interface

The problem under consideration is the reflection of light by a bulk metallic medium. Therefore a single interface between a dielectric and gold, a semi-infinite media, is located at $z = 0$, and the incoming light comes from a nonabsorbing, dielectric medium: the glass sample on which is deposited a very thick gold layer or simply air. The permittivities of the dielectric and the metal are $\varepsilon_d = 1$ and $\varepsilon_m = n_3^2$ (Table 11.1).

The incoming wave vector has, therefore, two components, $\mathbf{k}_i = k_0(u, 0, \sqrt{\varepsilon_d - u^2})$, with $k_0 = 2\pi/\lambda_0$, and $u = \sqrt{\varepsilon_d}\sin(\theta)$.[10] The transmitted (in metal) and reflected wave vectors are therefore:

$$\mathbf{k}_t = k_0(u, 0, w_2) = k_0(u, 0, \sqrt{\varepsilon_m - u^2}) \tag{11.1}$$

$$\mathbf{k}_r = k_0(u, 0, -w_2) = k_0(u, 0, -\sqrt{\varepsilon_d - u^2}) \tag{11.2}$$

The square root function of a complex number is multiple valued and requires a choice of the argument, usually the principal branch with the plane cut along the negative real axis. The z component of the wave vector is defined by a square root of a complex number and the orientation of the z-axis (Fig. 11.1.) implies the positivity of the real parts of w_i, with the above-mentioned definitions. On the other side, the sign of the imaginary parts of wave vectors must be coherent with the physically awaited decrease of the amplitude of the electromagnetic fields with the distance to the boundary. In this study, the electromagnetic fields involve $\exp(iw_i z)$.

The classical Fresnel's reflection coefficient can be defined for monochromatic illumination with (p) polarization, in the case of an absorbing material, as it results from the analytical computation of the amplitudes of the magnetic fields, satisfying Maxwell's equations and boundary conditions on the interface:

$$r_{12} = \frac{n}{d} = \frac{\varepsilon_m w_d - \varepsilon_d w_m}{\varepsilon_m w_d + \varepsilon_d w_m}. \tag{11.3}$$

Opening remarks deserve to be extracted from the above-mentioned paper[1]:

- The computation of the z complex component of the transmitted wave vector $\mathbf{k}_t$ must be computed with care of its sign, through the square root function. *Its real part must be positive, z being oriented in the direction of propagation of the incoming wave, and its imaginary part must have a coherent sign to decrease with increasing distance from the interface* ($z > 0$).

- The absorption of light is governed by the imaginary part of the permittivity, while the decrease of the amplitude of the electromagnetic field is due to the imaginary part of the wave vector.
- The plasmon resonance cannot be exactly excited practically even through Kretschman or Otto configurations. Actually the plasmon is described as a mathematical complex pole, and its fingerprint is only observed experimentally, thanks to a real wave vector (u is a real number associated to a real angle of incidence of light), while it would require a complex wave vector.[8,9,24,25] Therefore, the plasmon cannot be launched by classical illumination.
- The plasmon can be described as a mode that the metallic medium may sustain as a result of the presence of the surface, without an incoming field, or as a pole of the scattering matrix.[8,24]
- Drude's or the Drude-Lorentz models are commonly used to draw dispersion curves; nevertheless, they are not relevant for gold in a visible domain (especially around 550 nm), where the interband transitions must be taken into account.

Figure 11.2 illustrates the reflected intensity $R = |r_{12}|^2$, with $\lambda_0 = 670$ nm, $\varepsilon_d = 1$, $\varepsilon_m = -12.0 + 1.13i$. The plasmon cannot be launched under realistic illumination; u must be greater than 1, which is practically impossible since $u = \sin(\theta)$ in this case.

Nevertheless, the sign of the imaginary part of w_i must be controlled as it has a physical meaning[1]: $\Im(w_1) < 0$ and $\Im(w_2) > 0$ to pursue the exponential decrease of the amplitude of waves when one moves away from the interface in both media. Actually, the amplitudes of the fields are governed by $\exp(iw_i z)$. Particular attention should be focused on this numerical aspect as this degree of freedom for w_1 is related to the fact that the excitation of the plasmon corresponds to a virtual case, where θ is complex. To illustrate this point, let us consider a specific value of u corresponding to the plasmon excitation:[1]

$$u_{\mathrm{p}} = \sqrt{\frac{\varepsilon_{\mathrm{d}}\varepsilon_{\mathrm{m}}}{\varepsilon_{\mathrm{d}} + \varepsilon_{\mathrm{m}}}} = 1.044 + 0.00442i, \tag{11.4}$$

Therefore, the numerical values are $w_1 = 0.0154 - 0.300_i$, $w_2 = 0.155 + 3.62_i$, and $r_{12} = 0$, and the dispersion relation with the opposite sign, as in Agarwal:[1]

$$\varepsilon_{\mathrm{d}} w_2 - \varepsilon_{\mathrm{m}} w_1 = 0 \quad \text{or} \quad \mathbf{k}_{\mathrm{r}} \cdot \mathbf{k}_{\mathrm{t}} = 0 \tag{11.5}$$

This dispersion relation of the plasmon corresponds to the vanishing of the reflection coefficient $r_{12} = 0$ and therefore to the cancelation of its **numerator** instead of the classical cancelation of its denominator. Two arguments can be put to the front to support the coherence of this assertion.

(1) The transfer of illumination energy to the gold plate is the expression of the plasmon resonance. Therefore, the reflected intensity must vanish. This can only occur if its numerator of Eq. 11.3 vanishes:

$$\varepsilon_{\mathrm{m}}\sqrt{\varepsilon_{\mathrm{d}} - u^2} - \varepsilon_{\mathrm{d}}\sqrt{\varepsilon_{\mathrm{m}} - u^2} = 0 \quad \Leftrightarrow \quad u = \pm\sqrt{\frac{\varepsilon_{\mathrm{d}}\varepsilon_{\mathrm{m}}}{\varepsilon_{\mathrm{d}} + \varepsilon_{\mathrm{m}}}} \tag{11.6}$$

The resolution of this equation involves a squaring and therefore a loss of sign. *This fact explains the fact that it is commonly admitted that the above solution corresponds to the cancellation of the denominator of Eq. 11.3, leading to a misunderstanding of resonance.*

(2) Any resonance is considered to be based on "making something from nothing."[1] Therefore, the problem can be reformulated as a pole of a transfer function of the system. The transfer function is defined as the ratio of the "output" of the system to its "input." In plasmonics, the output is the field in the metal and the input is the reflected field, which must vanish (*instead of the incoming field*). This interpretation is consistent with that by Agarwal.[1] Therefore, the transfer function H follows:

$$E_2 = HE_{\mathrm{r}} = t_{12} . r_{12}^{-1} E_i = \frac{2\varepsilon_{\mathrm{m}} w_1}{\varepsilon_{\mathrm{m}} w_1 + \varepsilon_{\mathrm{m}} w_2} \frac{\varepsilon_{\mathrm{m}} w_1 + \varepsilon_{\mathrm{d}} w_2}{\varepsilon_{\mathrm{m}} w_1 - \varepsilon_{\mathrm{d}} w_2} = \frac{2\varepsilon_{\mathrm{m}} w_1}{\varepsilon_{\mathrm{m}} w_1 + \varepsilon_{\mathrm{d}} w_2} \tag{11.7}$$

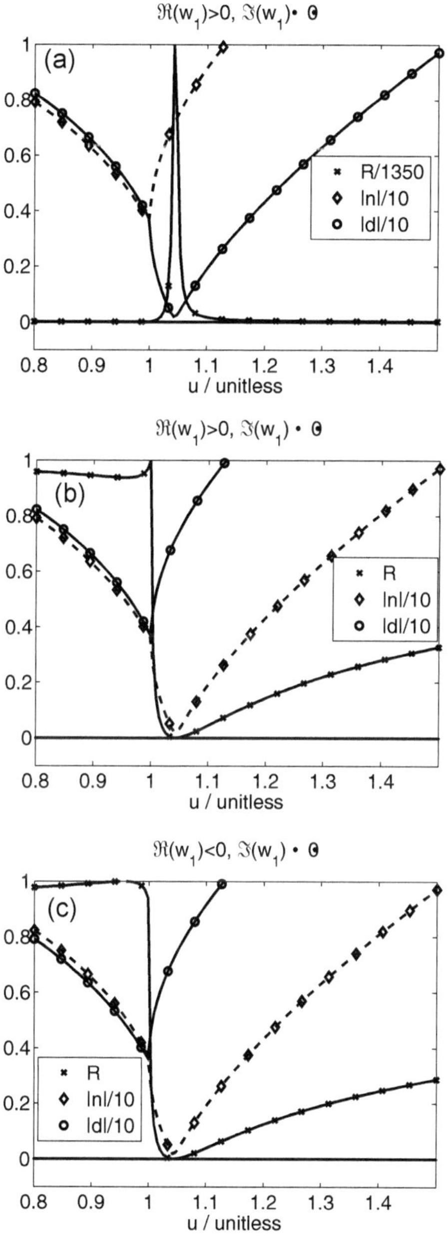

Figure 11.2 Reflection by a single interface air/gold (λ_0 = 670 nm) as a function of the dimensionless parallel to the interface component of the wave vector.

The pole of the transfer function is the one found above (Eq. 11.6). Remember that we have assessed that this solution cancels the numerator but not the denominator of the reflection coefficient r_{12}.

Figure 11.2 shows the erroneous plot computed directly without control of the sign of the imaginary part of w_1 (a) and the imaginary part accurate results (b–c), (c) being fully coherent with the resonator concept as well as the Agarwal[1] approach. Finally the above formalism is compatible with the definition of the Brewster angle, for which the reflection also vanishes but for the real angle, and the real ε_m. The reader is encouraged to compute the above-mentioned quantities for various properties to handle the subtle definition of the square root of a complex number. The discrimination between Brewster and plasmon angles is mainly related to the sign of the real part of the permittivity. Its imaginary part contributes to properly moving away the solution from the real axis.

11.3 Plasmon Resonance in a Thin Film

To explain the basic principle of the SPR biosensor, a simple model is proposed. The sensitive part of the biosensor is supposed to be only a thin gold layer deposited on a glass substrate; to simplify the discussion, no stick layer and no functionalization are considered (Fig. 11.1). Therefore, in this section, the thickness of gold is simply denoted by e, and $\varepsilon_1 = n_1^2$, $\varepsilon_2 = n_3^2$, and $\varepsilon_3 = n_5^2$ (Table 11.1). The plasmon results from the excitation of this metal layer by a laser with wavelength λ_0, under angle θ. The gold layer is illuminated in reflection, through the dielectric sample, with a p-polarized wave, in the Kretschmann configuration.[35] The thickness of the gold layer can be viewed as an additional degree of freedom with regard to the infinite medium studied in the previous section. Therefore, the thickness is a parameter that can be used to tune the plasmon resonance. Basically, the launch of the plasmon is detected through an attenuation of the reflected intensity R. Indeed, the plasmon occurrence is a resonance of the interaction between light and matter. This resonance is highly sensitive to any variation of the setup parameters and specifically to the change of the optical index $n_5 = \sqrt{\varepsilon_3}$ of the above medium, and therefore, the presence of the biological molecule in the vicinity of the gold layer induces a change of the position of the plasmon resonance, which varies accordingly.

11.3.1 Computation of the Parameters Required to Launch the Plasmon

To understand the SPR biosensor, an electromagnetic modeling of the interaction between a plane wave and a metallic layer is proposed. Even if the illumination is incoming under angle θ, the problem can be considered scalar, and the resolution of Maxwell's equations for monochromatic waves amounts to solving a linear system, which, unknown, is the amplitude of the reflected magnetic field whose square modulus is R. The analytical calculation of the reflection coefficient $R = |r_{13}|^2 I_0$, where I_0, the illumination intensity, equal to unity in the following, is given by:

$$r_{13} = \frac{n}{d} = \frac{r_{12} + r_{23}\exp(2iew_2)}{1 + r_{12}r_{23}\exp(2iew_2)}$$

$$= \frac{(\varepsilon_2 w_1 - \varepsilon_1 w_2)(\varepsilon_3 w_2 + \varepsilon_2 w_3) + (\varepsilon_3 w_2 - \varepsilon_2 w_3)(\varepsilon_2 w_1 + \varepsilon_1 w_2)e^{2iw_2 e}}{(\varepsilon_2 w_1 + \varepsilon_1 w_2)(\varepsilon_3 w_2 + \varepsilon_2 w_3) + (\varepsilon_3 w_2 - \varepsilon_2 w_3)(\varepsilon_2 w_1 - \varepsilon_1 w_2)e^{2iw_2 e}}, \tag{11.8}$$

where r_{ij} are the Fresnel coefficients of the interface j:[31]

$$r_{ij} = \frac{\varepsilon_j w_i - \varepsilon_i w_j}{\varepsilon_j w_i + \varepsilon_i w_j}, \tag{11.9}$$

with $w_i = 2\pi\sqrt{\varepsilon_i - u^2}/\lambda_0$ and $u = \sqrt{\varepsilon_i}\sin(\theta)$, like in the previous section. Of course, R can be considered relative to the incoming intensity factor, and if $e = 0$ (no gold layer), $r_{13} = r_{12}$ and $r_{23} = 0$. The plasmon resonance is revealed by $R \to 0$ and corresponds to $r_{12} + r_{23}$ $\exp(2_i e w_2) \to 0$. This equation can be considered a coupling between the plasmon of the first interface (between glass and gold, through r_{12}) and the surface plasmon of the second one (between gold and the upper medium, through r_{23}). The rigorous condition is fulfilled if:

$$\frac{\varepsilon_2 w_1 - \varepsilon_1 w_2}{\varepsilon_2 w_1 + \varepsilon_1 w_2} + \frac{\varepsilon_3 w_2 - \varepsilon_2 w_3}{\varepsilon_3 w_2 + \varepsilon_2 w_3}\exp(2iew_2) = 0 \tag{11.10}$$

The cancelation of R is therefore reached only if its numerator vanishes, ensuring that we have reduced r_{13} to a common denominator.

Otherwise, the possible cancelation of each denominator of r_{ij} could distort the discussion of results. Moreover, the conditions of cancelation for the denominator are not the same as those for the numerator. Again, we simply assess that $R = 0$ occurs if its numerator vanishes. The arguments developed in the previous section remain valid, both in terms of the transfer function and for the signs of the coordinates of wave vectors. Nevertheless, the plasmon can only be experimentally launched with real values of θ ($|u| < n_1$).

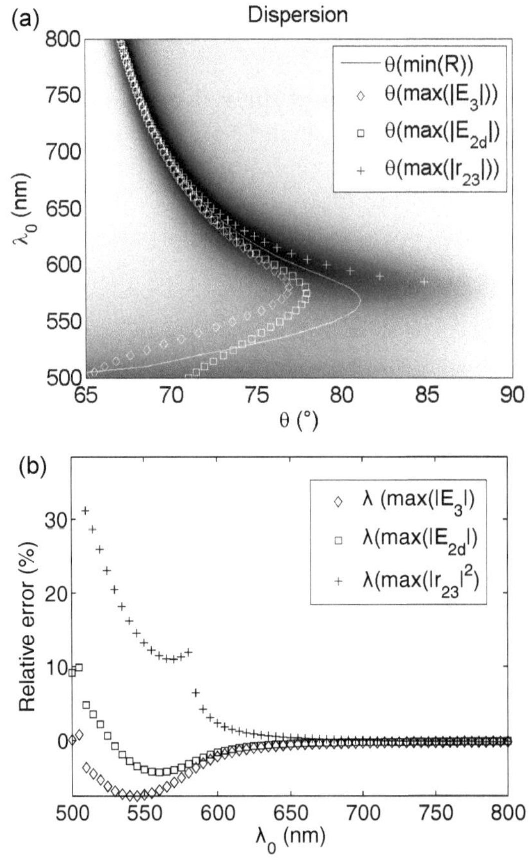

Figure 11.3 Dispersion of the plasmon in a gold layer of thickness $e = 50$ nm, deposited on a glass sample ($n_1 = 1.5$), the above medium being water $n_3 = 1.33$.

Figure 11.3a shows the dispersion curve, from $R(\theta, \lambda_0)$ for gold. The permittivities are those of Palik.[28] The white line delineates the minimum of R in a well, corresponding to the fingerprint of the plasmon resonance, that is, the near vanishing of R. The values of the maximum of $|r_{23}|$ (crosses) differs from those of min(R), contrary to what is generally claimed.[31,11] Some conclusions can be now drawn:

(1) The SPR biosensor cannot be described by neglecting the thickness of the gold plate.
(2) The maximum of the field reflected by the gold/water interface (squares) includes the thickness but is as far as not sufficient to find the plasmon. The field that is transmitted from glass in the gold layer E_2 contributes equally to the plasmon.
(3) The maximum of the amplitude of the transmitted field (argyles) does not correspond exactly to the plasmon excitation.

Figure 11.3b shows the relative error of all these angles, respectively, to the fingerprint of the plasmon angle, corresponding to min(R).

Therefore, the complete model (R) of the SPR biosensor must be used if optimization is required. In the next subsections, both the thickness and the sensitivity are consequently defined by using r_{13}.

11.3.2 A First Optimization of the Thickness

In principle, $R = 0$ could therefore be monitored by the adjustment of the thickness e of the gold layer:

$$e = \frac{1}{2iw_2}\log\left(-\frac{\varepsilon_2 w_1 - \varepsilon_1 w_2}{\varepsilon_2 w_1 + \varepsilon_1 w_2}\,\frac{\varepsilon_3 w_2 + \varepsilon_2 w_3}{\varepsilon_3 w_2 - \varepsilon_2 w_3}\right) \tag{11.11}$$

Equation 11.11 shows that the mathematical solution of the problem e is complex, whereas it is physically a thickness of the gold layer, the permittivity of gold $\varepsilon_2 = n_2^2$ being a complex number. Again, such a mathematical solution is a pole, in the complex plane, and cannot be used as is. Moreover, the logarithm is a multiple-valued function of its complex argument. As for the square root, the principal branch of the argument can be used: the plane cut is therefore along the negative real axis. This choice is governed by the required region of continuity of the function.

In this case, the real part of the mathematical solution for e is positive and its imaginary part is arbitrary. Nevertheless, the distance between the pole corresponding to the resonance and the real axis can help to choose the best-working wavelength, considering that the gold optical index is known. Hence, fixing θ enables to optimize the thickness e using Eq. 11.11, or Poincare's plot or phase plot of e. Figure 11.4 shows the walks of the plasmon as a function of the real part and the imaginary part of the thickness when the wavelength varies. This plot helps to choose a laser and the corresponding thickness, to decrease the sensitivity of the biosensor, on the thickness. Thus the best values are the intersects of the curves with the real axis. Some angles of incidence are therefore unsuitable if the corresponding imaginary part of e never vanishes. Nevertheless, the horizontal tangents correspond to zones of stability and to the tight dependence of the plasmon to the change of the optical index of gold. Of course, the permittivities are sensitive to the conditions of gold film preparation. But these first results of optimization remain significant.

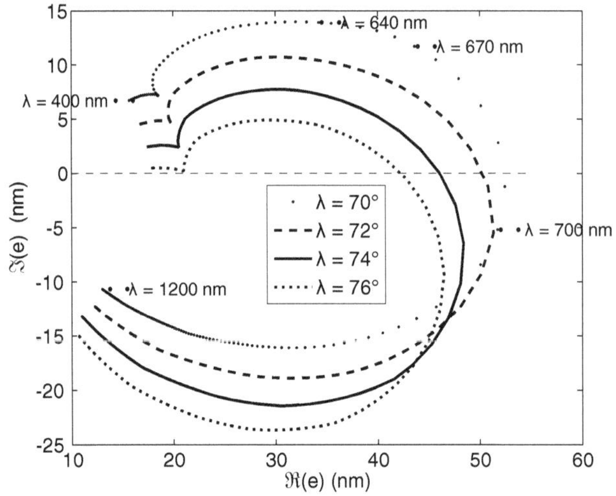

Figure 11.4 Poincare's plot of the plasmon walk in the complex plane of the thickness e, for various angles of incidence, when varying the illumination wavelength.

Figure 11.4 is also an illustration of the thickness as a degree of freedom of the plasmon launch. Actually, a mathematical solution exists for all wavelengths. Nevertheless, e must remain a real number. For λ_0 = 670 nm, the best angle of incidence is in [72, 76]° and the thickness e is in [43, 53] nm. This result could be more accurately determined with this abacus-type method, but it is unnecessary to devote more effort, the model being not accurate enough, as we will show in the next section.

11.3.3 The Sensitivity of the SPR Biosensor

To improve the model of the SPR biosensor, we have to define its sensitivity and, therefore, its mode of detection. The target is the detection of the presence of biomolecules of interest in the upper medium. For this, the shift of the plasmon resonance, induced by the absence/presence of the biomolecules, is measured (Fig. 11.5). For this, we consider the nanosensor as a single layer of gold on glass. The medium of detection is the above medium (water).

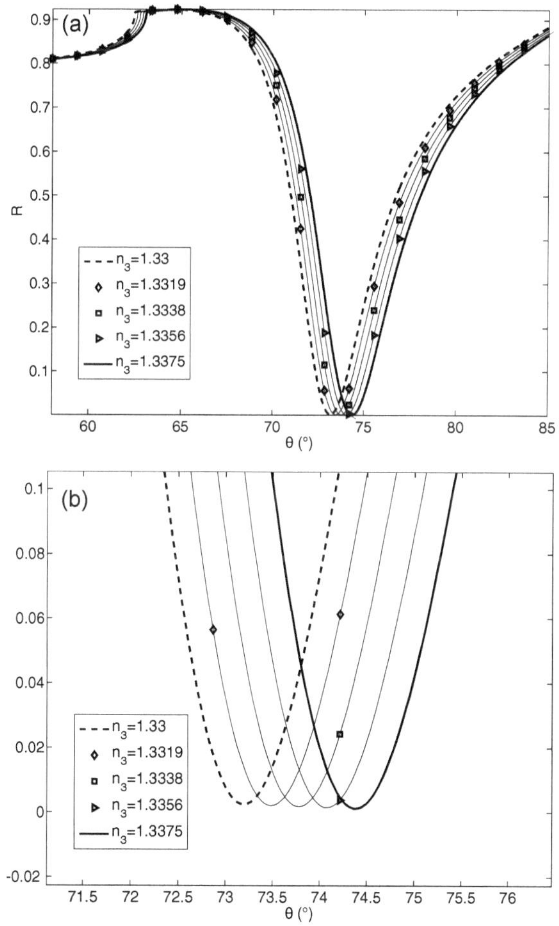

Figure 11.5 Computed reflected intensity $R = |r_{13}|^2$ as a function of the illumination angle for a small variation of the above medium. The thickness of the gold layer is 50 nm; $n_1 = 1.5$ (glass).

The shift of the plasmon resonance can be measured as the shift of the minimum or that of the maximum of the derivative of $R(\theta)$ (Fig. 11.6):

$$S = \theta(\min(R(n_3 + \Delta n_3))) - \theta(\min(R(n_3))) \tag{11.12}$$

$$S_d = \theta\left(\frac{dR}{d\theta}(n_3 + \Delta n_3)\right) - \theta\left(\frac{dR}{d\theta}(n_3)\right) \tag{11.13}$$

$$V = \frac{\max(R) - \min(R)}{\max(R) + \min(R)} \tag{11.14}$$

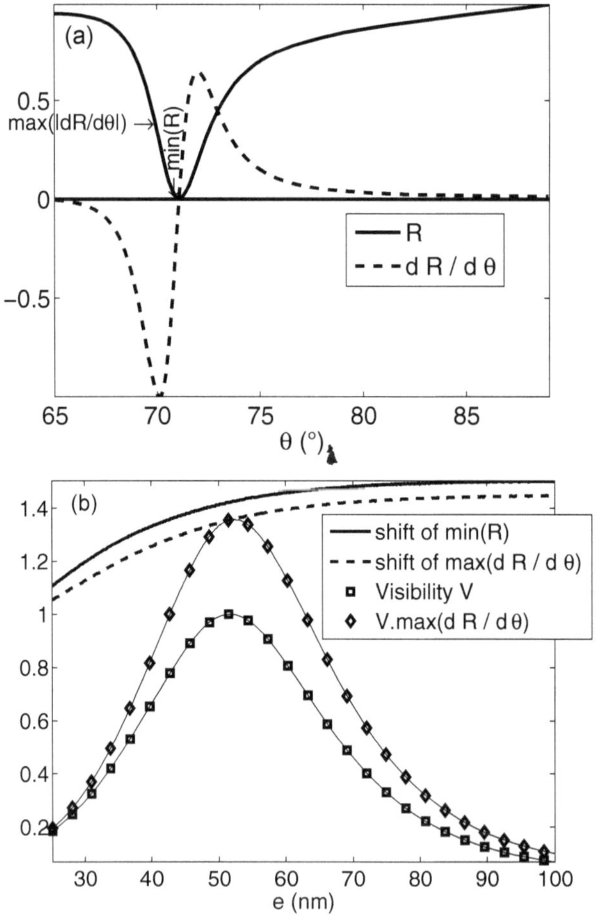

Figure 11.6 Sensitivities: Shift of the minimum S, shift of the maximum of derivative S_d, visibility V, and product $V \cdot S_d$ as functions of the thickness e, $\Delta n_3 = 0.01$.

The shift of the minimum of R or of the maximum of its derivative (Fig. 11.6a) are increasing functions of the thickness e. On the contrary, the visibility exhibits a maximum around 50 nm. The best compromise for detection is sufficient visibility and sensitivity. Therefore, Fig. 11.6b illustrates the possible targets of optimization. Finally the optimization of the plasmon fingerprint (minimum of R) seems to be the best choice in the following. Nevertheless, in numerical optimization, the model of SPR must be as accurate and realistic as possible.

11.4 Optimization of SPR

To optimize such sensors, the following conditions must be fulfilled. First, the numerical models of biosensors have to describe the sensitive part of the biosensors as well as the method of detection.[3,7,16] However, the more realistic the model is, the more computational time and memory it requires. Indeed, the optimization method requires the repeated evaluation of the model, and therefore a compromise must be struck between these two opposing constraints. Second, the model must be stable and accurate within the allowed domain of search of the best parameters. The control of exogenous parameters of the model (e.g., the discretization,[15] the number of terms in series,[14] the method of inversion, or the mathematical schemes[2,6,23,33]). A more accurate model is proposed in the following section by taking into account all the parameters mentioned in Fig. 11.1 and Table 11.1. Then, this model is used for optimization.

11.4.1 An Accurate Model of SPR

The model of SPR in the previous section neglects both the chromium stick layer of gold on glass and the functionalization layer (Fig. 11.1). For optimization purposes, we use the (S) formalism to compute accurately the intensity reflected by the full model of SPR, including the chromium and functionalization layers (Fig. 11.7).[23] This mathematical formulation has the advantage of numerical stability over a wide range of thicknesses and permittivities.[23]

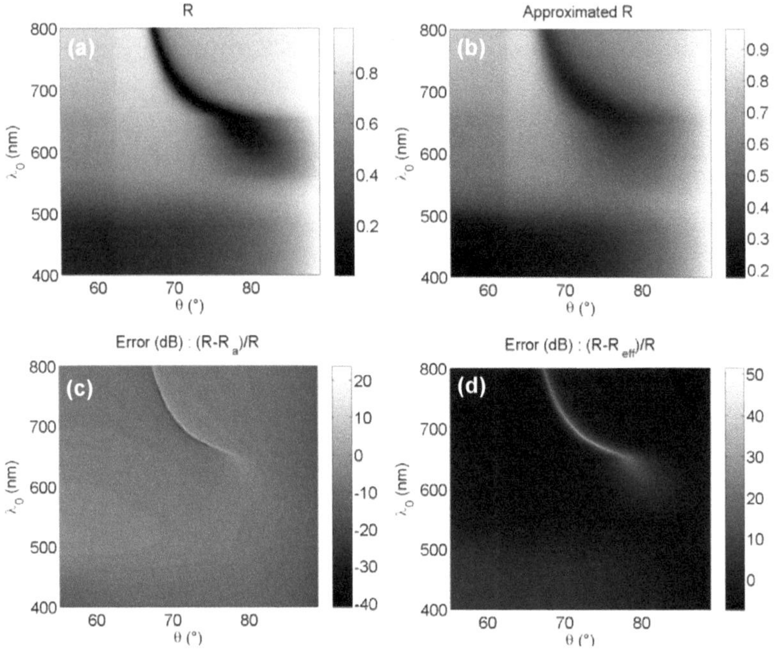

Figure 11.7 Dispersion of the plasmon in a gold layer of thickness e = 50 nm, deposited on a glass sample (n_1 = 1.5), the above medium being water n_3 = 1.33. A 2 nm thick layer of chromium separates gold and glass, and a 5 nm thick functionalization layer (optical index n_4 = 1.5) is included in the model. (a) $R(\theta, \lambda_0)$, (b) the approximated model using the effective index for the metal layer, (c) the relative error (dB) with regard to the approximated model of the single gold layer, and (d) the relative error (dB) with regard to the approximated model with metal effective index.

A simple visual comparison of the dispersion computed with the approximated model (Fig. 11.3a) and the realistic model (Fig. 11.7) seems to demonstrate the negligible influence of the chromium and functionalization layers. However, the relative error in dB is significant, particularly in the area of plasmon excitation. Indeed, it has been shown that a nanometer layer of grip can degrade and shift the plasmon resonance.[9,12,26,34] The relative error should be always used for scientific comparison of models that are close together. An illustration of the influence of two approximations can be observed in Fig. 11.7c,d. Figure 11.7c shows the error induced by the neglecting of the chromium layer and of the functionalization layer. The error can reach more than 20 dB, around the plasmon dispersion curve. The approximation whose relative error is shown in Fig. 11.7d is based on the effective index of the metal layer. Considering the thickness of the stick layer (Cr) is 4% of the gold one, one may consider it negligible, or better, the effective index of the bimetal layer can be calculated:

$$n_{\text{eff}} = \sqrt{\frac{\Re n_2^2 e_1 + \Re n_3^2 e_2}{e_1 + e_2}} \quad \text{and} \quad k_{\text{eff}} = \sqrt{\frac{\Im n_2^2 e_1 + \Im n_3^2 e_2}{e_1 + e_2}}, \tag{11.15}$$

and therefore, $n_2 = n_3 = n_{\text{eff}} + k_{\text{eff}}$. this case, the approximation is better, except for the plasmon location, where the relative error can reach 50 dB.

The sensitivity of the SPR biosensor, defined in Eq. 11.12, becomes:

$$S = \theta(\min(R), n_4 + \Delta n_4) - \theta(\min(R), n_4), \tag{11.16}$$

the optical index of the functionalization changing if it sticks the target biological molecules. Again, we verified that the target of the optimization is the quality of the plasmon resonance, which maximizes the sensitivity of the biosensor (visibility of the reflection curve $R(\theta)$ and angular shift of its minimum).

11.4.2 Optimization of SPR with Two Heuristic Methods

The evolutionary methods have been used to optimize the SPR, with a stick chromium layer[9] and have been improved to decrease the number of evaluations before convergence.[4] On the other hand, a collaborative method has been proposed[19] and also improved for plasmonics.[18] Pseudo codes and a full discussion of these methods and complementary physical discussions of SPR can be found in Barchiesi.[5] In this subsection, we describe both optimized algorithms and give some numerical results of the optimization of the SPR biosensor.

11.4.2.1 The adaptive nonuniform hyperelitist evolutionary method

The pseudo algorithm 11.1 presents the steps of computation for the evolutionary method.

Algorithm 11.1 Pseudo codes of the improved evolutionary method for plasmonics: the adaptive nonuniform hyperelitist method (ANUHEM). R is the fitness function that must be minimized.

Require: D domain of the N acceptable parameters e_i, $\theta, \ldots$

Require: Fixed parameters of the model R: ε_i if the wavelength is fixed.

Require: λ (number of parameter sets in the secondary population after mutation: the children), μ (number of parents), T (maximum of generations), and SC (stop criterion), p (number of elements to be recombined)

1. [Initial population with U: multidimensional uniform law in D]
   $p_{\mathrm{m}} = 1 \ldots \mu \leftarrow U(D)$
   $y_{\mathrm{m}} = 1 \ldots \mu \leftarrow R(p_{\mathrm{m}} = 1 \ldots \mu)$
2. **While** $|y_1 - y_\mu| / |y_1 + y_\mu| > SC$ **do**
3. [Hyperelitist crossover with random choice of integers n through $U(\{1, \ldots, \mu\})$
   $p'_{\mathrm{m}} = 1 \ldots \lambda \leftarrow \Sigma_p y_n^{-1} p_n / \Sigma_p y_n^{-1}$
4. [Adaptive nonuniform mutation using topology of solutions]
   $b \leftarrow std(p_{\mathrm{m}} = 1 \ldots \mu)/std(y_{\mathrm{m}} = 1 \ldots \mu)$
   $\sigma \leftarrow 1 - U([0; 1])^{(1-g/T)_b}$
   $p''_{\mathrm{m}} = 1 \ldots \lambda \leftarrow UB(D, \sigma)$ {[$UB(D, \sigma)$: multidimensional nonuniform law in D with constraints $p'' \in D$]}
   $y_\mu + 1 \ldots \mu + \lambda \leftarrow R(p''_j)$
5. [Selection (S) of the μ best elements among the λ for the initial population of the next step (elitist scheme[30])]
6. $y_{\mathrm{m}} = 1 \ldots \lambda + \mu \leftarrow F(\{p''_j, p'_j\})$
   $p_{\mathrm{m}} = 1 \ldots \mu \leftarrow S(\{p''_{\mathrm{m}} = 1 \ldots \lambda, p_m = 1 \ldots \mu\})$
7. **end while**

The proposed method uses a bounded generation of parameters within the initial domain D to prevent attraction toward trivial solutions. The plasmonic improvements of the classical Schwefel's method[30] consist of using nonuniform distribution of probability in the mutation step and weighted breeding. These two tricks enable us to increase the convergence of the method, maintaining the diversity of the search.[4] The elitist method of selection is more powerful than the nonelitist method in this case. In this case, the selection is made among the $\lambda + \mu$ parents and children.

11.4.2.2 The adaptive particle swarm optimization method

Particle swarm optimization (PSO) is basically a cooperative method where the set of parameters **p** at step t (corresponding to generation g in evolutionary methods) is considered a moving particle (or bee) in the N-dimensional space of search. The particles of a swarm communicate good positions to each other and adjust their own positions **p** and velocities v at each step t:

$$\mathbf{v}(t+1) = \omega\mathbf{v}(t) + r_1c_1(\mathbf{p}_b(t) - \mathbf{p}(t)) + r_2c_2(\mathbf{p}_b - \mathbf{p}(t)) \quad (11.17)$$

$$\mathbf{p}(t+1) = \mathbf{p}(t) + \mathbf{v}(t+1) \quad (11.18)$$

where r_1 and r_2 are random variables between 0 and 1, $\mathbf{p}_b(t)$ is the particle's best position, $\mathbf{p}_b$ is the global best, ω is the inertial weight, and c_1 and c_2 are the acceleration coefficients. The parameters ω, c_1, and c_2 could be constant or time dependent (i.e., updated at each step).

After computing the new velocity, the particle moves toward a new position following Eq. 11.18. The particle's new velocity $\mathbf{v}(t+1)$ combines its previous value and the distances between the particle's current position and its own best-found position, that is, its own best experience $\mathbf{p}_b(t)$ and the swarm global best $\mathbf{p}_b$. The first term of Eq. 11.18 is an inertia term, which prevents the only local search and therefore preserves diversity in the exploration of D.

The adaptive PSO (APSO) uses also the nonuniform distribution and can be compared with the method proposed by Zhan *et al.*,[36] where the authors outlined the necessity of updating the acceleration coefficients, following the three phases of optimization that are expected during the PSO loop (exploration, exploitation, or convergence phase). Through the value of $b(t)$:

- ω, the inertial coefficient, is close to 1 in the exploration phase, then decreases, and then increases in the exploitation state. This preserves the diversity of solutions and prevents the convergence to a local minimum. Finally it decreases toward 0 in the convergence state (when many particles are close to those for the global minimum).
- c_j, the acceleration coefficient, tends toward 0 when t reaches T.[37] Therefore T controls convergence and remains the only nonadaptive parameter of the method.

Then, the adaptation of the exogenous parameters is ensured through the nonuniform law, with $b(t)$ defined in algorithm 11.1:

$$\omega(t) = \max(b(t))^{-1} \text{ and } c_1 r_1 = 2U_b, c_2 r_2 = 2U_b \quad (11.19)$$

All the adaptive parameters are therefore monitored by the topology of the fitness function R. Actually, $b(t)$ can be seen as the slope of the solutions, that is, the spreading/dispersion of the population with regard to their quality, and can differ for each component of $\mathbf{p}(t)$ and therefore takes into account the sensitivity of R to the variations of all parameters of $\mathbf{p}$. The values of $b(t)$, for each parameter, govern the convergence speedup as follows:

- In the exploration state, $b(t)$ is close to 1 as well as $c_1(t)$ and $c_2(t)$. The value of $b(t)$ gives equal weight to the different contributions of the velocity. If the speed of the particle is too high, the particle (obtained at the generation t) leaves the search space and is therefore replaced by a random particle, through a uniform law, as in the initialization step. This contributes to the diversity at this stage.
- In the exploitation state, $b(t)$ increases and also the convergence speedup.[37]
- In the final state of convergence, $b(t)$ decreases again and the wells of the objective function R are more carefully exploited.

If velocity is too high, the particle can get out the domain D of allowed parameters. In this case, the particle is then freezed to its previous position in the space of search and therefore is useless for search. Kessentini and Barchiesi[18] solved this problem recently, but this improvement is not used in the following numerical study.

11.4.2.3 Numerical optimization of SPR

The chromium stick layer of thickness e_1 gives an additional degree of freedom to find $R(e_1, e_2, \theta) = 0$. Therefore, an infinity of solutions may be found using these free parameters. Assuming that chromium is only used to stick gold on glass, its thickness can be fixed to 2 nm.[9,17,20,26] Thus, the optimization involves only two parameters, e_2 and θ. Of course a double loop could be also used in this case, but the number of evaluations of the model, which would be required to reach the same precision on parameters, would be at least 10 times greater.[9,14] The fixed parameters are therefore the wavelength of the laser illumination (λ_0 = 670 nm), the optical indexes ($n_1 = 1.5$, $n_2(670\text{ nm}) = 3.08 + 3.35i$, $n_3(670\text{ nm}) = 0.163 + 3.46i$, $n_4 = 1.5$, $n_5 = 1.33$), the chromium thickness ($e_1 = 2$ nm), and the functionalization thickness ($e_3 = 5$ nm). The domain of research is $D = \{[1;\ 100]\text{ nm} \times [65;\ 90]°\}$.

Table 11.2 shows the results of both optimization methods. The stop criterion is $R < 10^{-3}$, and the maximum allowed generations (or iterations) is 200. The number of particles is $N = 30$ for APSO and $\lambda = 20$, $\mu = 10$, and $\rho = 2$ for ANUHEM; therefore, the number of evaluations of $R(\theta, e_2)$ are the same in both methods. The adjustment of the exogenous parameters of the optimization method has to be conducted carefully before its blind use. For heuristic methods, realizations of the same algorithm by varying only the feeds of the random generator must be made to characterize the convergence of the method and to check the low dependence of the software on the random number generation. We use 1,000 realizations of each optimization method. These realizations enable the numerical evaluation of the success rate, defined as a percentage of convergence before the allowed limit—$T = 100$, in this case. The mean number of each best parameter can also be computed as well as its standard deviation. If the standard deviation is small enough, the confidence level in the heuristic method increases.

Table 11.2 Optimization of SPR: ANUHEM and APSO. The success rate, the mean number of evaluations in the case of success, and the mean of e_2 and θ are indicated with the standard deviation over 1,000 realizations. The maximum number of generations is less than $T = 200$ to be considered successful. The results obtained for the approximated model (a single gold plate between glass and water) are also indicated

	ANUHEM	APSO
Success (%)	89.6%	99.8%
Mean number of evaluations	400	1,100
e_2	49.7 (0.6)	49.8 (0.5)
θ	74.55 (0.04)	74.56 (0.06)
Success (%)	90%	99.7%
Mean number of evaluations	400	1,100
e_2 (approximated model)	51.3 (0.7)	51.3 (0.5)
θ (approximated model)	73.26 (0.03)	73.27 (0.05)

The main constraint of this bench was the number of evaluations at each generation or iteration. This explains the difference between the success rate of the two methods. Actually, the selection pressure λ/μ is not optimal for ANUHEM. Nevertheless, both methods give about the same results and dispersion of the parameters. The number of evaluations is much less than that required by a double loop so as to obtain the same precision ($D = \{[1; 100]\text{ nm} \times [65; 90]°\}$), and therefore, a precision of 0.5 nm on e_2 and of 0.05° would require $198 \times 500 = 99.10^3$ evaluations.

Table 11.2 demonstrates also the usefulness of considering a complete model of SPR. Actually, the optimization using the approximated model gives $e_2 \approx 51$ nm and $\theta = 73.3°$. Instead, using the more complete model of SPR furnishes $e_2 \approx 50$ nm and $\theta = 74.5°$. The uncertainty intervals that could be constructed from the mean values and the uncertainties, even enlarged by a security factor 2, do not overlap, especially for θ. The SPR biosensor design is hopefully the most sensitive to this parameter. A more complete analysis of sensitivity can be found in Barchiesi.[5] Nevertheless, the use of the most complete model has been proven necessary if accurate results are awaited.

11.5 Conclusion

We revisited the plasmon resonance of a gold interface and a gold layer by introducing the concept of the plasmon fingerprint and by linking the concept of plasmon resonance to the general concept of resonators. We also gave some deterministic results on its optimization, and we discussed its sensitivity. Then, a more complete, rather numerically stable model of SPR was proposed to take into account the stick layer of gold on glass and the functionalization layer, and we implemented it to optimize the SPR. For this, we proposed two heuristic methods that have been improved for plasmonics. The obtained results show that the optimization requires stable, accurate models that are as realistic as possible. Indeed, only 2 nm of chromium and 5 nm of functionalization can shift significantly the plasmon resonance. The performances of the optimization methods may lead us to investigate the optimization of more complex biosensors, with confidence.

References

1. Agarwal, G. S. (1973). New method in the theory of surface polaritons, *Phys. Rev. B*, **8**(10), 4768–4779.
2. Barchiesi, D. (1996). A 3-D multilayer model of scattering by nanostructures, application to the optimisation of thin coated nano-sources, *Opt. Commun.*, **126**, 7–13.
3. Barchiesi, D. (2006). Scanning near-field optical data contrast measurement: a tomography-like near-field reconstruction, *Appl. Opt.*, **45**(29), 7597–7601.
4. Barchiesi, D. (2009). Adaptive non-uniform, hyperellitist evolutionary method for the optimization of plasmonic biosensors, *Int. Conf. Comput. Ind. Eng.*, 542–547, doi:10.1109/ICCIE.2009.5223795
5. Barchiesi, D. (2011). *Numerical Optimization of Plasmonic Biosensors*, Vol. 4, INTECH, Open Access Publisher.
6. Barchiesi, D., Grimault, A.-S., Grosges, T., Macias, D., and Vial, A. (2005). Apertureless scanning near-field optical microscopy: on the way to the optical metrology of nanostructures, *J. Korean Phys. Soc.*, **47**, S166–S174.
7. Barchiesi, D., Guizal, B., and Grosges, T. (2006). Accuracy of local field enhancement models: toward predictive models? *Appl. Phys.*, **584**, 55–60.
8. Barchiesi, D., Kremer, E., Mai, V., and Grosges, T. (2008a). A Poincare's approach for plasmonics: the plasmon localization, *J. Microsc.*, **229**, 525–532.
9. Barchiesi, D., Macias, D., Belmar-Letellier, L., Van Labeke, D., Lamy de la Chapelle, M., Toury, T., Kremer, E., Moreau, L., and Grosges, T. (2008b). Plasmonics: influence of the intermediate (or stick) layer on the efficiency of sensors, *Appl. Phys.*, **5**(93), 177–181.
10. Born, M., and Wolf, E. (1993). *Principle of Optics*, Pergamon Press, Oxford.
11. Cardona, M. (1971). Fresnel reflection and surface plasmon, *Am. J. Phys.*, **39**, 1277.
12. Ekgasit, S., Thammacharoen, C., Yu, F., and Knoll, W. (2005). Influence of the metal film thickness on the sensitivity of surface plasmon resonance biosensors, *Appl. Spectrosc.*, **59**, 661–667.
13. Friebel, M., and Meinke, M. (2006). Model function to calculate the refractive index of native hemoglobin in the wavelength range of 250-1100 nm dependent concentration, *Appl. Opt.*, **45**, 2838–2842.

14. Grosges, T., Barchiesi, D., Toury, T., and Grehan, G. (2008). Design of nanostructures for imaging and biomedical applications by plasmonic optimization, *Opt. Lett.*, **33**(23), 2812–2814.

15. Grosges, T., Borouchaki, H., and Barchiesi, D. (2007). Improved scheme for accurate computation of high electric near-field gradients, *Opt. Express*, **15**(3), 1307–1321.

16. Grosges, T., Vial, A., and Barchiesi, D. (2005). Models of near-field spectroscopic studies: comparison between finite-element and finite-difference methods, *Opt. Express*, **13**, 8483–8497.

17. Hoaa, X., Kirk, A., and Tabrizian, M. (2007). Towards integrated and sensitive surface plasmon resonance biosensors: a review of recent progress, *Bio sens. Bio electron.*, **23**, 151–160.

18. Kessentini, S., and Barchiesi, D. (2010). "A new strategy to improve particle swarm optimization exploration ability," in *Intelligent Systems (GCIS), 2010 Second WRI Global Congress on, Vol. 1* (IEEE), 27–30, doi:10.1109/GCIS.2010.147

19. Kessentini, S., Barchiesi, D., Grosges, T., Giraud-Moreau, L., and de la Chapelle, M. L. (2011). Adaptive non-uniform particle swarm application to plasmonic design, *Int. J. Appl. Metaheuristt. Comput. (IJAMC)*, **2**(1), 18–28.

20. Kolomenskii, A., Gershon, P., and Schuessler, H. (1997). Sensitivity and detection limit of concentration and absorption measurements by laser-induced surface-plasmon resonance, *Appl. Opt.*, **36**, 6539–6547.

21. Kretschman, E., and Raether, H. (1968). Radiative decay of nonradiative surface plasmons excited by light, *Z. Naturforsch. A*, **23**, 2135–2136.

22. Lecaruyer, P., Canva, M., and Rolland, J. (2006). Metallic film optimization in a surface plasmon resonance biosensor by the extended Rouard method, *Appl. Opt.*, **46**(12), 2361–2369.

23. Li, L. (1996). Formulation and comparison of two recursive matrix algorithms for modeling layered diffraction gratings, *J. Opt. Soc. Am. A*, **13**(5), 1024–1035.

24. Maystre, D., and Neviere, M. (1977). Sur une methode d'etude theorique quantitative des anomalies de Wood des reseaux de diffraction, application aux anomalies de plasmons, *J. Opt. (Paris)*, **8**, 165–174.

25. Maystre, D., and Neviere, M. (1982). *General Study of Grating Anomalies from Electromagnetic Surface Modes*, John Wiley & Sons, New York.

26. Neff, H., Zong, W., Lima, A., Borre, M., and Holzhiiter, G. (2006). Optical properties and instrumental performance of thin gold films near the surface plasmon resonance, *Thin Solid Films*, **496**, 688–697.

27. Otto, A. (1968). Excitation of nonradiative surface plasma waves in silver by the method of frustrated total reflection, *Z. Phys.*, **216**, 398–410.

28. Palik, E. D. (1985). *Handbook of Optical Constants*, Academic Press, San Diego, USA.

29. Raether, H. (1988). *Surface Plasmons on Smooth and Rough Surfaces and on Gratings*, Springer-Verlag, Berlin.

30. Schwefel, H. P. (1995). *Evolution and Optimum Seeking*, John Wiley & Sons, New York.

31. Simon, H. J., Mitchell, D. E., and Watson, J. G. (1975). Surface plasmons in silver films—a novel undergraduate experiment, *Am. J. Phys.*, **43**(7), 630–636.

32. Swalen, J. D., II, J. G. G., Philpott, M. R., Brillante, A., Pockrand, I., and Santo, R. (1980). Plasmon surface polariton dispersion by direct optical observation, *Am. J. Phys.*, **48**(8), 670–672.

33. Vial, A., Grimault, A.-S., Macias, D., Barchiesi, D., and dela Chapelle, M. L. (2005). Improved analytical fit of gold dispersion: application to the modeling of extinction spectra with a finite-difference time-domain method, *Phys. Rev. B*, **71**(8), 085416–085423.

34. Xilian, S., and Jianda, S. (2006). Influence of crinterlayer on the structure and optical properties of ag films on glass substrate by magnetron sputtering, *Appl. Surf. Sci.*, **59**(5), 661.

35. Yeatman, E. M., and Ash, E. A. (1987). Surface plasmon microscopy, *Electron. Lett.*, **23**, 1091–1092.

36. Zhan, Z.-H., Zhang, J., Li, Y., and Chung, H. S.-H. (2009). Adaptive particle swarm optimization, *IEEE Trans. Syst., Man, Cybernet. B: Cybernet.*, **39**, 1362–1381.

37. Zhao, X (2008). Convergent analysis on evolutionary algorithm with non-uniform mutation, *Evol. Comput. (IEEE)*, 940–944.

Chapter 12

Coherent Anti-Stokes Raman Scattering Microscopy

Nadia Djaker-Oudjhara

CSPBAT Laboratory, Paris XIII University, 74 rue Marcel Cachin, Bobigny 93017, France

nadia.oudjhara@univ-paris13.fr

Several techniques based on Raman scattering have been developed to address the diverse needs of and situations in which biomedical samples can be found. To study biomolecules, spontaneous Raman spectroscopy is mostly used in the near-infrared region of the optical spectrum where laser damage and excitation of autofluorescence is minimal; however, spontaneous Raman scattering produces a very weak signal, oftentimes orders of magnitude weaker than its fluorescence counterpart. Several techniques have been developed to address this shortcoming, including enhanced Raman scattering (SERS[a] and TERS[b]) and coherent anti-Stokes Raman scattering (CARS).

[a]surface-enhanced Raman scattering
[b]tip-enhanced Raman scattering

Nanoantenna: Plasmon-Enhanced Spectroscopies for Biotechnological Applications
Edited by Marc Lamy de la Chapelle and Annemarie Pucci

ISBN 978-981-4303-61-3 (Hardcover), 978-981-4303-62-0 (eBook)
www.panstanford.com

12.1 Introduction

The SERS process involves the use of Raman-active molecules very close to a metal surface, so surface plasmons can very efficiently couple the energy of the incoming laser light with the molecules, thereby amplifying the signal up to 10^{14} times.[1] Although the specific details leading to this huge amplification are still not entirely understood, Raman scattering even from single molecules has been observed, making SERS ideal for label-free molecular assays. The only limitations of SERS are that it does require the presence of a relatively large metal surface and that it decays rapidly with distance away from the surface. CARS amplifies Raman signals without adding any exogenous elements. A CARS signal is generated when two lasers are focused into the sample, one acting as the source, leading to Stokes-shifted Raman scattering, while the other one is tuned to the same frequency as a specific Raman peak so as to selectively scatter light at this selective Raman mode. The two lasers interact nonlinearly within their overlap volume, simultaneously exciting the characteristic chemical vibration and driving the amplification of the corresponding anti-Stokes peak. The resulting CARS signal is more than a thousand times stronger than the original Raman signal and can be used for selective imaging chemically without the use of fluorescent dyes.

12.1.1 Coherent Anti-Stokes Raman Scattering

Infrared and spontaneous Raman scattering imaging techniques are the two most common methods in vibrational microscopy. The first one is limited by low spatial resolution because of long wavelengths (infrared) that are used. Spontaneous Raman scattering microscopy with the excitation laser in the visible or near-infrared range avoids this problem; however, it is limited by the small cross section of Raman emission ($\sigma = 10^{-30}$ cm$^2 \cdot$s), which requires the use of high-power lasers (from tens to hundreds of milliwatts) focused on the sample. CARS microscopy also uses intrinsic molecular vibration properties; however, there are several subtle differences between spontaneous Raman and coherent Raman processes. The coherent signal is due to collective vibrational oscillations with a well-defined phase, where intensity grows quadratically with the number of molecules, whereas the spontaneous signal is linear with the number of molecules. While the spontaneous signal is emitted in all directions in space, the coherent signal is emitted only in the direction where the signal is constructive, a condition called *phase matching*. The anti-Stokes Raman scattering process is shown in Fig. 12.1. The sample is simultaneously illuminated by a *pump* wave at a frequency ω_P and a Stokes wave at a frequency ω_S. An anti-Stokes photon is resonantly generated at a frequency ω_{AS} ($\omega_{AS} = 2\omega_P - \omega_S$) when the frequency difference between the pump and the Stokes wave corresponds to the resonant vibrational frequency of the excited bond ω_R ($\omega_P - \omega_S = \Omega_R$).

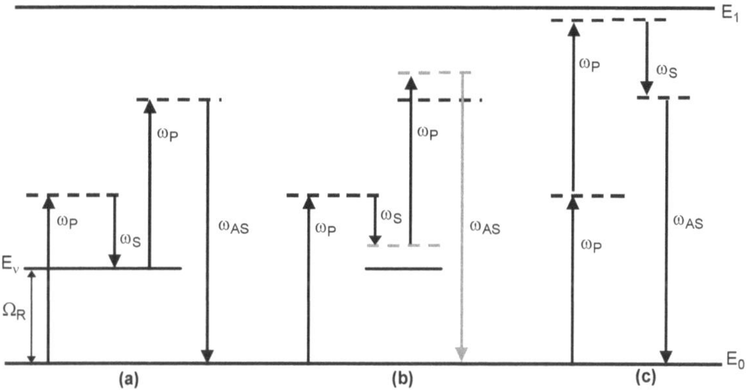

Figure 12.1 Energy diagram (CARS process). (a) Resonant anti-Stokes generation process ($\omega_P - \omega_S = \Omega_R$), (b) nonresonant process ($\omega_P - \omega_S \neq \Omega_R$), and (c) contribution of a nonresonant electronic state. E_0 and E_1 are electronic states of the molecule, and E_v is the resonant vibrational state with a resonance Raman frequency Ω_R.

12.1.2 CARS Analytical Techniques and Configurations Microscopy

Although the coherent scattering anti-Stokes Raman process was reported for the first time in 1965 by Maker and Terhune[2] as a new tool for chemical analysis, spectroscopy applied to gaseous and condensed states of materials.[3] It was in 1982, with the development of laser sources and electronics, that the coherent scattering anti-Stokes Raman process was combined with microscopy by Duncan *et al.*[4–6] Two visible lasers (pump and Stokes) in a collinear configuration and two-dimensional detectors for recording the

CARS signal were used. Visible lasers lead to a very strong noise due to the nonresonant two-photon absorption (Fig. 12.1c) in the chemical bonds, so in 1999, Zumbusch *et al.*[7] developed a CARS microscope with two infrared lasers, which greatly improved the efficiency of the process (improved the signal-to-noise ratio [SNR]).

Because of the coherent nature of the scattered radiations (the molecular bonds oscillate in phase and interfere constructively in the CARS process), the CARS signal is 10^6 times stronger than that of spontaneous Raman scattering. Unlike fluorescence or spontaneous Raman scattering imaging, the contrast mechanism in CARS imaging is produced by the sum of all coherent resonant and nonresonant radiations emitted by the sample (Fig. 12.1a,b).

CARS microscopy has many advantages:

(1) It does not require fluorescent labeling; the samples are directly imaged without no chemical or biological treatment.
(2) As there is no population in an excited electronic state of the molecule, there is no photo damage or photo bleaching.
(3) It is more sensitive than spontaneous Raman scattering imaging, while using average excitation powers tolerable for biological objects.
(4) As multiphoton microscopy, it offers opportunities for sectioning and three-dimensional confinement of the signal generation area without using confocal detection.
(5) The use of infrared excitation minimizes the heating of the sample and the nonresonant noise, specifically for biological samples because of the low absorption of water at these wavelengths, and increases penetration depth for imaging thick tissues and cells.

However, the noise in a CARS image comes mainly from contributions of the electronic third-order susceptibility of the sample and the solvent, in addition to the nonresonant CARS signal from the solvent. Different techniques have been developed to reduce the nonresonant CARS signal from the solvent. Among them (see Fig. 12.2) are the following:

- The backward CARS signal detection (epi-CARS or E-CARS)[8] configuration effectively reduces the nonresonant signal of the solvent by using the property of CARS generation, which is constructive in the direction of propagation of the exciting beams (forward directions— ee the next section, "Element of

Theory"), whereas it is destructive in the backward direction. This property increases considerably the sensitivity of CARS generation. Another configuration based on the same principle is the use of counterpropagating excitation beams (C-CARS). Numerical simulations have shown that the signal "forward" in this case is not very constructive; this greatly reduces the nonresonant contribution of the solvent.[9]

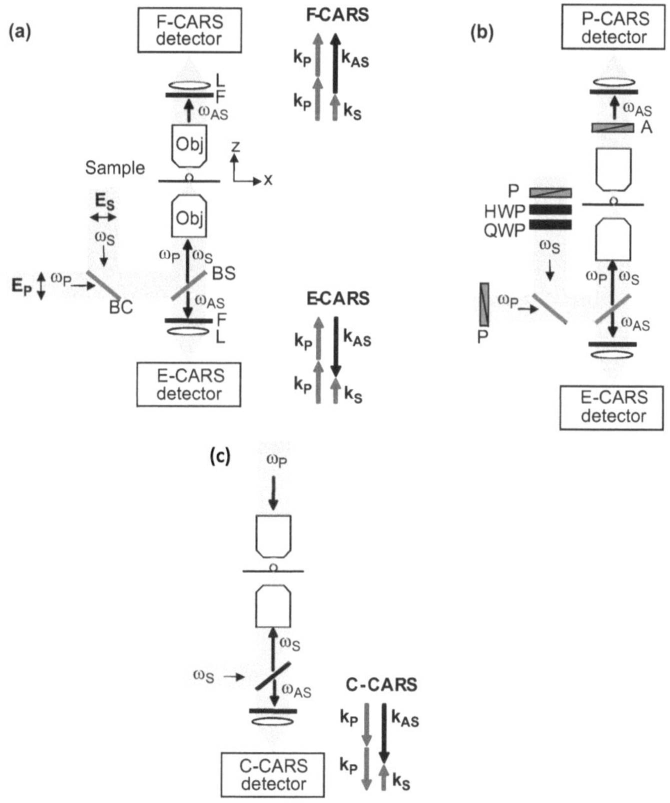

Obj: objective, BS: beam splitter, BC: 50/50 beam separator, F: filter, NA: numerical aperture, L: lens, P: polariser, A: analyser, HWP: (λ/2) wave-plate, QWP: (λ/4) wave-plate.

Figure 12.2 Different configurations of CARS microscopy (adapted from Ref. 9). The pump (ω_P) and Stokes (ω_S) lasers are horizontally polarized (E_P and E_S). The three wave vectors ($\mathbf{k}_P$, $\mathbf{k}_S$, $\mathbf{k}_{AS}$) of the involved fields are along the optical axis. (a) E-CARS and F-CARS microscopes with co-propagative excitation, (b) P-CARS microscope and (c) C-CARS microscope with counterpropagation excitation.

- The CARS signal is sensitive to the excitation field's polarization. It can be shown that the nonresonant signal is depolarized by a factor that depends on the third-order susceptibility of the observed sample. The CARS image contrast and sensitivity are significantly improved if one uses linear polarization states with an angle between the excitation fields and analyzes the CARS signal perpendicularly to the induced nonresonant CARS nonlinear polarization—this scheme is called P-CARS.[10]
- Other more sophisticated techniques can improve the sensitivityf— or example, time-resolved CARS (T-CARS),[11] where a temporal delay between the pair pump/Stokes lasers and a third laser (probe) is applied. This is based on the fact that the nonresonant process involving virtual energy states is faster than the resonant process involving molecule vibrational state lifetimes. Another technique is based on monitoring the spectral coherence of the exciting beams[12] using pulse shaping and spectral phase modulators (SLMs) with ultrashort laser pulses.

These systems control the interference between different excitation field components, which significantly reduces the nonresonant signal. Interferometry has also been used to retrieve the phase of the generated CARS signal and the associated pure Raman signal in homodyne detection[13] and more recently at interfaces.[14] While there are different possible configurations for CARS microscopy, the most commonly used are illustrated in Fig. 12.2. The excitation beams (pump and Stokes) are copropagative and horizontally polarized (Fig. 12.2a,b), and the CARS signal is recorded in both backward (E-CARS) and forward (F-CARS) directions of the optical axis. In P-CARS configuration (Fig. 12.2b), a system of polarizers is added to differentiate the resonant from the nonresonant CARS signal. Another configuration of CARS microscopy is shown in Fig. 12.2c using C-CARS. As mentioned previously, these last two techniques permit significantly to improve the sensitivity of the CARS signal. There are other approaches to improve the optical resolution and sensitivity, rather difficult and complicated to produce, for example, CARS microscopy combined with near-field microscopy.[15] Some metal samples can also be used to generate the CARS signal[16] (see the third section).

12.2 CARS Generation: Elements of Theory

12.2.1 Introduction

Laser development in the sixties permitted the exploration of most nonlinear optical phenomena and their integration in different analysis tools and imaging systems, including microscopy and spectroscopy. These sources of coherent light (monochromatic, directional, and high intensity) interact with the material, giving rise to what is called nonlinear optical phenomena. In this case, the response of an excited material with an intense optical field depends nonlinearly on the intensity of this field. In this case, the optical responses of the material depend strongly on the excitation intensity and the spectral, temporal, and spatial nature of the resulting electromagnetic field. The relationship between the induced polarization in the material and the applied electric field is generally modeled by a power series of the driving fields and can be expressed by the following relation:[17,18]

$$P_i = \varepsilon_0[\chi^{(1)}_{ij}E_j + \chi^{(2)}_{ijk}E_jE_k + \chi^{(3)}_{ijkl}E_jE_kE_l + \cdots] \quad (12.1)$$

where ε_0 is the vacuum permittivity and $\chi^{(i)}$ is the electric susceptibility tensor of order i ($i, j, k, \ldots = 1, 2, 3, \ldots$): The first term of this development, or linear susceptibility, defines the dielectric constant of the medium ($\varepsilon = 1 + \chi^{(1)}$). E_i is the field amplitude along the i-axis.

By using Maxwell's equations, we can write for each polarization frequency component $P(\omega)$:

$$P(\omega) = \varepsilon_0\chi^{(1)}E(\omega) + P^{\mathrm{NL}}(\omega) \quad (12.2)$$

where $\chi^{(1)}$ is a linear and purely real susceptibility (linear absorption is neglected) and P^{NL} is the induced nonlinear polarization in the medium. According to Maxwell's equations, we can write, then:

$$\nabla^2E(\omega) + n^2(\omega)\frac{\omega^2}{c^2}E(\omega) = -\frac{\omega^2}{\varepsilon_0 c^2}P^{\mathrm{NL}}(\omega) \quad (12.3)$$

where $n(\omega) = (1 + \chi^{(1)}(\omega))^{1/2}$ is the refraction index of the medium at frequency ω.

Equation 12.3 is called the *nonlinear wave equation*. It has the shape of the usual wave's equation in a medium of refractive index $n(\omega)$ but has a source term proportional to the nonlinear polarization. The fields' amplitudes' product together with the nonlinear susceptibility contained in $P^{NL}(\omega)$ describes the interactions between laser waves and the nonlinear medium. According to their spatial and temporal dependence, the polarization components $P^{(i)}(\omega_m)$ can indeed generate two types of processes in the nonlinear medium. The first corresponds to a different propagation of the incident laser waves (refractive index and intensity are modified), while the second represents the generation of new electric field Fourier components that were not present in the original incoming waves (frequency mixing, stimulated Raman scattering, phase conjugation, etc.).

In general, it is possible to solve Eq. 12.3 so that a number of approximations is used, leading to simplifications allowing analytical solutions of the propagation equation. A commonly used simplification is the slowly varying envelope approximation (SVEA), which assumes that the field's envelopes vary only slightly over a wavelength. Mathematically, this translates into:

$$\left|\frac{\partial^2 E}{\partial z^2}\right| << \left|2k\frac{\partial E}{\partial z}\right| \tag{12.4}$$

With this, one can neglect the second derivative in the propagation equation. In the case of time-dependent pulse propagation, envelopes are slowly varying in z and t, and the wave propagation equation in nonlinear optics becomes:

$$\left|\frac{\partial}{\partial z} + \frac{n}{c}\frac{\partial}{\partial t}\right| E = \frac{i}{2\varepsilon_0}\frac{\omega}{nc} P^{NL} e^{-ikz} \tag{12.5}$$

12.2.2 CARS Process

CARS is a four-wave mixing process, which involves, as in the case of the third-harmonic generation (THG), the susceptibility parameter $\chi^{(3)}$. CARS is therefore a third-order nonlinear process, based on Raman scattering, where pump and Stokes lasers at frequencies ω_P and ω_S, respectively ($\omega_P > \omega_S$), interact (see Fig. 12.3a). The pump and Stokes electromagnetic fields' amplitudes, E_P and E_S, respectively, with $\mathbf{k}_P$ and $\mathbf{k}_S$ wave vectors interact with the medium and induce a third-order polarization, which is described by:[3,19]

$$P^{(3)}(\omega_{AS}) = \varepsilon_0 \chi^{(3)}(\omega_{AS}; \omega_P; -\omega_S)\, E_P^2(\omega_P) E_S^*(\omega_S) \tag{12.6}$$

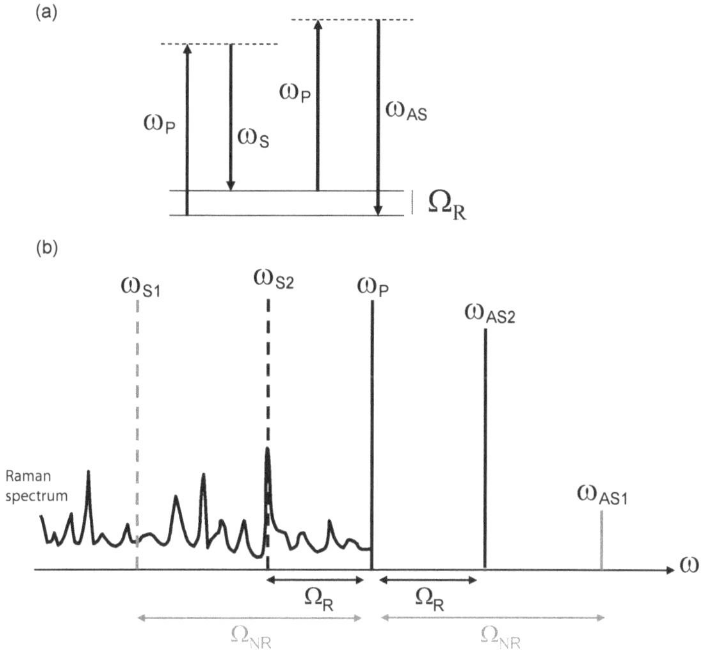

Figure 12.3 (a) Resonant CARS process ($\omega_P - \omega_S = \Omega_R$) and (b) CARS generation enhanced for a resonant Raman band of an excited molecule ($\omega_{AS2} = 2\omega_P - \omega_{S2}$).

Using the wave equation, we can write, in the case of anti-Stokes wave generation:

$$\frac{\partial}{\partial z} E_{AS} = i \frac{\omega_{AS}}{n_a c} \chi^{(3)} E_P^2 E_S^* e^{-i\Delta k z} \tag{12.7}$$

By integrating the last equation as a function of z (distance along the optical axis), with the initial condition $E_{AS}(0) = 0$ and taking the distance D representing the interaction CARS distance, or the size of the sample, as the intensity is proportional to the square of the electric field, one obtains:

$$I_{AS} \propto \left|E_{AS}\right|^2 = \left(\frac{\omega_{AS}}{n_a c}\right)^2 \cdot \left|\chi^{(3)}\right|^2 \cdot I_P^2 \cdot I_S \cdot D^2 \cdot \left(\frac{\sin\left(\Delta k \cdot \frac{D}{2}\right)}{\Delta k \cdot \frac{D}{2}}\right)^2 \tag{12.8}$$

The parameter Δk expresses the relationship between the fields' wave vectors involved in the CARS generation (represented by Fig. 12.2b). An important condition to achieve the generation of coherent waves is called the *phase-matchin g condt i n*. This condition appears in Eq. 12.8 in the term sin $c(\Delta k \cdot D/2)$, which is nonzero when $\Delta k \cdot D << \pi$. This condition is achieved when $\Delta k = 0$, or $D \approx 0$. For a sample size D smaller than the wavelength, the phase-matching condition is achieved regardless of Δk. In the opposite case, when $D >> \lambda_P$, the condition $\Delta k = 0$ is required to obtained a CARS signal. In the case of CARS microscopy, the latter relation is always true; this is due to the use of immersion objectives with a very high numerical aperture allowing wave vector couples to carry always the phase-matching condition.[20]

12.2.3 CARS Spectrum

The third-order CARS polarization generates a coherent radiation at frequency ω_{AS} ($\omega_{AS} = 2\omega_P - \omega_S$) when the two excitation frequencies' difference corresponds to the resonant Raman frequency of the medium ($\omega_P - \omega_S = \Omega_R$) (see Fig. 12.3). Consider that these induced polarizations are modeled by a set of harmonic oscillators; the classical equation describing the oscillators movement is given by:

$$\ddot{x} + 2\Gamma \dot{x} + \Omega_R^2 x = \frac{F(x,t)}{m} \tag{12.9}$$

where x is the extension from the equilibrium position and (2Γ) is the Raman vibrational level width. The resonant frequency of this oscillator is given by $\Omega_R = (k/m)^{1/2}$. In a normal Raman process $F(x, t) = 0$, so how is the driving force $F(x, t)$ for the vibration in the molecule for the CARS process obtained?

In a molecule, light interacts with the dipole moment, which contains an induced part written as:

$$\vec{\mu}_{in} = \varepsilon_0 \left(\alpha^{ij} + \frac{d\alpha^{ij}}{dx} x + \cdots \right) E_j \tag{12.10}$$

This induced part depends on the applied electric field E_j, and α is the polarizability. The parameter χ is the susceptibility used in the case of a large number of atoms and molecules, while α is used for a single atom or molecule.

- For Rayleigh scattering, we write $\vec{\mu}_0 = \varepsilon_0 \alpha^{ij} E_j$.
- For the Raman process, we write $\vec{\mu}_1 = \varepsilon_0 \frac{d\alpha^{ij}}{dx} E_j$.

In this last case, x and E_j are harmonic functions, and their combination gives rise to new frequency components as $(\omega_P \pm \Omega_R)$. In the case of CA RS scattering, the force F depends on the applied electric fields E_P and E_S:

$$F(x,t) = \frac{\varepsilon_0}{2}\left(\frac{d\alpha^{ij}}{dx}E_P E_S^* + \cdots\right) \tag{12.11}$$

Combining Eqs. 12.9 and 12.11, and considering that the polarization is $P^{(NL)} = N \cdot \mu_{in} = \varepsilon_0 \chi^{(3)} E_P^2 E_S^*$, we can deduce that:

$$\chi^{(3)} = \frac{\varepsilon_0 \left(\frac{d\alpha^{ij}}{dx}\right)^2}{m[(\Omega_R^2 - \omega^2) + i(2\Gamma\omega)]} \tag{12.12}$$

When the oscillator vibration frequency is close to the vibration frequency of the Raman band $(\omega \approx \Omega_R)$, we obtain:

$$\Omega_R^2 - \omega^2 + 2i\Gamma\omega = (\Omega_R + \omega)(\Omega_R - \omega) + 2i\Gamma\omega \approx 2\Omega_R(\Omega_R - \omega + i\Gamma) \tag{12.13}$$

We define a parameter a for resonant Raman scattering $\chi^{(3)} = \chi_R^{(3)}$ (when: $\omega \approx \Omega_R$) as:

$$a = \frac{\varepsilon_0}{2m\Omega_R}\left(\frac{d\alpha^{ij}}{dx}\right)^2 \tag{12.14}$$

We can write then:

$$\chi_R^{(3)} = \frac{a}{\delta + i\Gamma} \tag{12.15}$$

where δ is the frequency difference defined by $\delta = \Omega_R - (\omega_P - \omega_S)$.

In addition to the resonant part of third-order susceptibility, one must also consider a nonresonant part $\chi_{NR}^{(3)}$ (real), which is usually caused by the electronic gain of the medium. Finally, we obtain:

$$\chi^{(3)} = \chi_R^{(3)} + \chi_{NR}^{(3)} \tag{12.16}$$

The induced third-order polarization in the CARS process is expressed with a first part depending on the resonant Raman response and a nonresonant response independent from the excitation frequencies. These two parts of polarization are described by:

$$P^{(3)} = (P_R^{(3)}\ P_{NR}^{(3)}) \propto (\chi_R^{(3)} + \chi_{NR}^{(3)})\ E_P^2 E_S^* \tag{12.17}$$

The parameter $\chi_R^{(3)}$ is a complex ($\chi_R^{(3)} = \chi_R' + i\chi_R''$), which represents the specific Raman response of each chemical bond in CARS imaging. These real and imaginary resonant parameters can be expressed as a function of Γ and the frequency difference δ; these two parameters are shown in Fig. 12.4:

$$\chi_R' = \frac{a\delta}{\delta^2 + \Gamma^2} \tag{12.18}$$

$$\chi_R'' = -\frac{a\Gamma}{\delta^2 + \Gamma^2} \tag{12.19}$$

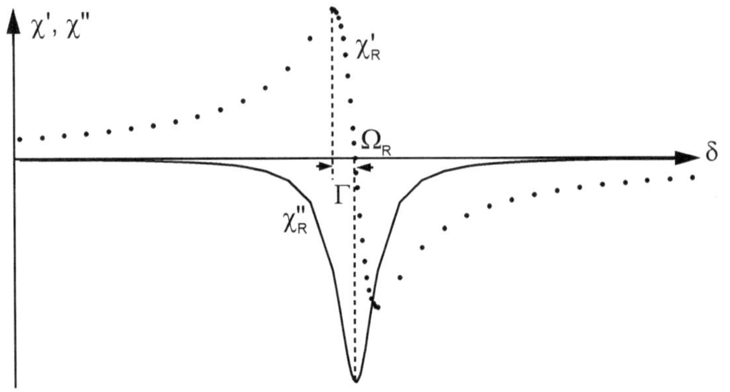

Figure 12.4 The real part (χ_R') and the imaginary (χ_R'') part of third-order susceptibility, depending on the frequency difference δ.

However, the nonresonant term $\chi_{NR}^{(3)}$ is a real parameter, which is independent of excitation frequencies and represents part of the CARS signal, which is not specifically associated to the bond of interest. This nonresonant contribution limits the contrast and CARS detection sensitivity and is to be minimized in the experiment (see Fig. 12.3b; $\omega_P - \omega_{S1} = \Omega_{NR}$).

The intensity of the CARS signal (I_{CARS}) depends quadratically on the third-order susceptibility $\chi^{(3)}$. The curve ($|\chi^{(3)}|^2$) versus the difference frequency δ is shown in Fig. 12.5. This curve depicts the general shape of the CARS spectrum, which appears, from the spontaneous Raman spectrum, shifted to lower frequencies due to nonresonant contribution ($\chi_{NR}^{(3)}$). This offset is written as:

$$\delta_{max/min} = \Omega_R + \frac{1}{2}\left(-\frac{\chi_R}{\chi_{NR}} \pm \sqrt{\left(\frac{\chi_R}{\chi_{NR}}\right) + 4\Gamma^2}\right) \tag{12.20}$$

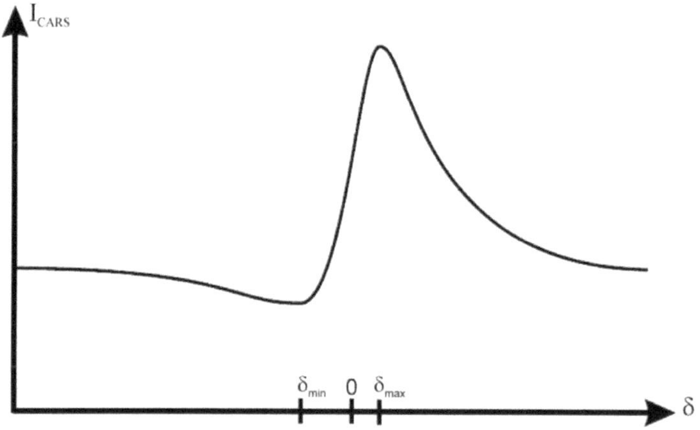

Figure 12.5 The CARS spectrum is shifted compared with the spontaneous Raman scattering spectrum because of the interference between both parameters ($\chi_R^{(3)}$) and ($\chi_{NR}^{(3)}$).[21]

In case the resonant signal is much stronger than the non-resonant signal,[c] the extreme values of δ can be written as:

$$\delta_{max/min} = \Omega_R + \frac{1}{2}\left(-\frac{\chi_R}{\chi_{NR}} \pm \frac{\chi_R}{\chi_{NR}}\right) \tag{12.21}$$

With $\delta_{max} \approx \Omega_R$ and $\delta_{min} \approx \Omega_R - (\chi_R/\chi_{NR})$, this shift is negligible when the resonant CARS signal is much stronger than the non-resonant CARS signal.

The molecular specificity of CARS microscopy is to produce images with a high contrast (coherent process) and spectral specificity, without any staining; this allows to perform molecular imaging, which is of particular importance in the field of cell and tissue imaging.

12.3 CARS Imaging

One of the key applications of CARS microscopy is noninvasive cellular imaging using molecular bonds that are present naturally in the cell. The use of infrared lasers with moderate excitation powers (a few hundred megawatts) decreases cell damage[22] and can produce images with good spatial resolution (~ microns) without

[c] $\chi_R >> \chi_{NR} \Rightarrow |\chi_R/\chi_{NR}| >> \Gamma$

the need for any labeling or staining. Image-contrast CARS builds up from the coherent emission of excited molecular bonds and their constructive interference in the forward direction, giving rise to the F-CARS image.

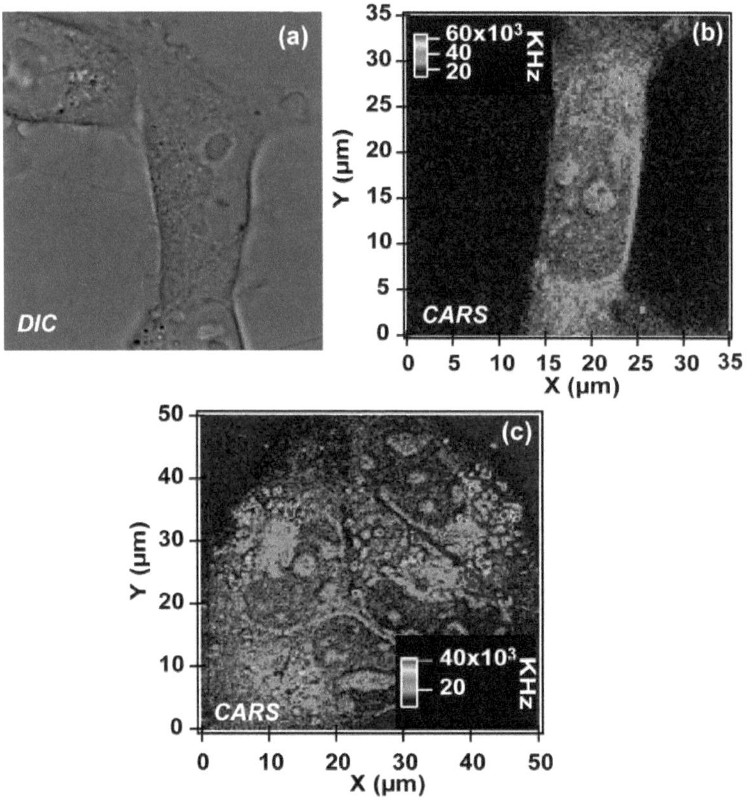

Figure 12.6 Images of COS-7 cells obtained by DIC microscopy (a) and CARS microscopy (b, c). (F-CARS images). The two excitation laser beam pulses ω_P (13,699 cm^{-1}) (λ_P = 730 nm) and ω_S (10,854 cm^{-1}) (λ_S = 921 nm) produce an anti-Stokes wave ω_{AS} to 16,544 cm^{-1} (λ_{AS} = 604 nm) corresponding to the vibration frequency of the Raman symmetric aliphatic CH bond present mainly in plasma lipid membranes and intracellular membranes.[32]

Figure 12.6 shows a comparison of live cell images obtained by differential interference contrast (DIC) microscopy (Fig. 12.6a) and CARS microscopy (Fig. 12.6b,c; F-CARS images). CARS contrast is obtained in this case by the symmetric aliphatic CH bond, which is present mainly in cell plasma membranes, intracellular lipids, and intracellular organelles such as the Golgi apparatus (the nucleus). Small vesicles are clearly visible in CARS (although it is not visible by DIC microscopy). Both pump and Stokes laser beams of frequency ω_P (13,513 cm^{-1}) and ω_S (10,668 cm^{-1}) produce an anti-Stokes wave at ω_{AS} (16,358 cm^{-1}) corresponding to the resonant frequency of the CH bond (2,845 cm^{-1}). E-CARS images are similar to Fig. 12.6a,b and highlight the vesicles with a diameter smaller than the wavelength.[24,25]

Because of its specific spectral imaging, CARS offers a significant advantage over other optical microscopy techniques since it allows direct physical and chemical imaging of the material. This specificity allows for probing of different molecular bonds in different cellular systems. For example, the CH bond has been imaged in lipid membranes and nucleic, and the PO bond (1,090 cm^{-1}) has been detected in chromosomes. It is also possible to image the medium surrounding the cell, as done for the OH bond.[22,26,27] The CARS microscopy technique was also used for imaging artificial lipid membranes[28–30] and, more recently, for *in vivo* imaging in tissues.[31]

12.3.1 Deuterium Labeling

The resonant CARS signal provided by a specific excited chemical bond is not free of the solvent background.[d] The nonresonant CARS signal from the solvent is independent of the Raman shift ($\omega_P - \omega_S$) and can be stronger than the resonant CARS signal. This situation decreases the contrast image considerably. To maximize the resonant signal, picosecond (ps) laser pulses (2–5 ps) are used to optimize the overlap between the spectral width of the Raman band (from 10 to 20 cm^{-1}) and excitation laser pulses. However, this strategy may be insufficient to follow a specific molecular bond in a congested spectral region (as the one lying below 1,500 cm^{-1}) or for many molecular components present in biological samples, such as cell membranes. For example, this is the case for lipids that contain many CH bonds—CH_2 and CH_3 with various Raman resonance frequencies ranging from 2,800 cm^{-1} to 3,500 cm^{-1}.[33] Differentiating

[d]The CARS process always occurs, even if the vibrational resonance of a chemical bond is different from the frequency difference between the pump and Stokes lasers. When $\omega_P - \omega_S = \Omega_R$, the CARS signal is strongly enhanced.

lipids is hard because their vibrational frequencies overlap and/or are quite similar.

This obstacle can be cleared up in CARS microscopy, using minimally invasive staining, by switching CH_2 groups by CD_2 groups. This does not disturb the cellular environment because the hydrogen and deuterium atoms are chemically very close. Aliphatic or aromatic CD_2 bonds have a Raman resonance frequency at 2,100 cm^{-1}, located in a spectral region between 1,800 cm^{-1} and 2,700, which is described as the "silent region" of the cell because of not overlapping with other cellular endogenous Raman resonances.[32,34] Several applications of this specific noninvasive labeling have been reported, either for live cell imaging[35] or investigation of segregation of domains in model systems (i.e., GUVs[e]).[36]

12.3.2 Recent Developments

12.3.2.1 Multiplex-CARS microscopy

CARS microscopy is combined with various spectroscopic techniques, including multiplex-CARS (M-CARS) spectroscopy, which has been used to study the physicochemical structures of different biological samples.[37] Often CARS microscopy experiments employ 2 ps lasers that serve both to drive a particular vibrational mode and to probe the induced coherence. In this case, only a single point of the CARS spectrum is recorded at each image position. To explore the CARS spectrum further, the Stokes laser has to be tuned to different wavelengths. The SNR of such "single-point" CARS spectra is inherently limited by laser-induced fluctuations. However, detailed chemical identification or characterization of the physical state of the sample often requires the analysis of spectral line shapes, relative peak heights of different vibrational features, etc.

This requires measurement of the complete CARS spectrum, which is readily achieved with a high SNR using M-CARS microscopy.[29] Here, the Stokes laser is abroad broadband femtosecond laser, whereas the pump laser is a narrow ps laser; the anti-Stokes signal is recorded simultaneously over a significant part of the vibrational spectrum at each position in the image. As an additional advantage,

[e]giant unilamellar vesicles

this experimental approach, in which the M-CARS signal is measured relative to the nonresonant background, can provide quantitative measurement of the local density (concentration 0 of the vibrational mode of interest[38] (see Fig. 12.7).

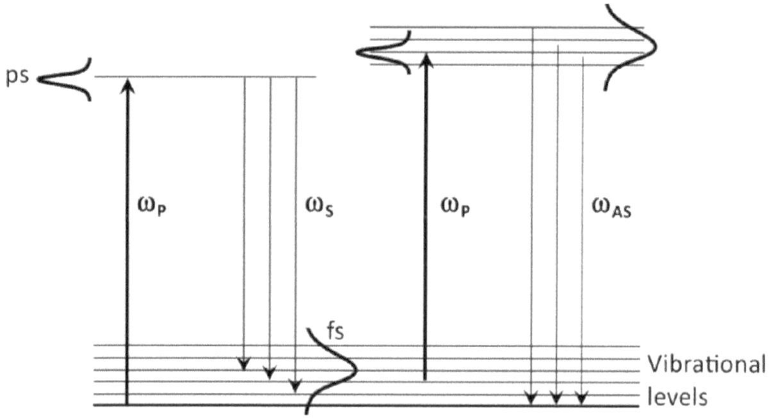

Figure 12.7 Energy diagram of resonant Raman M-CARS. The pump beam is ω_P (ps pulse), while the Stokes beam is ω_S (femtosecond pulse).[29]

In biology, M-CARS microscopy is specifically used for lipids (e.g., DOPC, DSPC, POPS, etc.)[f] and proteins (e.g., AMP, ADP, ATP, and NAD^+)[g] characterization[39] and for studying the organization and thermodynamic state of lipid bilayers systems.[40,41]

As fluorescence correlation microscopy (FCS), CARS correlation microscopy (CS-CARS) was used for investigating dynamic processes by analyzing the fluctuations of the CARS signal in small observation volumes. This technique allowed analysis of dynamic three-dimensional distribution of beads or molecular clusters with high sensitivity.[42,43]

12.3.2.2 Microscopy/interferometry CARS

One of the major complications in CARS microscopy comes from the fact that the CARS signal is not background free. As explained in the section "CARS Spectrum," the total CARS signal originates from

[f] DOPC, 1,2-dioleoyl-*sn*-glycero-3-phosphocholine; POPS, 1-palmitoyl-2-oleoyl-*sn*-glycero-3-phosphocholine; DSPS, 1,2-distearoyl-*sn*-glycero-3-phosphocholine

[g] AMP, adenosine monophosphate; ADP, adenosine diphosphate; ATP, adenosine triphosphate; NAD^+, nicotinamide adenine dinucleotide

the third-order nonlinear susceptibility of the material, which is a sum of a resonant $\chi_{R}^{(3)}$ and a nonresonant electronic contribution $\chi_{NR}^{(3)}$. In biological samples, the aqueous environment gives rise to a substantial nonresonant response, which can overwhelm weak resonant signals and significantly reduce imaging contrast. Improving the vibrational contrast in CARS imaging requires suppression of the nonresonant background, while the resonant vibrational signal is retained. Although the use of pulses in the ps range significantly optimizes the signal-to-background ratio,[26] additional background suppression is needed for observing weak signals. Polarization-sensitive detection and T-CARS have also been used for background rejection, but these techniques also attenuate the resonant contribution and are therefore less suitable for the detection of weak signals. Another solution is to use techniques based on CARS interferometry,[44,45] where the generated CARS field is mixed with a local oscillator field (nonresonant CARS). This detection method is called *optical heterodyne detection* (OHD) and offers a number of attractive qualities. First, it allows a direct comparison with Raman spectral profiles without the complication of a nonresonant background. Second, the heterodyne signal is linearly proportional to the concentration, which permits a straightforward quantitative interpretation of images. Last, heterodyne mixing provides the possibility of interferometric CARS signal enhancement by mixing with a strong local oscillator field.[13,46]

This technique is exploited using detection in either a time domain[47,48] or a frequency domain.[13] Several applications in biology take advantage of heterodyne detection in CARS microscopy for lipid multilayers and cell imaging.[46,49]

Recently, a different approach[12,50] based on coherent control techniques by controlling the phase of the CARS excitation beams has been proposed. The great advantage of this approach is that the excitation comes from a single ultrashort pulsed laser (tens of femtoseconds). This laser is spectrally broad enough to contain both pump and Stokes frequencies. The laser spectral phase is controlled using a liquid crystal spatial modulator (SLM), the spectral resolution in this case being limited by the pixel size of the SLM. By varying this phase, the different spectral components of the pulse interfere in the sample more or less constructively, thereby increasing the contrast in the image, while reducing the nonresonant signal. Several demonstrations have been reported, where the nonresonant CARS signal is greatly reduced.[51,52]

Another approach has recently used the coherent addition of nonresonant and resonant CARS signals when the active focal spot is located at an interface between a resonant and a nonresonant medium. Thanks to a simple analysis of the CARS signal in angular space, it has been possible to completely removed the nonresonant background in CARS imaging.[14,53]

12.4 Surface-Enhanced Cars

12.4.1 Introduction

Surface-enhanced Raman scattering (SERS) and surface-enhanced coherent anti-Stokes Raman scattering (SECARS) are techniques strongly dependant on the surface plasmon excitations of metal nanostructures. With local fields that can be several orders of magnitude higher than the incident fields, the SERS/SECARS response is strongly sensitive to (1) the *surface morphology* of the nanostructure and (2) the *polarization state* of the exciting field. Since CARS is a four-wave mixing process in which the anti-Stokes signal results from the parametric coupling of an incident pump laser (ω_p) and a Stokes laser (ω_s), with $\omega_p - \omega_s = \Omega_R$, and the CARS generation is dependent quadratically on the pump field and linearly on the Stokes field, a local stronger enhancement is expected for CARS than for conventional SERS.

Beyond the fact that CARS is more sensitive than spontaneous Raman spectroscopy, it remains less sensitive compared with fluorescence for which single-molecule detection is now possible. To overcome this drawback, resonators are used, such as microdroplets[54,55] or nanoparticles,[56] and recently Koo *et al.*[57] claimed to reach single-molecule sensitivity with a surface-enhanced CARS setup.

Enhancement of scattering processes such as linear[58] and non-linear[59] Raman scattering, second-harmonic generation (SHG)[60] or THG,[61] and nonresonant four-wave mixing[62] by Fabry–Perot cavities was also successfully demonstrated. Spontaneous emission radiation pattern modification has been used to improve the extraction of light in semiconductor devices.[63] In all these experiments, the basic effect of the cavity is to *tailor* the electromagnetic environment of the emitters. In particular, the cavity restricts the number of available radiative modes (compared with the free space) to which the field radiated by the emitters can couple. The modification of the CARS far-field radiation pattern has been clearly demonstrated in planar dielectric microcavities.[64] Such a strategy permits us to confine the CARS emission power into a narrow, angular cone.

12.4.2 SECARS in Metallic Nanostructures

SECARS in silver colloids was first observed by Liang *et al.*[65] in 1994. In their experiment, to satisfy the phase-matching condition, two beams were crossed on a sample at a small angle that resulted in a 10 μm coherence length. The density of the silver colloid was adjusted such that several particles lay within the coherence length. Recently it was verified that a tightly focused spot, which was produced with collinear illumination optics using a high-numerical-aperture (NA) objective lens, allows the relaxation of the phase-matching condition and accordingly achieves high spatial resolution.[7,66,67] Under a CARS microscope using such a high-NA lens, CARS can be generated locally at the proximity of an isolated single metallic particle. Owing to the cubic property of the CARS process, the enhanced CARS signal can be further localized compared with spontaneous SERS.

With local fields several orders of magnitude higher than the incident driving field, the SERS response is typically 10^6 higher compared with conventional spontaneous Raman scattering. Combined with electronic resonances in the molecule, the total signal enhancement is high enough to detect the Raman scattering from single molecules.[68] Since the strength of the local fields crucially depends on the geometry of the nanoparticles, a careful design of the metallic structure is required for optimizing the efficiency of SERS molecular detection assays. Surface plasmon–enhanced fields are also the candidates of choice for improving the nonlinear optical response from molecules. SECARS is of particular interest because it offers a route for exploring the coherent vibrational response of molecules in very dilute samples. Initial SECARS measurements have indicated that enhancement of molecular optical nonlinearities can indeed be achieved at surfaces of noble metal nanostructures.[16,57,69]

So far, however, a straightforward implementation of a SECARS assay has not been demonstrated yet. An important hurdle in developing nanostructures suitable for SECARS is the intrinsic emission of the metal substrate. While the emissive response of the metal nanostructure is often negligible in SERS experiments, radiative contributions of the substrate itself are prominent under conditions of ultrafast laser irradiation.[70–73]

Coherent anti-Stokes scattering signals have been observed from coupled pairs of gold nanospheres in near-field four-wave mixing experiments,[74] indicating that plasmon resonances enable the generation of radiation at shifted frequencies when driven by two optical fields of different colors simultaneously. Nonetheless, it is unclear how the coherent anti-Stokes emission scales with the dimension and shape of the nanostructure, information that will be essential in potential SECARS experiments. Also, intrinsic coherent anti-Stokes emission is observed in lithographically patterned gold nanozigzag wires. Polarization-dependent measurements reveal that the nanostructure's anti-Stokes response is polarized in the direction of the transverse surface plasmon resonance of the wire.[75]

12.4.3 Tip-Enhanced CARS

The CARS technique has been translated to the near field with the adoption of near-field fiber-based signal collection with far-field excitation, CARS-SNOM,[15] which has achieved a resolution of 128 nm. Tip-enhanced CARS, TECARS, was shown to be feasible by Ichimura *et al.*[16] with the imaging of a biomacromolecule cluster (polyalanine) with a resolution of 10–20 nm and sensing a vibrational spectroscopic signal from a remarkably small subzeptolitre volume. For TECARS, the laser source should be flexible enough to be able to achieve near-optimal excitation of tip plasmon resonance and Raman-mode vibrational resonance. Ideally, the pump wavelength ought to be matched to the wavelength peak of tip scattering/enhancement. There is also the need to reduce the average laser power to avoid tip damage. The metal tip will produce fluorescence (four-wave mixing) with ps excitation, so there is a need to use a spectrometer or a narrow bandpass filter to remove most of the fluorescence from the tip. Neither tip-enhanced two-photon excitation fluorescence (PEF) (a three-photon process) nor TECARS (a four-photon process) needs far-field background removal.[76]

12.5 Conclusion

Electromagnetic field enhancement by surface plasmons is a technique that allows strong amplification of different physical processes. Both linear (fluorescence, Raman, etc.) and nonlinear (SHG,[77] THG,[78] CARS, etc.) physical processes generated from biological molecules near a metal surface are enhanced; the final objective is to achieve single-molecule detection.

In the case of Raman processes, CARS generation is stronger than spontaneous Raman. Since CARS is a four-wave mixing process in which the anti-Stokes signal results from the parametric coupling of an incident pump and a Stokes lasers,[h] a local stronger enhancement is expected for CARS than for conventional SERS and makes the SECARS technique more challenging than SERS. However, strong field energies are needed in CARS generation, and CARS is not background free. This makes single-molecule detection hard.

Tip-enhancement-based techniques (Raman and CARS) are powerful tools that acquire chemical information on molecular species and conformation of nanodevices with nanoscaled spatial resolution but are still complex and expensive techniques to be applied to biomedical diagnosis.

Acknowledgments

The author acknowledges useful discussions of CARS microscopy with Dr. Hervé Rigneault (Institut Fresnel Mosaic, Marseille, France) before and during the writing of the manuscript.

References

1. Kneipp, K., Wang, Y., Kneipp, H., Perelman, L. T., and Itzkan, I. (1997). Single molecule detection using surface-enhanced Raman scattering (SERS), *Phys. Rev. Lett.*, **78**, 1667–1670.

---

[h]CARS signal depends quadratically with the pump field and linearly with the Stokes field.

2. Maker, P. D., and Terhune, R. W. (1965). Study of optical effects due to an induced polarization third order in the electric field strength, *Phys. Rev.*, **137**, A801–A818.

3. Beglay, R. F., Harvey, A. B., and Bayer, R. L. (1974). Coherent anti-Stokes Raman spectroscopy, *Appl. Phys. Lett.*, **25**, 387–390.

4. Duncan, M. D. (1984). Molecular discrimination and contrast enhancement using a scanning coherent anti-stokes raman microscope, *Opt. Commun.*, **50**, 307–312.

5. Duncan, M. D., Reintjes, J., and Manuccia, T. J. (1982). Scanning coherent anti-stokes raman microscope, *Opt. Lett.*, **7**, 350–352.

6. Duncan, M. D., Reintjes, J., and Manuccia, T. J. (1984). Imaging biological compounds using the cars microscope, *Proc. Soc. Photo-Opt. Instrum. Eng.*, **482**, 46–52.

7. Zumbusch, A., Holtom, G. R., and Xie, X. S. (1999). Three-dimensional vibrational imaging by coherent anti-Stokes Raman scattering, *Phys. Rev. Lett.*, **82**, 4142–4145.

8. Volkmer, A., Cheng, J. X., and Xie, X. S. (2001). Vibrational imaging with high sensitivity via epidetected coherent anti-Stokes Raman scattering microscopy, *Phys. Rev. Lett.*, **8702**, 023901.

9. Cheng, J. X., Volkmer, A., and Xie, X. S. (2002). Theoretical and experimental characterization of coherent anti-Stokes Raman scattering microscopy, *J. Opt. Soc. Am. B-Opt. Phys.*, **19**, 1363–1375.

10. Cheng, J. X., Book, L. D., and Xie, X. S. (2001). Polarization coherent anti-Stokes Raman scattering microscopy, *Opt. Lett.*, **26**, 1341–1343.

11. Volkmer, A., Book, L. D., and Xie, X. S. (2002). Time-resolved coherent anti-Stokes Raman scattering microscopy: imaging based on Raman free induction decay, *Appl. Phys. Lett.*, **80**, 1505–1507.

12. Dudovich, N., Oron, D., and Silberberg, Y. (2002). Single-pulse coherently controlled nonlinear Raman spectroscopy and microscopy, *Nature*, **418**, 512–514.

13. Evans, C. L., Potma, E. O., and Xie, X. S. N. (2004). Coherent anti-Stokes Raman scattering spectral interferometry: determination of the real and imaginary components of nonlinear susceptibility chi((3)) for vibrational microscopy, *Opt. Lett.*, **29**, 2923–2925.

14. Gachet, D., Brustlein, S., and Rigneault, H. (2010). Revisiting the Young's double slit experiment for background-free nonlinear Raman spectroscopy and microscopy, *Phys. Rev. Lett.*, **104**, 213905.

15. Schaller, R. D., Ziegelbauer, J., Lee, L. F., Haber, L. H., and Saykally, R. J. (2002). Chemically selective imaging of subcellular structure in human hepatocytes with coherent anti-stokes Raman scattering (CARS) near-field scanning optical microscopy (NSOM), *J. Phys. Chem. B*, **106**, 8489–8492.

16. Ichimura, T., Hayazawaa, N., Hashimotoa, M., Inouyea, Y., and Kawataa, S. (2004). Tip-enhanced coherent anti-Stokes Raman scattering for vibrational nanoimaging, *Phys. Rev. Lett.*, **92**, 220801.

17. Bloembergen, N. (1965). *Nonlinear Optics*, Benjamin.

18. Boyd, R. (1992). *Nonlinear Optics*, Academic, New York.

19. Lotem, H., Lynch, R. T., and Bloembergen, N. (1976). Interference between Raman resonances in four-wave difference mixing, *Phys. Rev. A*, **14**, 1748–1755.

20. Djaker, N., Sandeau, N., Gachet, D., Lenne, P. F., and Rigneault, H. (2006). Refractive effects in coherent anti-Stokes Raman scattering (CARS) microscopy, *Appl. Opt.*, **45**, 7005–7011.

21. Hellerer, T. (2004). *CARS-Mikroskopie: Entwicklung und Anwendung*, Ludwig-Maximilians-Universität München.

22. Cheng, J. X., Jia, Y. K., Zheng, G. F., and Xie, X. S. (2002). Laser-scanning coherent anti-stokes Raman scattering microscopy and applications to cell biology, *Biophys. J.*, **83**, 502–509.

23. Volkmer, A., Cheng, J. X., and Xie, X. S. (2001). Vibrational imaging with high sensitivity via epidetected coherent anti-Stokes Raman scattering microscopy, *Phys. Rev. Lett.*, **87**, 23901–23904.

24. Djaker, N., Marguet, D., and Rigneault, H. (2006). Coherent anti-Stokes Raman scattering microscopy, *Med. Sci.*, **22**, 853–859.

25. Djaker, N., Lenne, P. F., and Rigneault, H. (2004). Vibrational imaging by CARS microscopy, *Proc. SPIE Int. Soc. Opt. Eng.*, **5463**, 133–139.

26. Cheng, J. X., Volkmer, A., Book, L. D., and Xie, X. S. (2001). An epi-detected coherent anti-Stokes Raman scattering (E-CARS) microscope with high spectral resolution and high sensitivity, *J. Phys. Chem. B*, **105**, 1277–1280.

27. Nan, X. L., Cheng, J. X., and Xie, X. S. (2003). Vibrational imaging of lipid droplets in live fibroblast cells with coherent anti-Stokes Raman scattering microscopy, *J. Lipid Res.*, **44**, 2202–2208.

28. Cheng, J. X., Volkmer, A., Book, L. D., and Xie, X. S. (2002). Multiplex coherent anti-Stokes Raman scattering microspectroscopy and study of lipid vesicles, *J. Phys. Chem. B*, **106**, 8493–8498.

29. Muller, M., and Schins, J. M. (2002). Imaging the thermodynamic state of lipid membranes with multiplex CARS microscopy, *J. Phys. Chem. B*, **106**, 3715–3723.

30. Potma, E. O., Jones, D. J., Cheng, J.-X., Xie, X. S., and Ye, J. (2002). High-sensitivity coherent anti-Stokes Raman scattering microscopy with two tightly synchronized picosecond lasers, *Opt. Lett.*, **27**, 1168–1170.

31. Wang, H. F., Fu, Y., Zickmund, P., Shi, R. Y., and Cheng, J. X. (2005). Coherent anti-stokes Raman scattering imaging of axonal myelin in live spinal tissues, *Biophys. J.*, **89**, 581–591.

32. Djaker, N., Lenne, P. F., Marguet, D., Colonna, A., Hadjur, C., and Rigneault, H. (2007). Coherent anti-Stokes Raman scattering microscopy (CARS): instrumentation and applications, *Nucl. Instrum. Methods Phys. Res.*, **571**, 177–181.

33. Nan, X., Yang, W. Y., and Xie, X. S. (2004). CARS microscopy: lights up lipids in living cells, *Biophoton. Int.*, **11**, 44–47.

34. Potma, E. O., and Xie, X. S. (2005). Direct visualization of lipid phase segregation in single lipid bilayers with coherent anti-Stokes Raman scattering microscopy, *ChemPhysChem*, **6**, 77–79.

35. Holtom, G. R., Thrall, B. D., Chin, B. Y., Wiley, H. S., and Colson, S. D. (2001). Achieving molecular selectivity in imaging using multiphoton Raman spectroscopy techniques, *Traffic*, **2**, 781–788.

36. Li, L., Wang, H. F., and Cheng, J. X. (2005). Quantitative coherent anti-Stokes Raman scattering imaging of lipid distribution in coexisting domains, *Biophys. J.*, **89**, 3480–3490.

37. Tahara, T., Toleutaev, B. N., and Hamaguchi, H. (1994). Picosecond time-resolved multiplex coherent anti-stokes-raman scattering spectroscopy by using a streak camera isomerization dynamics of all-trans and 9-cis retinal in the lowest excited triplet-state, *J. Chem. Phys.*, **100**, 786–796.

38. Wurpel, G. W. H., Schins, J. M., and Müller, M. (2002). Chemical specificity in three-dimensional imaging with multiplex coherent anti-Stokes Raman scattering microscopy, *Opt. Lett.*, **27**, 1093–1095.

39. Rinia, H. A., Bonn, M., and Muller, M. (2006). Quantitative multiplex CARS spectroscopy in congested spectral regions, *J. Phys. Chem. B*, **110**, 4472–4479.

40. Wurpel, G. W. H., Rinia, H. A., and Muller, M. (2005). Imaging orientational order and lipid density in multilamellar vesicles with multiplex CARS microscopy, *J. Microsc. (Oxford)*, **218**, 37–45.

41. Wurpel, G. W. H., Schins, J. M., and Muller, M. (2004). Direct measurement of chain order in single phospholipid mono- and bilayers with multiplex CARS, *J. Phys. Chem. B*, **108**, 3400–3403.

42. Cheng, J. X., Potma, E. O., and Xie, X. S. (2002). Coherent anti-Stokes Raman scattering correlation spectroscopy: probing dynamical processes with chemical selectivity, *J. Phys. Chem. A*, **106**, 8561–8568.

43. Hellerer, T., Schiller, A., Jung, G., and Zumbusch, A. (2002). Coherent anti-Stokes Raman scattering (CARS) correlation spectroscopy, *ChemPhysChem*, **7**, 630–633.

44. Marowsky, G., and Lupke, G. (1990). Cars-background suppression by phase-controlled nonlinear interferometry, *Appl. Phys. BP– hotophys. Laser Chem.*, **51**, 49–51.

45. Yacoby, Y., Fitzgibbon, R., and Lax, B. (1980). Coherent cancellation of background in 4-wave mixing spectroscopy, *J. Appl. Phys.*, **51**, 3072–3077.

46. Potma, E. O., Evans, C. L., and Xie, X. S. (2006). Heterodyne coherent anti-Stokes Raman scattering (CARS) imaging, *Opt. Lett.*, **31**, 241–243.

47. Marks, D. L., and Boppart, S. A. (2004). Nonlinear interferometric vibrational imaging, *Phys. Rev. Lett.*, **92**, 123905.

48. Vinegoni, C., Bredfeldt, J. S., Marks, D. L., and Boppart, S. A. (2004). Nonlinear optical contrast enhancement for optical coherence tomography, *Opt. Express*, **12**, 331–341.

49. Greve, M., Bodermann, B., Telle, H. R., Baum, P., and Riedle, E. (2005). High-contrast chemical imaging with gated heterodyne coherent anti-Stokes Raman scattering microscopy, *Appl. Phys. B*, **81**, 875–879.

50. Dudovich, N., Oron, D., and Silberberg, Y. (2003). Single-pulse coherent anti-Stokes Raman spectroscopy in the fingerprint spectral region, *J. Chem. Phys.*, **118**, 9208–9215.

51. Oron, D., Dudovich, N., and Silberberg, Y. (2003). Femtosecond phase-and-polarization control for background-free coherent anti-Stokes Raman spectroscopy, *Phys. Rev. Lett.*, **90**, 1–4.

52. Oron, D., Dudovich, N., and Silberberg, Y. (2002). Single-pulse phase-contrast nonlinear Raman spectroscopy, *Phys. Rev. Lett.*, **89**, 1–4.

53. Gachet, D., Billard, F., and Rigneault, H. (2008). Focused field symmetries for background-free coherent anti-Stokes Raman spectroscopy, *Phys. Rev. A*, **77**, 061802.

54. Qian, S. X., Snow, J. B., and Chang, R. K. (1985). Coherent Raman mixing and coherent anti-Stokes Raman scattering from individual micrometer-size droplets, *Opt. Lett.*, **10**, 499–501.

55. Cooney, J., and Gross, A. (1982). Coherent anti-Stokes Raman scattering by droplets in the Mie size range, *Opt. Lett.*, **7**, 218–220.

56. Chew, H., Wang, D. S., and Kerker, M. (1984). Surface enhancement of coherent anti-Stokes Raman scattering by colloidal spheres, *J. Opt. Soc. Am. B*, **1**, 56–66.

57. Koo, T. W., Chan, S., and Berlin, A. A. (2005). Single-molecule detection of biomolecules by surface-enhanced coherent anti-Stokes Raman scattering, *Opt. Lett.*, **30**, 1024–1026.

58. Cairo, F., Martini, F., and Murra, D. (1993). QED-vacuum confinement of inelastic quantum scattering at optical frequencies: a new perspective in Raman spectroscopy, *Phys. Rev. Lett.*, **70**, 1413–1416.

59. Fainstein, A., Jusserand, B., and Thierry-Mieg, V. (1995). Raman scattering enhancement by optical confinement in a semiconductor planar microcavity, *Phys. Rev. Lett.*, **75**, 3764–3767.

60. Dumeige, Y., Sagnes, I., Monnier, P., Vidakovic, P., Abram, I., Mériadec, C., and Levenson, A. (2002). Phase-matched frequency doubling at photonic band edges: efficiency scaling as the fifth power of the length, *Phys. Rev. Lett.*, **89**, 043901.

61. Soboleva, I. V., Murchikova, E. M., Fedyanin, A. A., and Aktsipetrov, O. A. (2005). Second- and third-harmonic generation in birefringent photonic crystals and microcavities based on anisotropic porous silicon, *Appl. Phys. Lett.*, **87**, 241110.

62. Becker, C., Wegener, M., Wong, A. S., and Freymann, G. V. (2006). Phase-matched nondegenerate four-wave mixing in onedimensional photonic crystals, *Appl. Phys. Lett.*, **89**, 131122.

63. Benisty, H., Neve, H. D., and Weisbuch, C. (1998). Impact of planar microcavity effects on light extraction—part I: basic concepts and analytical trend, *IEEE J. Quantum Electron.*, **34**, 1612–1631.

64. Gachet, D., Billard, G., and Rigneault, H. (2009). Coherent anti-Stokes Raman scattering in a microcavity, *Opt. Lett.*, **34**, 1789–1791.

65. Liang, E. J., Weippert, A., Funk, J. M., Materny, A., and Kiefer, W. (1994). Experimental-observation of surface-enhanced coherent anti-Stokes Raman scattering, *Chem. Phys. Lett.*, **227**, 115–120.

66. Hashimoto, M., and Araki, T. (2000). Molecular vibration imaging in the fingerprint region by use of coherent anti-Stokes Raman scattering microscopy with a collinear configuration, *Opt. Lett.*, **25**, 1768–1770.

67. Hashimoto, M., and Araki, T. (2001). Three-dimensional transfer functions of coherent anti-Stokes Raman scattering microscopy, *J. Opt. Soc. Am. A Opt. Image Sci. Vision*, **18**, 771–776.

68. Nie, S. M., and Emory, S. R. (1997). Probing single molecules and single nanoparticles by surface-enhanced Raman scattering, *Scė nce*, **275**, 1102–1106.

69. Ichimura, T. H., N., Hashimotoa, M., Inouyea, Y., and Kawataa, S. (2003). Local enhancement of coherent anti-Stokes Raman scattering by isolated gold nanoparticles, *J. Raman Spectrosc.*, **34**, 651–654.

70. Renger, J., Quidant, R., Van Hulst, N., Palomba, S., and Novotny, L. (2009). Free-space excitation of propagating surface plasmon polaritons by nonlinear four-wave mixing, *Phys. Rev. Lett.*, **103**, 266802.

71. Renger, J., Quidant, R., Van Hulst, N., and Novotny, L. (2010). Surface-enhanced nonlinear four-wave mixing, *Phys. Rev. Lett.*, **104**, 059903.

72. Imura, K., Nagahara, T., and Okamoto, H. (2005). Near-field two-photon induced photoluminscence from single gold nanorods and imaging of plasmon modes, *J. Phys. Chem. B*, **109**, 13214–13220.

73. Beversluis, M. R., Bouhelier, A., and Novotny, L. (2003). Continuum generation from single gold nanostructures through near-field mediated intraband transitions, *Phys. Rev. B*, **68**, 115433.

74. Danckwerts, M., and Novotny, L. (2007). Optical frequency mixing at coupled gold nanoparticles, *Phys. Rev. Lett.*, **98**, 026101–026104.

75. Kim, H. M., Xiang, C. X., Guell, A. G., Penner, R. M., and Potma, E. O. (2008). Tunable two-photon excited luminescence in single gold nanowires fabricated by lithographically patterned nanowire electrodeposition, *J. Phys. Chem. C*, **112**, 12721–12727.

76. Elfick, A. P., Downes, A. R., and Mouras, R. (2010). Development of tip-enhanced optical spectroscopy for biological applications, *Anal. Bb anal. Chem.*, **396**, 45–52.

77. Bouhelier, A., Beversluis, M., Hartschuh, A., and Novotny, L. (2003). Near-field second-harmonic generation induced by local field enhancement, *Phys. Rev. Lett.*, **90**, 13903.

78. Lippitz, M., Dijk, M. A. V., and Orrit, M. (2005). Third-harmonic generation from single gold nanoparticles, *Nano Lett.*, **5**, 799–802.

Chapter 13

Multi-Signal Processing Biosensors and Bioactuators Based on Biocomputing Concept

Evgeny Katz and Arcan Güven

Department of Chemistry and Biomolecular Science and NanoBio Laboratory (NABLAB), Clarkson University, Potsdam, New York 13699-5810, USA

ekatz@clarkson.edu

13.1 Introduction

Molecular[1–7] and biomolecular[8] logic gates and their network processing chemical input signals, similar to computers, received a great deal of attention and were rapidly developed in the first decade of this century. A subarea of unconventional computing,[9,10] logic gates process chemical information in a way that mimics Boolean logic operations, using binary definitions (**1**,**0**; **YES**/**NO**) for concentrations of reacting species. Using this approach, chemical reactions can be reformulated as information processing steps with built-in logic operations. Then the chemical processes are programmed, like computer programming, yielding networks that perform logic operations.

Nanoantenna: Plasmon-Enhanced Spectroscopies for Biotechnological Applications
Edited by Marc Lamy de la Chapelle and Annemarie Pucci

ISBN 978-981-4303-61-3 (Hardcover), 978-981-4303-62-0 (eBook)
www.panstanford.com

As a network that processes information becomes large, and the information is processed in greater quantities and at higher complexity, noise inevitably builds up and can ultimately degrade the useful signal that is the intended result of the computation. We then have to develop approaches to achieve what is known as *fault-tolerant ṅ formatȯ n processṅ g*, which involves noise control and noise suppression. One notable example of of fault-tolerant information processing is the robustness of many complex processes in cell functions.

Presently, we are aware of three primary fault-tolerant information processing paradigms. The first is the analog/digital electronics paradigm of the silicon-chip technology in modern computers. We know how to design such devices and they have been successfully built. Living organisms are the second paradigm: While we obviously know that this paradigm leads to scalability, we do not yet fully understand it, though significant strides have been made in exploring the structure and functioning of biological networks. The third, most recent paradigm, involves massive parallelism: quantum (quantum computing) or ensemble (variants of DNA computing), both in the preliminary research stages.

Biochemical computing based on enzymatic reactions[8,11–13] attempts to process information with biomolecules and biological objects.[14–23] However, the information processing paradigm assumed, has, in most cases, been that of ordinary electronics. Indeed, most biochemical computing studies attempt to realize and, most recently, network gates that mimic Boolean digital logic.[8,24]

Networks with computational steps that solely involve biochemical processes[25,26] are being researched for new technological capabilities: multi-input biosensors with new functionalities,[27–29] as well as approaches that allow removing the batteries from, and generally reducing the need for, inorganic leads and the electrical power supply for those stages of information processing that occur during biomedical testing, that are used in implantable devices, and for other fast decision-making steps in applications. Futuristic ideas for applications of biochemical logic include more direct brain-computer and body-computer interfacing for both reading out and inputting information, and generally erasing the barrier between inorganic and organic information processing in computer device functioning.

The primary challenge in the field of biochemical computing at present can be summarized as follows. Recent studies took the field somewhat beyond the earlier, simple gates that mimic two-input, one-output Boolean functions. Specifically, some gate networking has been accomplished,[25,26] the first steps in the analysis of network scalability have been reported,[30] and the first attempts at smart interfacing with ordinary electronics have been initiated.[31-34] In these studies, biocomputing based on enzymatic reactions has emerged as an appealing approach for information processing due to the specificity and other useful chemical-kinetics properties of enzyme reactions.

These advances have also set the stage for new challenges. In particular, it has been realized that large-scale networking and fault tolerance cannot be achieved without the development of a toolbox of new network elements, such as filters, signal splitters (for copying), signal balancers, and resetting functions. These non-Boolean network elements for biochemical computing might not follow the analogy with ordinary electronic devices too closely. Instead, concepts borrowed from nature, including delayed identity— part of the feed-forward network motif,[35] or memory, properties[36]— have recently received attention in information processing network studies.

13.2 Chemical Logic Gates and Networks

Chemical-computing studies have used molecular or supramolecular systems[1-7] to mimic processes typical for electronic computing devices, including simple Boolean logic operations such as **AND**, **OR**, and **XOR**, as well as more complex systems like molecular comparators, digital demultiplexers, encoder-decoders, keypad locks, write-read-erase memory units, half adders/half subtractors, and full adders/full subtractors. Chemical systems are, in principle, capable of performing computations at the level of a single molecule,[37] resulting in nanoscaling of the computing units[38] and allowing parallel computation performed by molecules involved in various chemical reactions.[39] However, the most important challenges in chemical computing are networking and fault tolerance. Complex, multicomponent chemical logic systems usually require ingenious molecular assemblies to allow compatibility between information processing subunits just to perform rather simple functions, despite their extreme synthetic complexity.[40]

Biomolecular systems, on the other hand, offer the advantage of specificity (of being very selective) in their chemical functions and therefore being more usable in complex chemical soup environments. As a result, complex enzyme-based information processing units could be more easily scaled up, giving rise to artificial biocomputing networks that can perform various logic functions and mimicking natural biochemical pathways.[8,41] An added advantage of biomolecular computing systems is their ability to process biochemical information received in the form of chemical signals directly from biological systems and their ability to operate in a biological environment.[42] This is important for interfacing the resulting biochemical-logic devices with processes in living organisms[43] in biomedical applications.

Recent experimental advances in enzyme-based biocomputing have included not only experimental demonstrations of several single Boolean gates, such as **AND**, **OR**, **XOR**, and **Inhibit**,[12,13] but also networking of up to five gates.[8,25,26] Similar logic operations were also realized using nonbiological chemical systems.[1-7] However, the biochemical systems offered the relative simplicity of assembled logic schemes. Ultimately, the increasing complexity of enzyme-based logic networks will require exploration of noise suppression approaches in biochemical logic-gate networks. Fault tolerance within the analog/digital information processing paradigm is accomplished by gate optimization for suppression of the analog noise amplification[8,44] as well as by network design and/or network topology.[30] For larger networks, another, digital, mechanism[45] of noise amplification may emerge. This, however, necessitates redundancy in network design and truly digital information processing with appropriate network elements for filtering, signal splitting, and so on.

The present sizes of biochemical computing networks have already allowed exploration of aspects of design and optimization issues related to suppression of analog noise amplification. Work on this topic has been discussed in recent publications.[8,44,45] The field is presently wide open for research into practical applications.

13.3 Biosensing Applications of Biocomputing Systems

Despite the great success and rapid development of chemical[1–7] and biochemical[8] computing systems, most systems represent only proof of the concept. Molecular systems that can perform computing/logic operations are not yet practical. Indeed, the unconventional chemical computing systems that do exist are capable of solving only basic arithmetic or logic operations in minutes or even hours. On the other hand, application of molecular logic systems for analytical purposes could yield a novel class of sensors that are able to accept many input signals and produce binary outputs in YES/NO form, which is particularly important for biomedical applications. This approach has already been successfully used to analyze protein libraries associated with multiple sclerosis.[46]

Logically processed feedback between a drug delivery application and physiological conditions can significantly improve drug targeting and efficiency.[47] The well-developed field of DNA biocomputing[18] spins out from solving combinatorial problems[48] to analyzing biomedical, multiparameter physiological conditions.[49] Programmable and autonomous DNA computing systems operating *in vitro* have demonstrated logic multiparameter analysis of disease-related biomolecular markers and could be applied in future for *in situ* medical diagnosis and cure.[49,50] For example, biosensor systems for detecting genetic modifications in avian influenza[51] were developed based on DNA computing principles, when various oligonucleotide signals were logically processed by a DNA logic network. Coupling enzyme logic systems with controlled self-assembling of nanoparticles has allowed logic **AND/OR** responses to cancer markers (matrix-metalloproteinase: MMP2 and MMP7).[52] The logically controlled aggregation of superparamagnetic Fe_3O_4 nanoparticles has been detected by magnetic resonance imaging (MRI), promising easy adaptation of the method to future *in vivo* medical applications.

The results of logically-processed biomolecular signals can be stored in enzyme-controlled set-reset flip-flop memory units.[53] The terminal memory units can be connected to various biocatalytic pathways processing multiple biochemical signals. Programmable memory units based on two different protein kinases as information input signals have allowed **AND**, **OR,** and **NOR** logic operations prior to information storage, thus providing information processing diversity.[54]

It should be still noted that most of the presently developed biocomputing systems are commonly activated by signals that are not relevant to biomedical applications.[55] Even when signals do have biomedical significance, they are usually used at nonphysiologically relevant concentrations.[51,52] This limits immediate use of the developed biocomputing systems for biomedical applications. Extensive research, combining efforts in biochemistry, computer science, and medicine is necessary to develop real biocomputing systems for biomedical applications.

Common biosensing devices are based on a single input (analyte). Multisignal logic gates rely on multiple target analytes (inputs) to yield YES/NO responses. Such digital biosensors are expected to offer higher fidelity biosensing than traditional single-input biosensors. By using various biomarkers as inputs for the enzyme gates and automatically processing physiological information, such biochemical logic systems can provide rapid and reliable assessment of overall physiological conditions and would initiate optimal, timely therapeutic intervention. The main challenge in further development of multiparameter biosensing systems with digital logic information processing is scaling up their complexity by assembling individual logic gates in complex logic networks aiming at intelligent medical diagnostics via biomolecular logic.

As an example, let's consider in detail a multi-input biocatalytic cascade analyzing two kinds of injuries.[56] The assessment of two common battlefield injuries, soft tissue injury (STI) and traumatic brain injury (TBI), was performed by a multi-enzyme biocatalytic cascade, shown in Fig. 13.1, that can be described as a comprehensive concatenated logic system, shown in Fig. 13.2. The logic network represented in Fig. 13.2 is actually identical to the biocatalytic cascade depicted in Fig. 13.1, but it offers a description that is more convenient for discussing system operation. The system architecture includes eight networked logic gates of the AND type. The biocatalytic system operated in two different modes, one for the analysis of TBI and another for the analysis of

STI, with the possibility of switching between them. The system was designed to analyze five biomarkers appearing in different combinations and to lead to a logic conclusion about the presence or absence of TBI or STI conditions. Three biomarkers— creatine kinase (CK), lactate dehydrogenase (LDH), and lactate (Lac)—corresponded to the physiological conditions characteristic of STI, while two other biomarkers—enolase (EN) and glutamate (Glu)—were reporting on the TBI diagnosis.

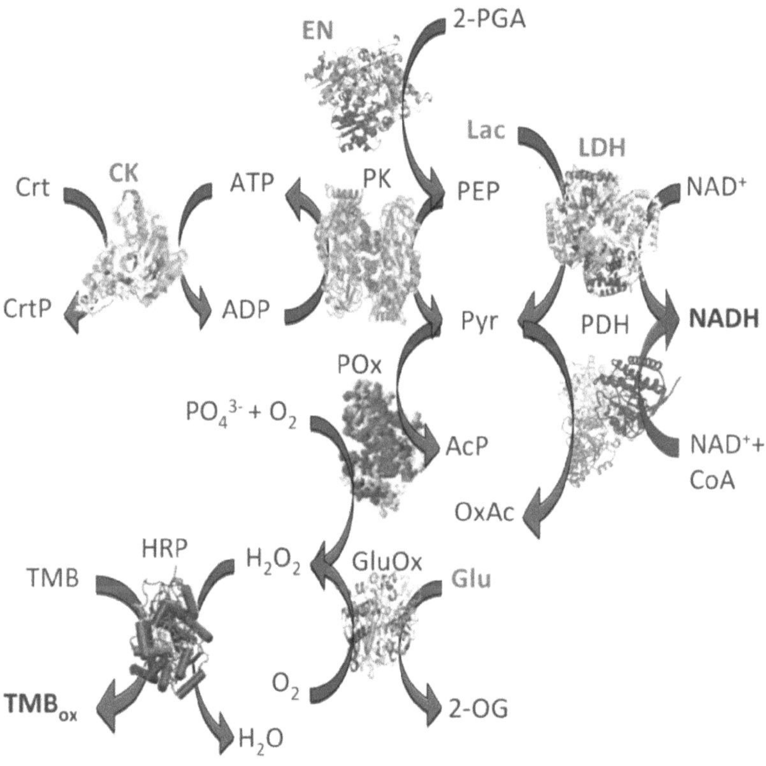

Figure 13.1 Multi-enzyme biocatalytic cascade for the analysis of STI and TBI. Biomarker inputs were used for STI (CK, Lac, LDH) and for TBI (EN and Glu). Output signals for STI and TBI are NADH and TMB_{ox}, respectively. Other products of the biocatalytic cascade are acetyl phosphate (AcP), oxalacetate (OxAc), 2-oxoglutarate (2-OG), and creatine-phosphate (CrtP). Note that for simplicity the scheme does not include some reacting cofactors, promoters, and byproducts.

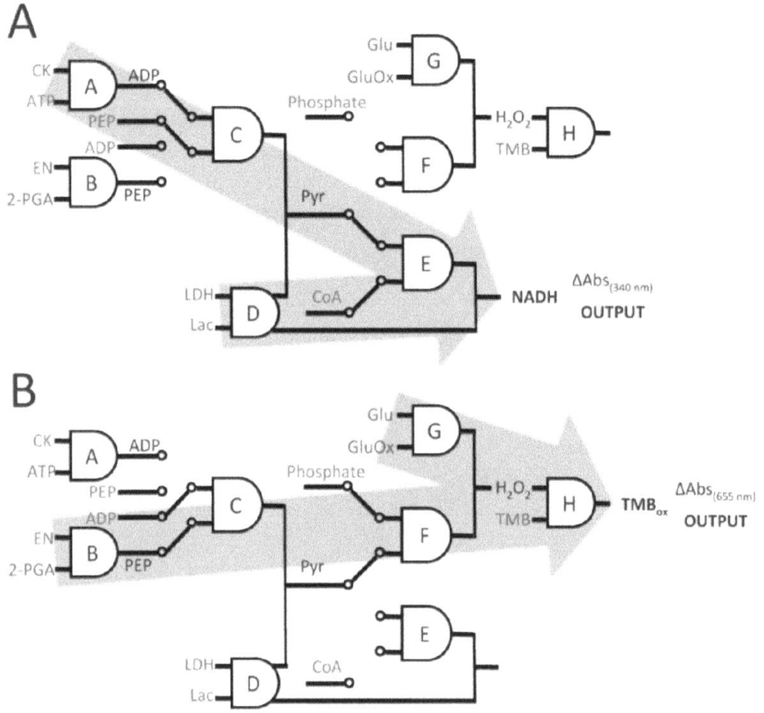

Figure 13.2 Equivalent logic schemes for the concatenated logic gates analyzing STI and TBI corresponding to the biocatalytic cascade shown in Fig. 13.1. (A) The system switched to the STI analysis mode. (B) The system switched to the TBI analysis mode. The switch regulated pathways for the STI and TBI operational modes are indicated by arrows.

All five biomarkers were applied as biochemical input signals activating the logic network at two different concentration levels: logic **0** corresponded to normal physiological concentrations of the biomarkers, while logic **1** was selected at elevated pathophysiological concentrations corresponding to the respective injuries. Simultaneous processing of many biomarkers through the complex biocatalytic cascade required optimization of the biocatalytic reactions by tuning the reaction rates in order to get comparable output signals for various combinations of biomarker inputs. Optimization was achieved through careful selection of the auxiliary inputs: adenine 5′-triphosphate (ATP), D(+)2-phosphoglyceric acid (2-PGA), glutamate oxidase (GluOx), and 3,3′,5,5′-tetramethylbenzidine dihydrochloride (TMB). In order to digitize the logic network operation, the auxiliary inputs were applied at two levels: logic **0** corresponded to the physical zero concentration, while logic-**1** was selected experimentally on optimization of the system. The auxiliary inputs did not provide any information about physiological conditions related to the injuries, but they were needed for optimal processing of the biomarker inputs. When they were applied at **0** levels, the system was mute and insensitive to the biomarker inputs, but their application at **1** levels provided optimized processing of the biomarker inputs applied in different combinations of **0** and **1** logic levels. Four additional inputs—phosphoenolpyruvic acid (PEP), adenine 5′-diphosphate (ADP), phosphate, and coenzyme A (CoA)—were used to switch the system operation between TBI and STI modes. These inputs were applied at logic **0** levels corresponding to the physical zero concentrations and logic **1** levels being experimentally optimized.

Since none of the used biomarkers was specific enough for the STI diagnosis, only simultaneous appearance of all three STI-related biomarkers (CK, LDH, and Lac) at logic 1 values would provide a reliable conclusion about the STI conditions. The system was operating in the following way (see Fig. 13.1 for the biochemical representation and Fig. 13.2A for the logic equivalent circuitry—the STI pathway is highlighted): The biocatalytic reaction of CK (STI biomarker) and ATP (gate A) resulted in the production of ADP. Further reaction of ADP with PEP biocatalyzed by pyruvate kinase (PK) (gate C) resulted in the formation of pyruvic acid (Pyr), which then reacted in the presence of CoA and pyruvate dehydrogenase (PDH) (gate E) to yield the reduced NADH, the output signal. Note that the pathway composed of A-C-E gates was activated only when the biomarker input CK, auxiliary input ATP, and switching inputs PEP and CoA appeared at logic 1 values. Simultaneous application of LDH and Lac inputs (STI biomarkers) at logic 1 (gate D) resulted in a reduction of NAD^+ and further increase of the NADH output. It should be noted that this pathway resulted in the concomitant production of Pyr, which was passing through gate E producing one more equivalent of NADH, thus further amplifying the output signal. Fig. 13.3a shows the optical changes in the system measured at λ = 340 nm (NADH absorbance) for different combinations of biomarker inputs. Only simultaneous application of all three biomarker inputs at logic 1 values (input combination 1, 1, 1) resulted in the high optical absorbance changes that allowed an unambiguous conclusion about the STI condition. The experimentally derived threshold level of 0.5 O.D. allowed perfect separation of the logic 0 and 1 levels for the output signal being different at least by two-fold, as shown in Fig. 13.3b. It should be noted that in all the measurements shown in Fig. 13.3 the auxiliary (ATP) and switching (PEP and CoA) inputs were applied at logic 1 values to allow optimal performance of the analytical pathway. At the same time the switching inputs ADP and phosphate were applied at logic 0 levels to inhibit the alternative TBI pathway.

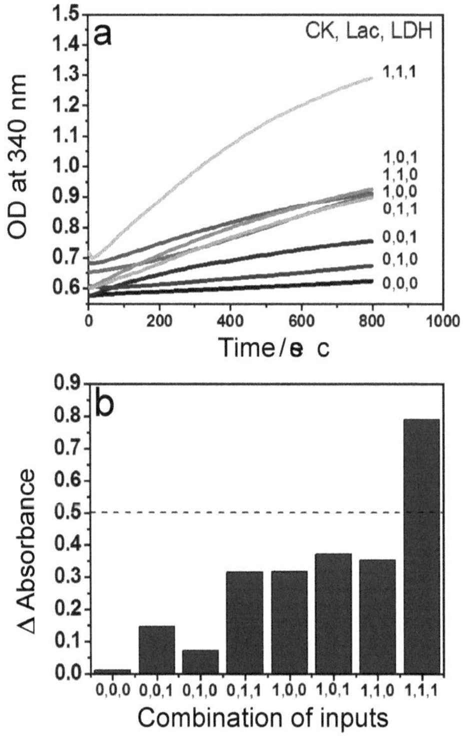

Figure 13.3 (a) Optical detection of the output signal (NADH) generated by the logic system operating for the STI analysis obtained upon different combinations of the injury biomarker input signals (CK, Lac, LDH). (b) Bar chart for the output signals generated by the enzyme logic system for the analysis of STI at 800 seconds. The STI diagnosis corresponds to the output signal above the decision threshold (dashed line). The auxiliary (ATP) and switching (PEP and CoA) inputs were at logic-**1** for all combinations of the biomarker inputs.

Unlike STI biomarkers, the TBI biomarkers, EN and Glu, are rather specific and can report on the presence of an injury even when they appear alone. The system operated in the following way (see Fig. 13.1 for the traditional biochemical representation and Fig. 13.2B for the logic equivalent circuitry—the TBI pathway is highlighted): The biocatalytic reaction of EN (TBI biomarker) and 2-PGA (gate B) resulted in the formation of PEP. In the next step PEP reacted with ADP (switching input) biocatalyzed by PK (gate C), resulting in the production of Pyr. The reaction of the Pyr with phosphate (switching input) biocatalyzed by pyruvate oxidase (POx) (gate F) yielded H_2O_2, which reacted with TMB in the presence of HRP (gate H). This reaction resulted in the oxidation of TMB and produced the absorbance increase at λ = 655 nm characteristic of TMB_{ox}, which was considered the final output signal from the pathway composed of B–C–F–H gates.

The same signal was generated through another pathway composed of gates G and H: The biocatalytic reaction of Glu (TBI biomarker) and glutamate GluOx resulted in O_2 reduction and yielded H_2O_2 (gate G). Then H_2O_2 reacted with TMB in the presence of HRP to yield TMB_{ox} with the respective absorbance changes (gate H).

Careful optimization of the system (by tuning the concentrations of the auxiliary inputs) allowed comparable output signals produced in the both pathways. Figure 13.4a shows the optical changes in the system measured at λ = 655 nm (TMBox absorbance) for different combinations of the biomarker inputs. Any or both biomarkers appearing at logic 1 level resulted in high absorbance changes, signaling TBI conditions. The experimentally derived threshold level of 0.02 O.D. allowed perfect separation of the logic 0 and 1 levels for the output signal, which was significantly different for the input combination 0,0 and all other combinations (0,1; 1,0; 1,1) (see Fig. 13.4b). It should be noted that in all the measurements shown in Fig. 13.4 the auxiliary (2-PGA, TMB, and GluOx) and switching (ADP and phosphate) inputs were applied at logic 1 values to allow optimized performance of the analytical pathway. At the same time, the switching inputs PEP and CoA were applied at logic 0 levels to inhibit the alternative STI pathway.

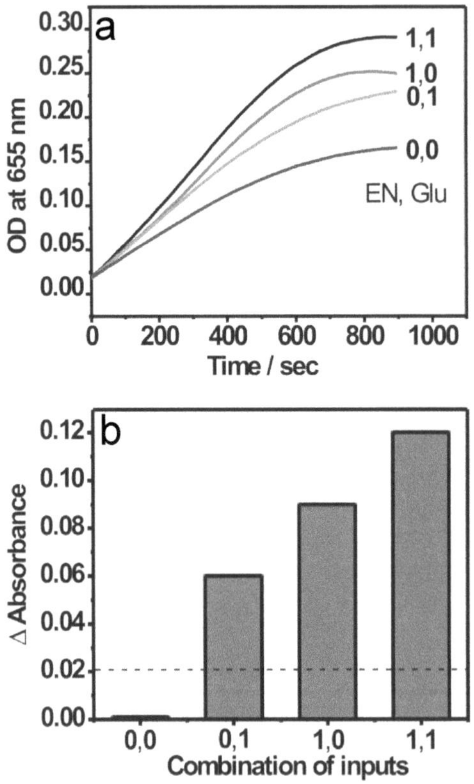

Figure 13.4 (a) Optical detection of the output signal (TMB_{ox}) generated by the logic system operating for the TBI analysis obtained on different combinations of the biomarker input signals (EN, Glu). (b) Bar chart for the output signals generated by the enzyme logic system for the analysis of TBI at 800 seconds. The TBI diagnosis corresponds to the output signals above the decision threshold (dashed line). The auxiliary (2-PGA, GluOx, and TMB) and switching (ADP and phosphate) inputs were at logic-**1** for all combinations of the biomarker inputs.

The complex biocomputing-bioanalytic system operated in two different modes on activation of the appropriate pathways. Operation of the logic system in the STI mode results in production of ADP by the biocatalytic reaction at gate A. Further reaction of ADP in gate C requires the presence of PEP, which is not produced by the system when it operates in the STI mode (note that gate B is mute because of the absence of EN input). Therefore PEP should be added artificially in order to activate gate C. On the other hand, when the system operates in TBI mode, PEP is produced *in situ* in gate B, while ADP which, is also needed for the operation of gate C, is missing (note that gate A is mute because of the absence of CK). Therefore, in this case ADP should be added artificially in order to activate gate C. Finally, to activate the network in the STI mode, the switching inputs PEP and ADP should be applied at the logic **1** and **0** values, respectively, while for the TBI mode they should be at the opposite **0** and **1** values. This switch allows the use of gates C and F in two different modes of operation. Similarly, in order to switch between two operational modes, gates F and E should be selectively activated by the correct concentrations of phosphate and CoA. Specifically, phosphate and CoA were applied at logic **0** and **1** levels in the STI mode and at logic **1** and **0** levels in the TBI mode.

13.4 Conclusion

A very sophisticated, multi-enzyme/multistep biocatalytic cascade can provide reliable diagnostics of physiological conditions using logic analysis of complex patterns of various biomarkers. The designed system exemplifies the novel approach to multisignal processing biosensors that mimick natural biochemical pathways and that operate according to the biocomputing concept.

Novel biocomputing sensor systems show promise for a wide range of analytical applications, ranging from biomedical and environmental monitoring to national defense and food safety. Biocomputing sensor systems of even moderate complexity could be used for closed-loop (Sense/Act/Treat) systems with biomedical applications such as patient-tailored, timely therapy. Such smart biosensing-bioactuating systems could also benefit other fields, from biotechnology to homeland security.

Integrated medical systems based on a sense/delivery feedback-loop are expected to revolutionize patient monitoring by enabling personalized medicine in connection to specific diseases and conditions. Such fully autonomous systems can be created by coupling sensing and delivery devices that would monitor changes in the level of biomarkers and release an appropriate drug in response. Most work around such feedback-loop systems is currently being devoted to the management of diabetes through the integration of an electrochemical glucose sensing element that can deliver an optimal dose of insulin. However, applications of sense-and-act systems extend beyond chronic disease management and could be extremely important in a medical emergency where rapid intervention is crucial for the survival of the patient. The successful integration of drug delivery feedback loops depends on the availability of "smart" interfaces for automated decision making and initiating proper therapeutic interventions. New algorithms and methods are needed in view of the limited experience with real-time control of drug delivery. Such interface of the sensing and delivery elements is thus a critical component of integrated medical systems.

Acknowledgment

This research was supported by the National Science Foundation (Grants DMR-0706209, CCF-1015983), by ONR (Grant N00014-08-1-1202), and by the Semiconductor Research Corporation (Award 2008-RJ-1839G).

References

1. de Silva, A. P., Uchiyama, S., Vance, T. P., and Wannalerse, B. (2007). A supramolecular chemistry basis for molecular logic and computation, *Coord. Chem. Rev.*, **251**, 1623–1632.
2. P. de Silva, A., Uchiyama, S. (2007). Molecular logic and computing, *Nat. Nanotechnol.*, **2**, 399–410.
3. Szacilowski, K. (2008). Digital information processing in molecular systems, *Chem. Rev.*, **108**, 3481–3548.
4. Credi, A. (2007). Molecules that make decisions, *Angew. Chem. Int. Ed.*, **46**, 5472–5475.
5. Pischel, U. (2007). Chemical approaches to molecular logic elements for addition and subtraction, *Angew. Chem. Int. Ed.*, **46**, 4026–4040.
6. Pischel, U. (2010). Digital operations with molecules — advances, challenges, and perspectives, *Austral. J. Chem.*, **63**, 148–164.
7. Andreasson, J., and Pischel, U. (2010). Smart molecules at work — mimicking advanced logic operations, *Chem. Soc. Rev.*, **39**, 174–188.

8. Katz, E., and Privman, V. (2010). Enzyme-based logic systems for information processing, *Chem. Soc. Rev.*, **39**, 1835–1857.
9. Calude, C. S., Costa, J. F., Dershowitz, N., Freire, E., and Rozenberg, G. (eds) (2009). *Unconventional Computation: Lecture Notes in Computer Science*, vol. 5715, Springer, Berlin.
10. Adamatzky, A., Bull, L., De Lacy Costello, B., Stepney, S., and Teuscher, C. (eds) (2007). *Unconventional Computing*, Luniver Press, Frome, UK.
11. Privman, V., Strack, G., Solenov, D., Pita, M., and Katz, E. (2008). Optimization of enzymatic biochemical logic for noise reduction and Scalability: How many biocomputing gates can be interconnected in a circuit? *J. Phys. Chem. B*, **112**, 11777–11784.
12. Strack, G., Pita, M., Ornatska, M., and Katz, E. (2008). Boolean logic gates that use enzymes as input signals, *Chem. Bio. Chem*, **9**, 1260–1266.
13. (a) Baron, R., Lioubashevski, O., Katz, E., Niazov, T., and Willner, I. (2006). Two coupled enzymes perform in parallel the "AND" and "InhibAND" logic gates operations, *Org. Biomol. Chem.*, **4**, 989–991; (b) Baron, R., Lioubashevski, O., Katz, E., Niazov, T., and Willner, I. (2006). Logic gates and elementary computing by enzymes, *J. Phys. Chem. A*, **110**, 8548–8553.
14. (a) Sivan, S., and Lotan, N. (1999). A biochemical logic gate using an enzyme and its inhibitor. 1. The inhibitor as switching element, *Biotechnol. Prog.*, **15**, 964-970; (b) Sivan, S., Tuchman, S., and Lotan, N. (2003). A biochemical logic gate using an enzyme and its inhibitor. Part II: The logic gate, *Biosystems*, **70**, 21–33.
15. Deonarine, A. S., Clark, S. M., and Konermann, L. (2003). Implementation of a multifunctional logic gate based on folding/unfolding transitions of a protein, *Future Gener. Comp. Sym.*, **19**, 87–97.
16. Ashkenazi, G., Ripoll, D. R., Lotan, N., and Scheraga, H. A. (1997). *Biosens. Bioelectron.* **12**, 85–95.
17. Unger, R., and Moult, J. (2006). Towards computing with proteins, *Proteins*, **63**, 53–64.
18. Stojanovic, M. N., Stefanovic, D., LaBean, T., and Yan, H. (2005). Computing with Nucleic Acids, in *Bioelectronics: From Theory to Applications*, (ed. Willner, I., and Katz, E.). Wiley-VCH, Weinheim, Germany, pp. 427–455.
19. Saghatelian, A., Volcker, N. H., Guckian, K. M., Lin, V. S. Y., and Ghadiri, M. R. (2003). DNA-Based photonic logic gates: AND, NAND, and INHIBIT, towards biosensing strategies based on biochemical logic systems, *J. Am. Chem. Soc.*, **125**, 346–347.

20. Ashkenasy, G., and Ghadiri, M. R., (2004). Boolean logic functions of a synthetic peptide network, *J. Am. Chem. Soc.*, **126**, 11140–11141.

21. Win, M. N., and Smolke, C. D. (2008). Higher-order cellular information processing with synthetic rna devices, *Scė nce*, **322**, 456–460.

22. Simpson, M. L., Sayler, G. S., Fleming, J. T., and Applegate, B. (2001). Whole-cell biocomputing, *Trends Bὁ technol.*, **19**, 317–323.

23. Gunton, J. D., Toral, R., Mirasso, C., and Gracheva, M. E. (2003). The role of noise in some physical and biological systems, *Recent Res. Devel. Appl. Phys.*, 6, 497–514.

24. Baron, R., Lioubashevski, O., Katz, E., Niazov, T., and Willner, I. (2006). Elementary arithmetic operations by enzymes: A model for metabolic pathway based computing, *Angew. Chem. Int. Ed.*, **45**, 1572–1576.

25. Niazov, T., Baron, R., Katz, E., Lioubashevski, O., and Willner, I. (2006). Concatenated logic gates using four coupled biocatalysts operating in series, *Proc. Natl. Acad. USA*, **103**, 17160–17163.

26. Strack, G., Ornatska, M., Pita, M., and Katz, E. (2008). Biocomputing security system: concatenated enzyme-based logic gates operating as a biomolecular keypad lock, *J. Am. Chem. Soc.*, **130**, 4234–4235.

27. Katz, E., Privman, V., and Wang, J. (2010). Towards biosensing strategies based on biochemical logic systems, *Proc. Conf. ICQNM 2010*, IEEE Comp. Soc. Conf. Publ. Serv., 1–9.

28. Manesh, K. M., Halámek, J., Pita, M., Zhou, J., Tam, T. K., Santhosh, P., Chuang, M.-C., Windmiller, J. R., Abidin, D., Katz, E., and Wang, J. (2009). Enzyme logic gates for the digital analysis of physiological level upon injury, *Bὁ sens. Bὁ electron.*, **24**, 3569–3574.

29. Pita, M., Zhou, J., Manesh, K. M., Halámek, J., Katz, E., and Wang, J. (2009). Enzyme logic gates for assessing physiological conditions during an injury: Towards digital sensors and actuators, *Sens. Actuat. B*, **139**, 631–636.

30. Privman, V., Arugula, M. A., Halámek, J., Pita, M., and Katz, E. (2009). Network analysis of biochemical logic for noise reduction and stability: a system of three coupled enzymatic AND gates, *J. Phys. Chem. B*, **113**, 5301–5310.

31. Tam, T. K., Zhou, J., Pita, M., Ornatska, M., Minko, S., and Katz, E. (2008). Biochemically controlled bioelectrocatalytic interface, *J. Am. Chem. Soc.*, **130**, 10888–10889.

32. Zhou, J., Tam, T. K., Pita, M., Ornatska, M., Minko, S., and Katz, E. (2009). Bioelectrocatalytic system coupled with enzyme-based biocomputing

ensembles performing boolean logic operations: Approaching "Smart" physiologically controlled biointerfaces, *ACS Appl. Mater. Interf.*, **1**, 144–149.

33. Privman, M., Tam, T. K., Pita, M., and Katz, E. (2009). Switchable electrode controlled by enzyme logic network system: Approaching physiologically regulated bioelectronics, *J. Am. Chem. Soc.*, **131**, 1314–1321.

34. Krämer, M., Pita, M., Zhou, J., Ornatska, M., Poghossian, A., Schöning, M. J., and Katz, E. (2009). Coupling of biocomputing systems with electronic chips: electronic interface for transduction of biochemical information, *J. Phys. Chem. B*, **113**, 2573–2579.

35. Alon, U. (2007). *An Introduction to Systems Biology: Design Principles of Biological Circuits*, CRC Press, Boca Raton, FL.

36. (a) Strukov, D. B., Snider, G. S., Stewart, D. R., and Williams, R. S. (2008). The missing memristor found, *Nature*, **453**, 80–83; (b) Yang, J. J., Pickett, M. D., Li, X, Ohlberg, D. A. A., Stewart, D. R., and Williams, R. S. (2008). Memristive switching mechanism for metal/oxide/metal nanodevices, *Nat. Nanotechnol.*, **3**, 429–433; (c) Cooper, L. N. (2000). Memories and Memory: A Physicist's Approach to the Brain, *Int. J. Mod. Phys. A*, **15**, 4069–4082; (d) Munakata, T. (2008). *Fundamentals of the New Artificial Intelligence: Neural, Evolutionary, Fuzzy and More*, 2nd edn, Springer, London; (e) Ventra, M. D., Pershin, Y. V., and Chua, L. O. (2009). Circuit Elements With Memory: Memristors, Memcapacitors, and Meminductors, *Proc. of the IEEE*, **97**, 1717-1724; (f) Pershin, Y. V., La Fontaine, S., and Ventra, M. D. (2009). Memristive model of amoeba learning, *Phys. Rev. E*, **80**, 021926; (h) Pershin, Y. V., and Ventra, M. D. (2008). Spin memristive systems: Spin memory effects in semiconductor spintronics, *Phys. Rev. B*, **78**, 113309.

37. (a) Stadler, R., Ami, S., Joachim, C., and Forshaw, M. (2004). Integrating logic functions inside a single molecule, *Nanotechnology*, **15**, S115–S121; (b) Raymo, F. M., and Giordani, S. (2001). Signal Processing at the Molecular Level, *J. Am. Chem. Soc.*, **123**, 4651–4652.

38. De Silva, A. P., Leydet, Y., Lincheneau, C., and McClenaghan, N. D. (2006). Chemical approaches to nanometre-scale logic gates, *J. Phys. Cond. Matter*, **18**, S1847–S1872.

39. Adamatzky, A. (2004). Computing with waves in chemical media: Massively parallel reaction-diffusion on processors, *IEICE Trans. Electronics*, **E87C**, 1748–1756.

40. Flood, A. H., Ramirez, R. J. A., Deng, W. Q., Muller, R. P., Goddard, W. A., and Stoddart, J. F. (2004). Meccano on the nanoscale — a blueprint

for making some of the world's tiniest machines, *Austr. J. Chem.*, **57**, 301–322.

41. Xu, J., and Tan, G. J., (2007). A review on DNA computing models, *J. Comput. Theor. Nanosci.*, **4**, 1219–1230.
42. Kahan, M., Gil, B., Adar, R., and Shapiro, E. (2008). Towards molecular computers that operate in a biological environment, *Physċ a D*, **237**, 1165–1172.
43. Shao, X.-G., Jiang, H.-Y., and Cai, W.-S. (2002). Advances in biomolecular computing, *Prog. Chem.*, **14**, 37–46.
44. Melnikov, D., Strack, G., Pita, M., Privman, V., and Katz, E. (2009). Analog noise reduction in enzymatic logic gates, *J. Phys. Chem. B*, **113**, 10472–10479.
45. Fedichkin, L., Katz, E., and Privman, V. (2008). Error correction and digitalization concepts in biochemical computing, *J. Comput. Theor. Nanosci*, **5**, 36–43.
46. Margulies, D., and Hamilton, A. D. (2009). Digital analysis of protein properties by an ensemble of DNA quadruplexes, *J. Am. Chem. Soc.*, **131**, 9142–9143.
47. Szacilowski, K. (2007). Biomedical implications of information processing in chemical systems: Non-classical approach to photochemistry of coordination compounds, *Bȯ systems*, **90**, 738–749.
48. Ezziane, Z. (2006). *Nanotechnology*, **17**, R27–R39.
49. Adar, R., Benenson, Y., Linshiz, G., Rosner, A., Tishby, N., and Shapiro, E. (2004). Stochastic computing with biomolecular automata, DNA computing: applications and challenges, *Proc. Natl. Acad. USA*, **101**, 9960–9965.
50. Simmel, F. C. (2007). Towards biomedical applications for nucleic acid nanodevices, *Nanomedċ ṅ e*, **2**, 817–830.
51. May, E. E., Dolan, P. L., Crozier, P. S., Brozik, S., Manginell, M. (2008). Towards de novo design of deoxyribozyme biosensors for GMO Detection, *IEEE Sens. J.*, **8**, 1011–1019.
52. Maltzahn, G. von, Harris, T. J., Park, J.-H., Min, D.-H., Schmidt, A. J., Sailor, M. J., and Bhatia, S. N. (2007). Nanoparticle self-assembly gated by logical proteolytic triggers, *J. Am. Chem. Soc.*, **129**, 6064–6065.
53. Pita, M., Strack, G., MacVittie, K., Zhou, J., and Katz, E. (2009). Set–reset flip-flop memory based on enzyme reactions: toward memory systems controlled by biochemical pathways, *J. Phys. Chem. B*, **113**, 16071–16076.

54. Tomizaki, K., and Mihara, H. (2007). Phosphate-mediated molecular memory driven by two different protein kinases as information input elements, *J. Am. Chem. Soc.*, **129**, 8345–8352.

55. Xia, F., Zuo, X., Yang, R., White, R. J., Xiao, Y., Kang, D., Gong, X., Lubin, A. A., Vallée-Bélisle, A., Yuen, J. D., Hsu, B. Y. B., and Plaxco, K. W. (2010). Label-free, dual-analyte electrochemical biosensors: A new class of molecular-electronic logic gates, *J. Am. Chem. Soc.*, **132**, 8557-8559.

56. Halámek, J., Bocharova, V., Chinnapareddy, S., Windmiller, J. R., Strack, G., Chuang, M.-C., Zhou, J., Santhosh, P., Ramirez, G. V., Arugula, M. A., Wang, J., and Katz, E. (2009). Multi-enzyme logic network architectures for assessing injuries: digital processing of biomarkers, *Molecular Bio systems*, **6**, 2554–2560.

Index

Color Insert

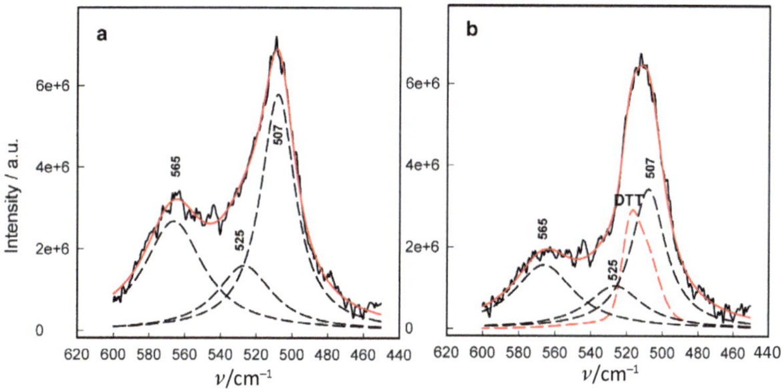

Figure 1.9

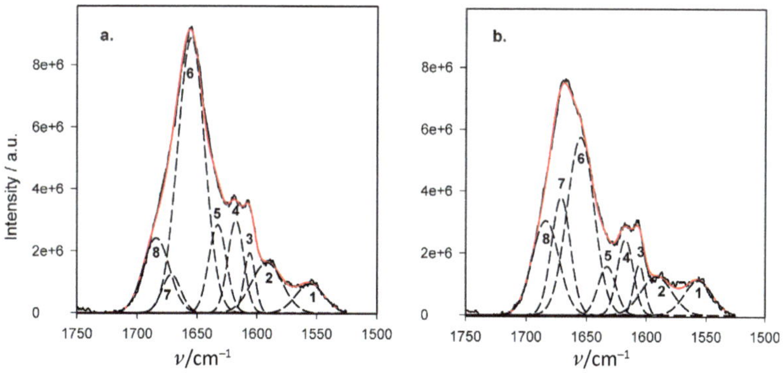

Figure 1.10

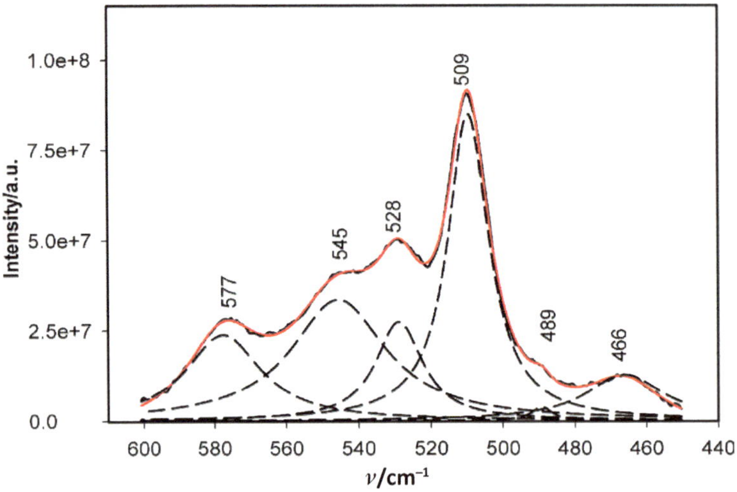

Figure 1.11

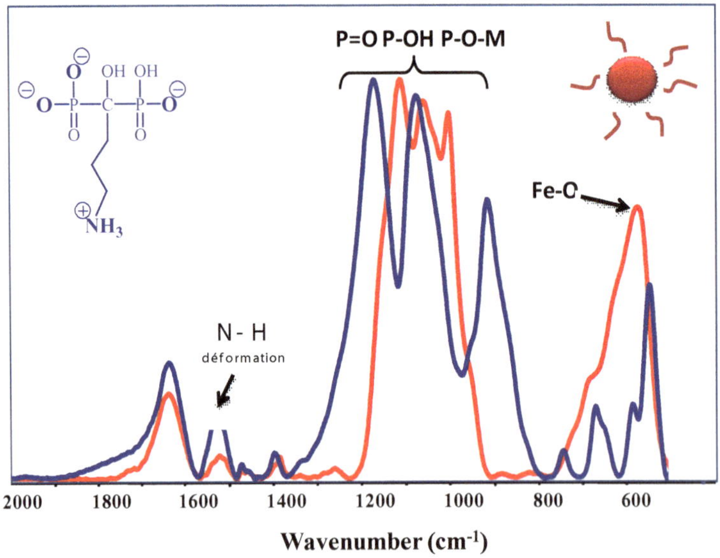

Figure 4.5

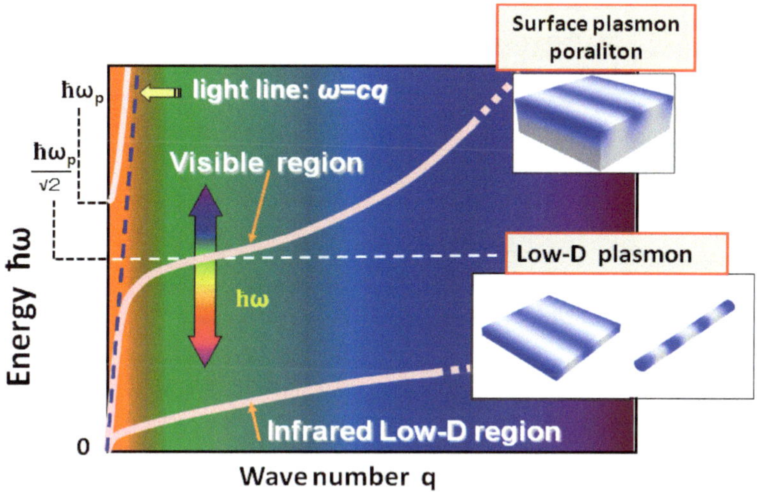

Figure 5.2

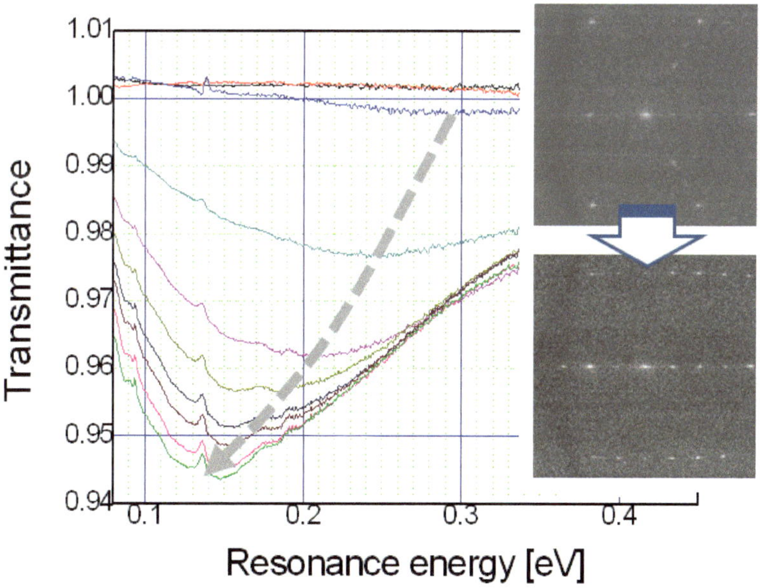

Figure 5.9

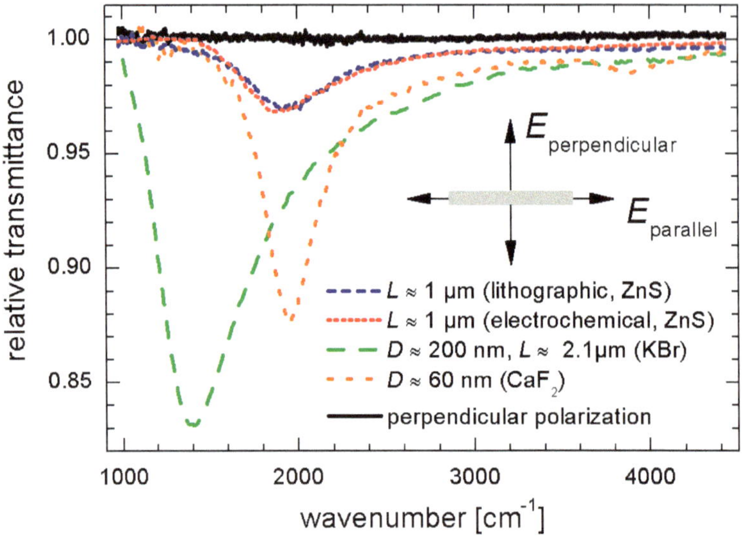

Figure 6.1

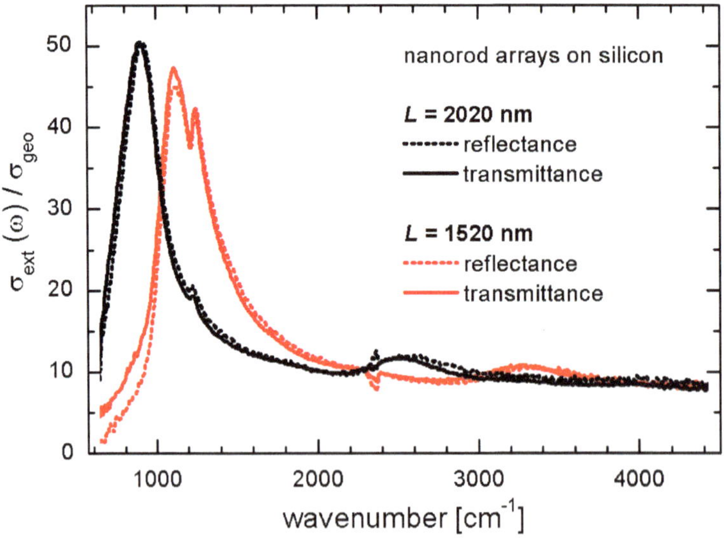

Figure 6.3

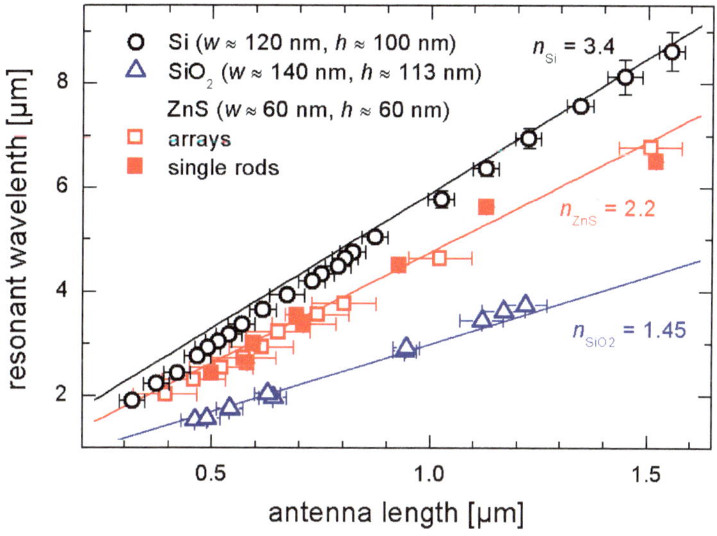

Figure 6.4

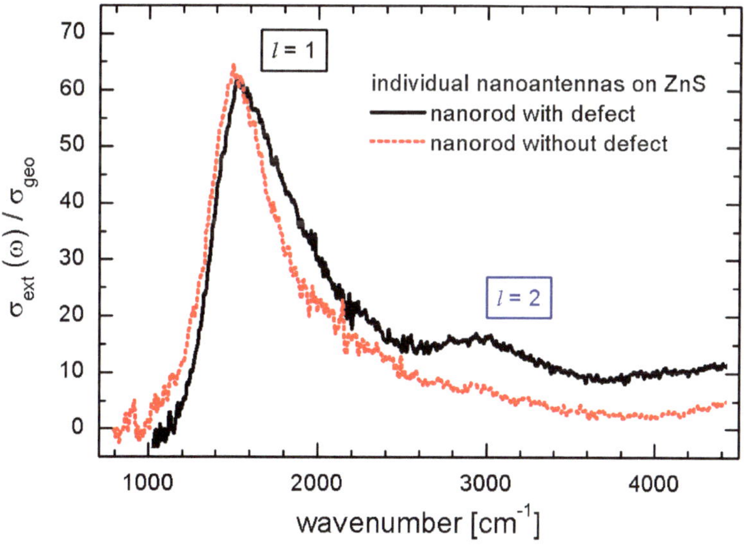

Figure 6.6

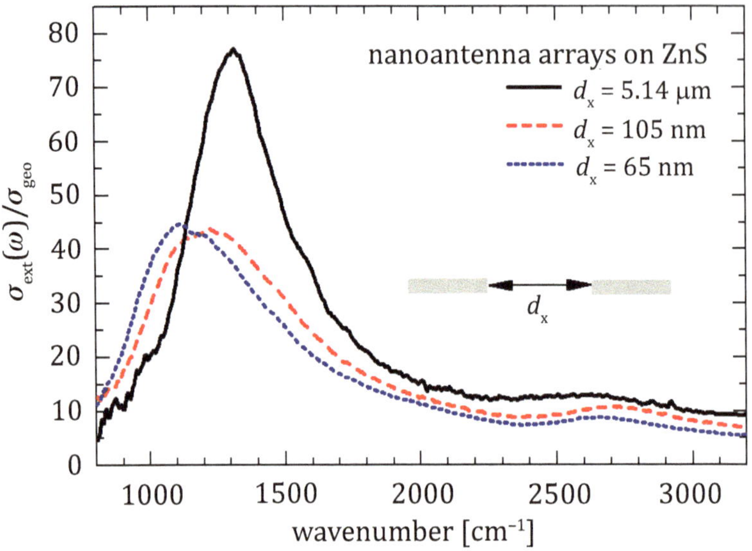

Figure 6.7

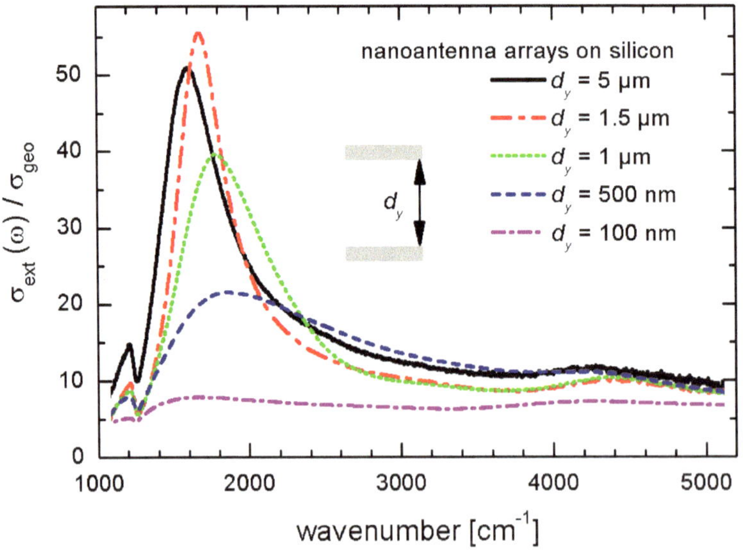

Figure 6.8

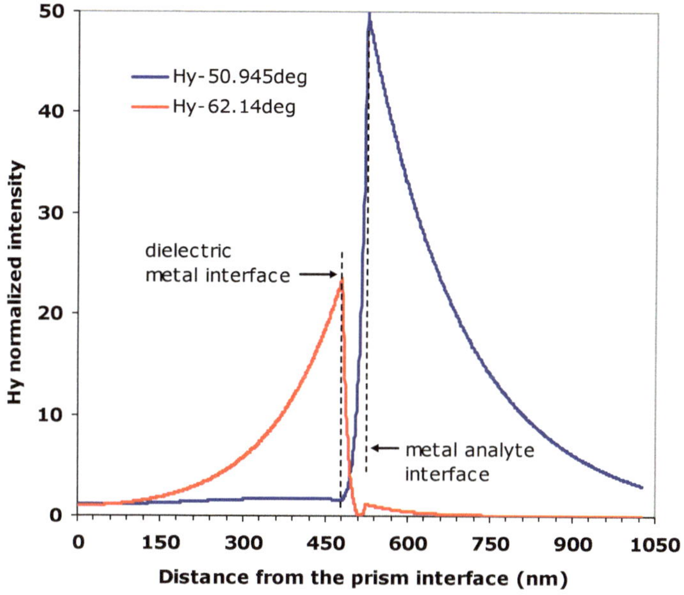

Figure 7.8

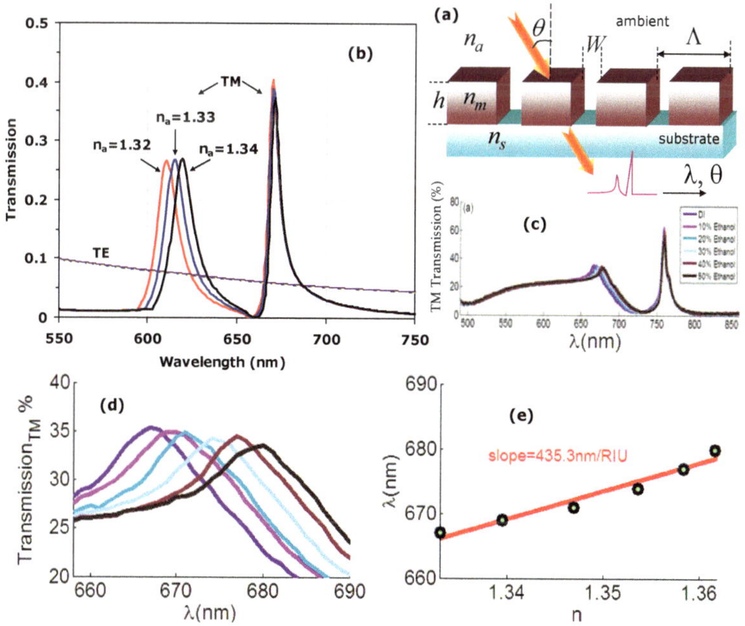

Figure 7.10

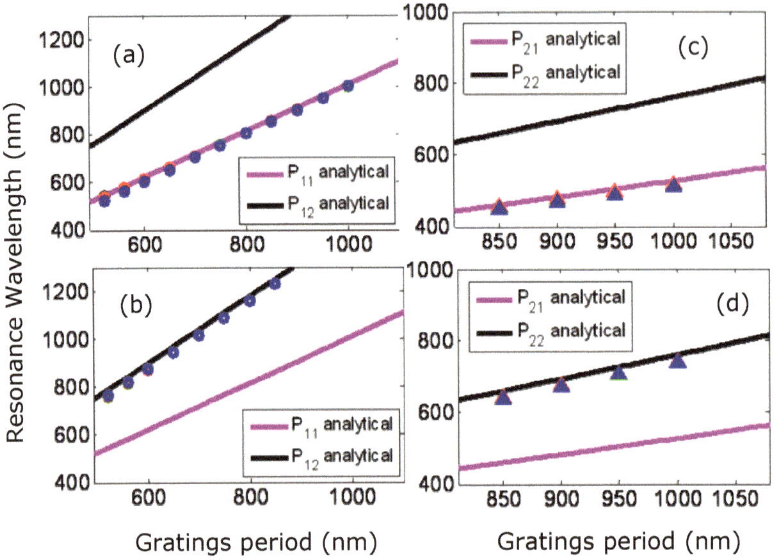

Figure 7.11

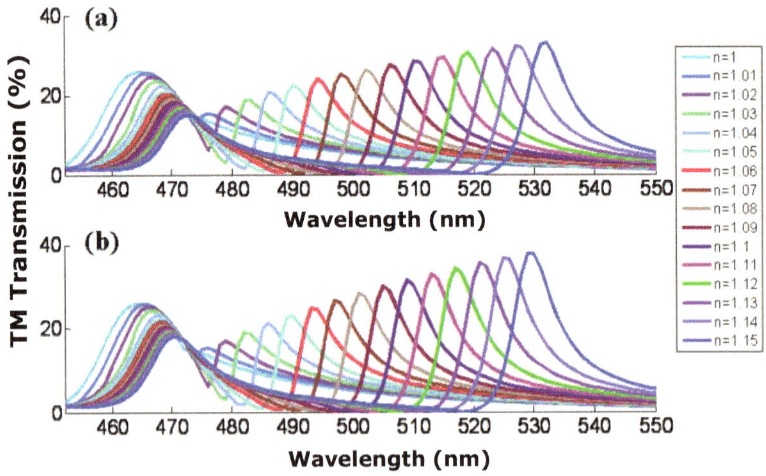

Figure 7.12

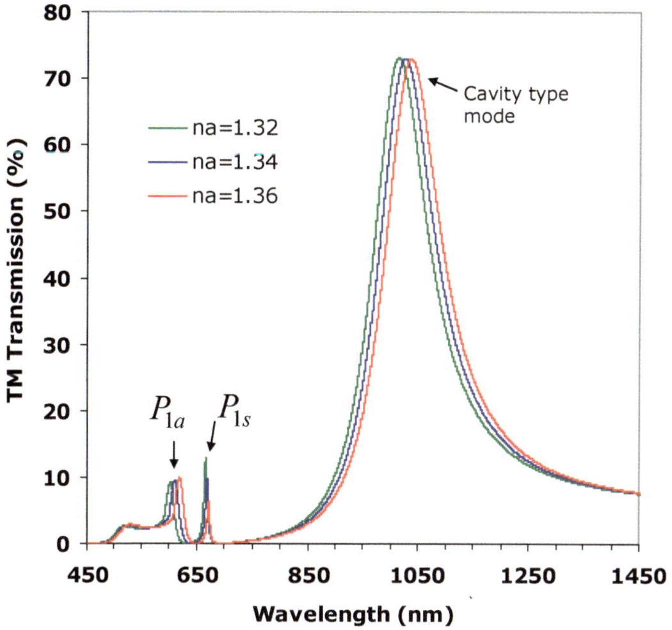

Figure 7.14

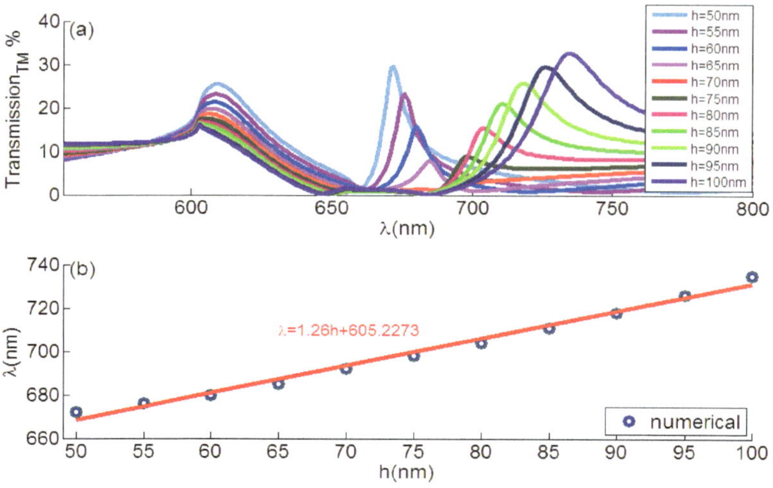

Figure 7.15

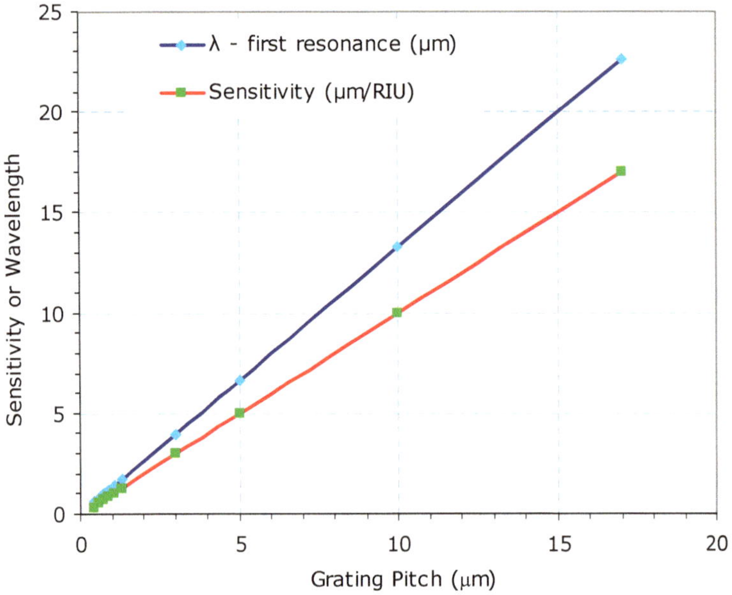

Figure 7.16

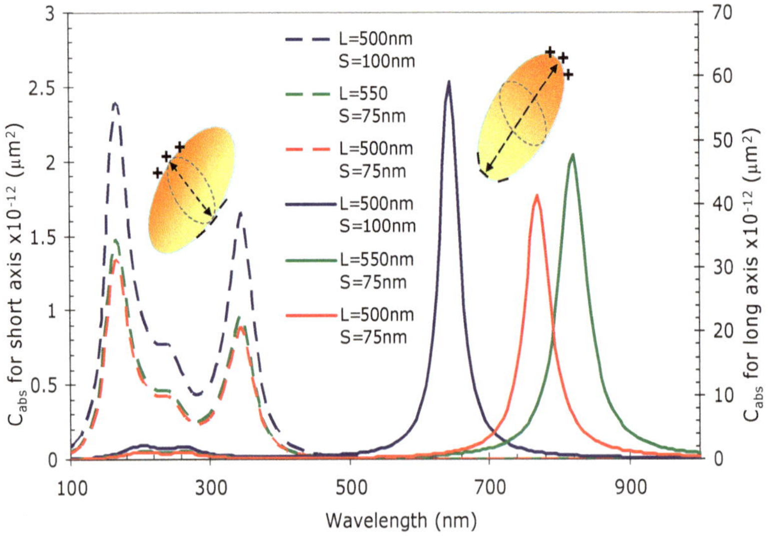

Figure 7.20

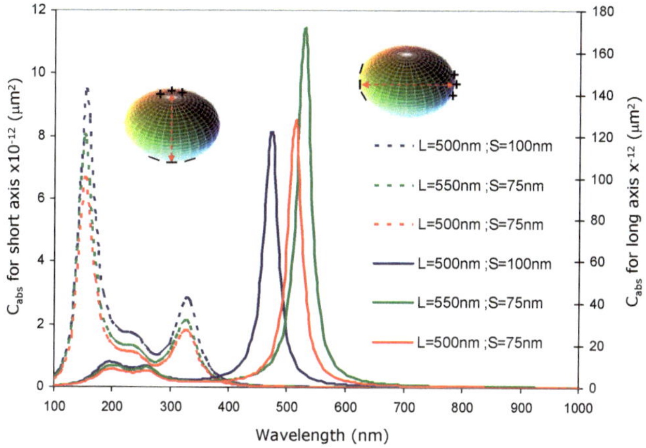

Figure 7.21

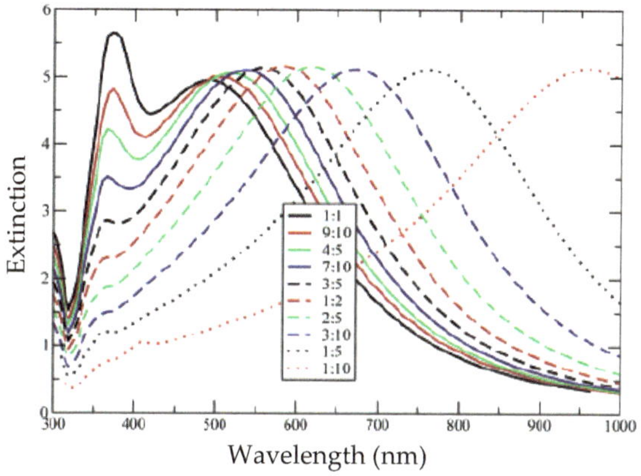

Figure 7.22

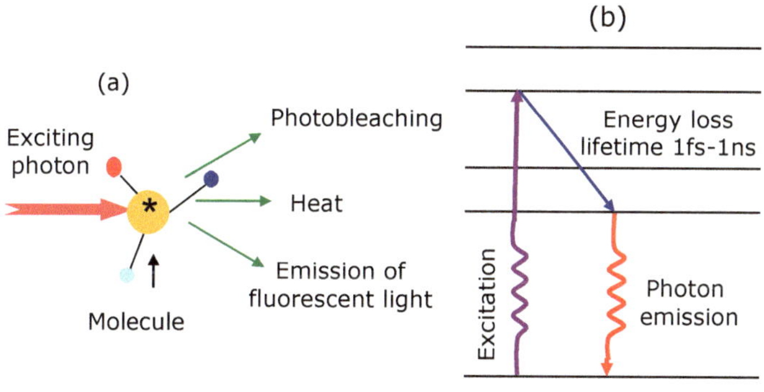

Figure 7.25

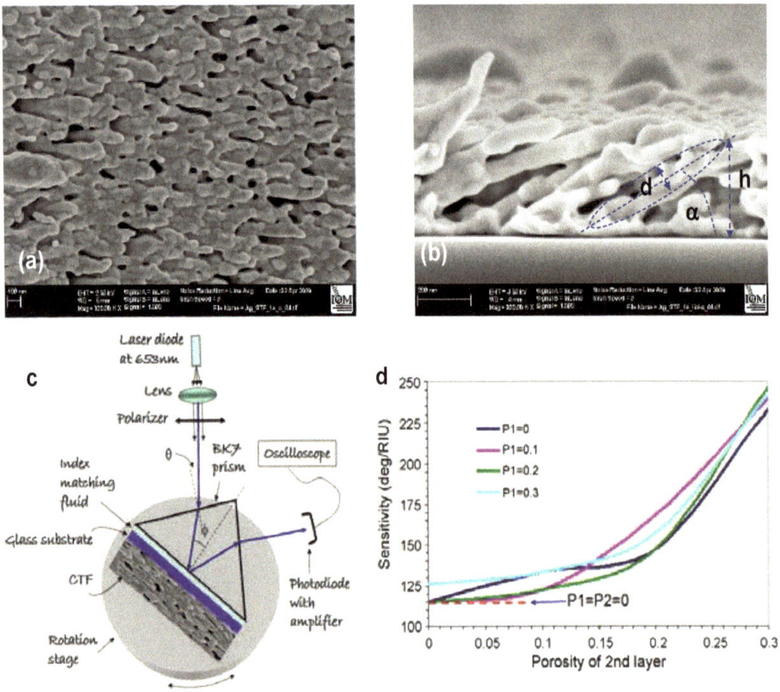

Figure 7.30

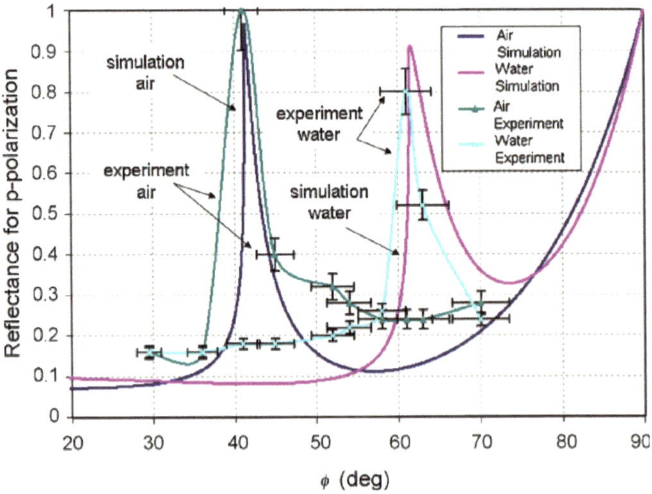

Figure 7.31

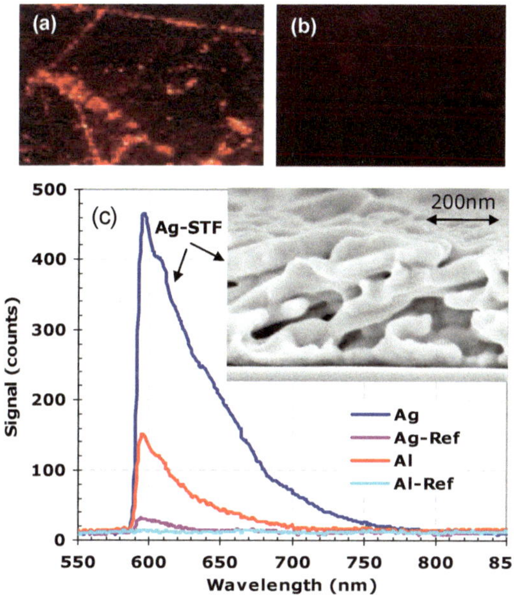

Figure 7.33

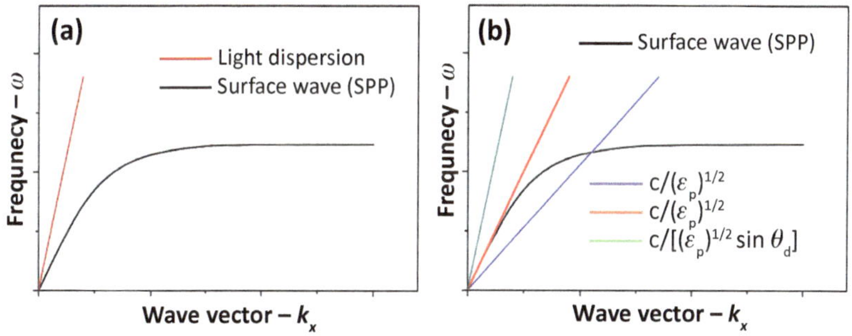

Figure 8.2

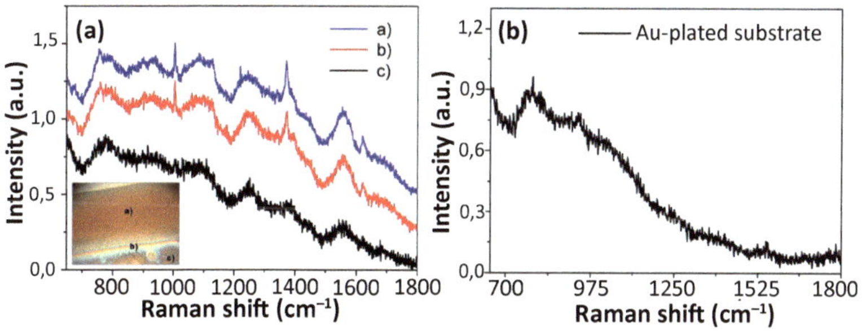

Figure 8.3

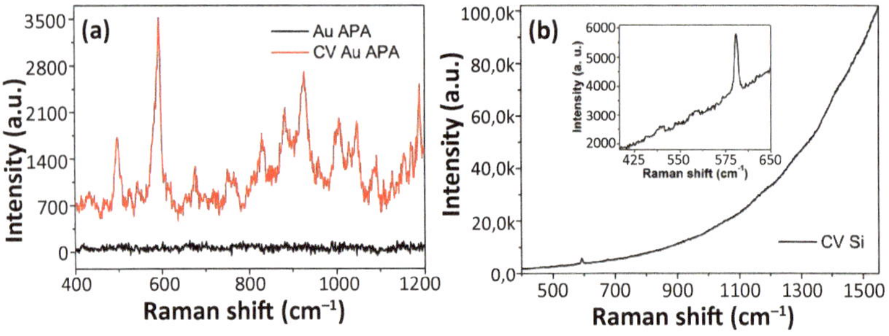

Figure 8.6

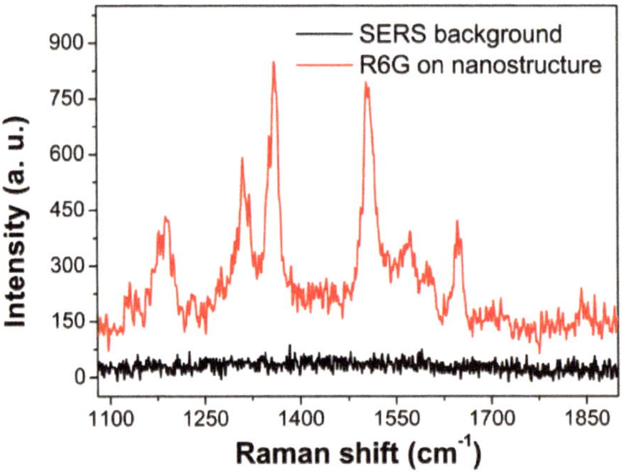

Figure 8.9

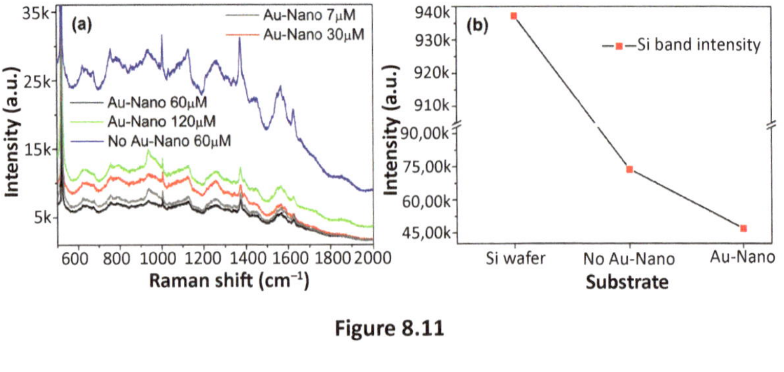

Figure 8.11

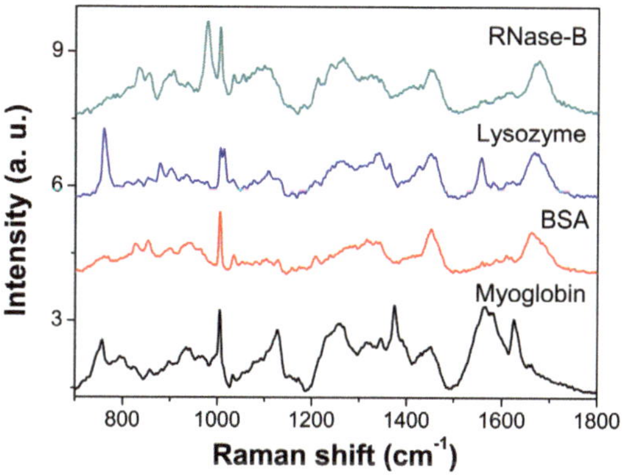

Figure 8.12

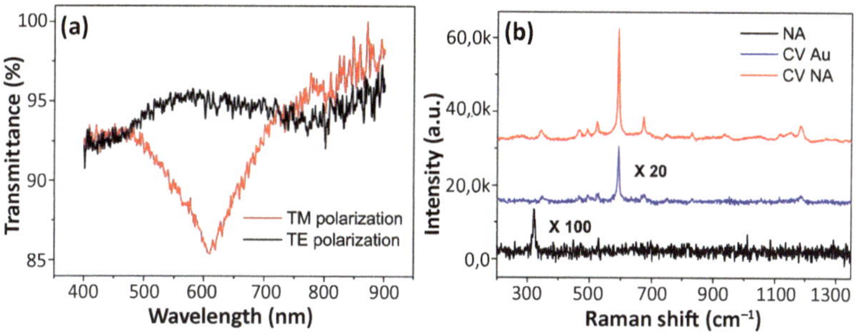

Figure 8.15

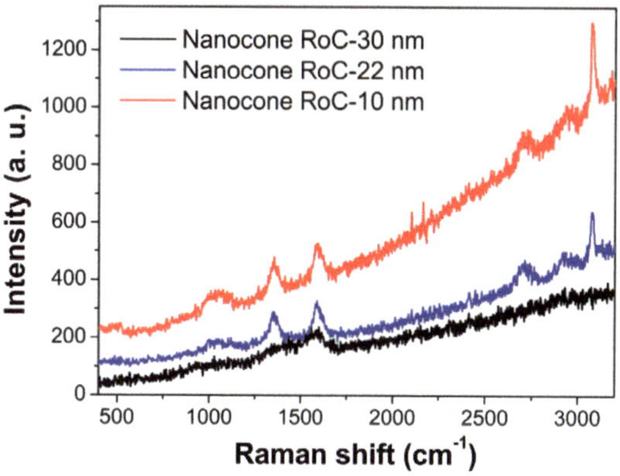

Figure 8.18

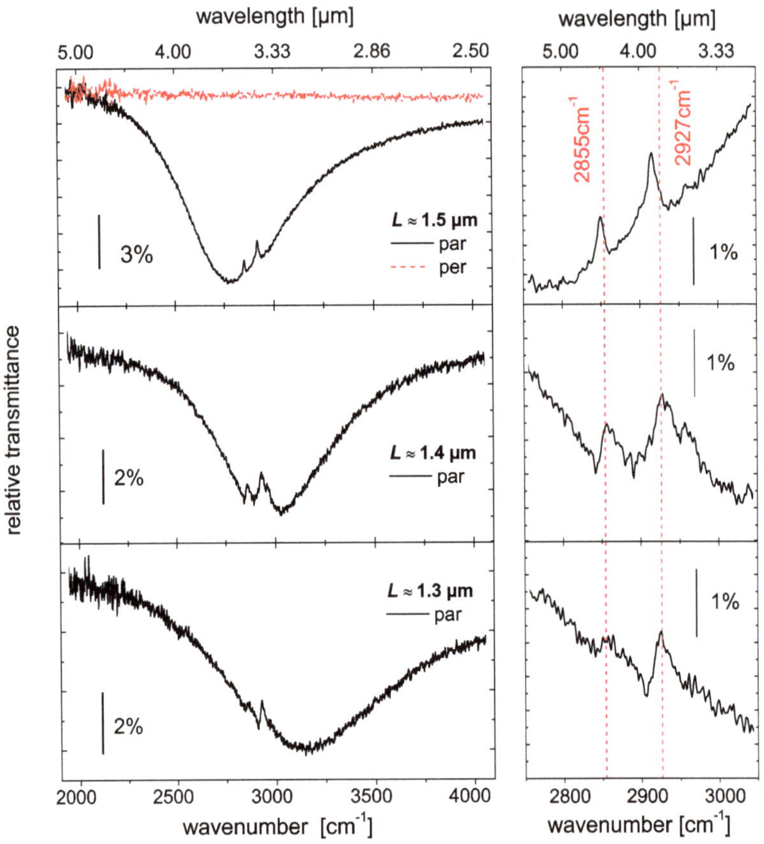

Figure 9.2

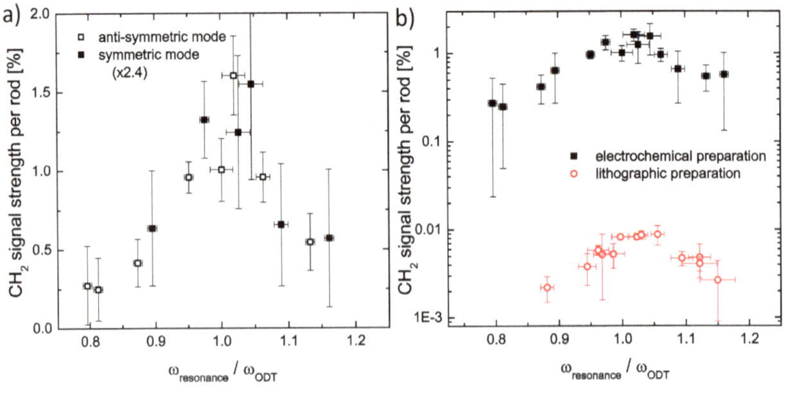

Figure 9.3